Progress in Systems and Control Theory
Volume 17

Series Editor
Christopher I. Byrnes, Washington University

Advances in Nonlinear Dynamics and Control: *A Report from Russia*

Alexander B. Kurzhanski

Editor

1993

Birkhäuser

Boston Basel Berlin

Alexander B. Kurzhanski
Faculty of Computational Mathematics
and Cybernetics (VMK)
Moscow State University
119899 Moscow, Russia

Library of Congress Cataloging In-Publication Data

Advances in nonlinear dynamics and control : a report from Russia /
Alexander B. Kurzhanski, editor.
p. cm. -- (Progress in systems and control theory ; v. 17)
Includes bibliographical references.
ISBN 0-8176-3736-2
1. Control theory. 2. Dynamics. 3. Nonlinear theories.
I. Kurzhanski, A. B. II. Series.
QA 402.3.A365 1993 93-39908
003'.85--dc20 CIP

Printed on acid-free paper

Birkhäuser

ISBN 0-8176-3736-2
ISBN 3-7643-3736-2
Typeset by the Authors in LATEX.

9 8 7 6 5 4 3 2 1

Contents

Preface

The purpose of this volume is to present a coherent collection of overviews of recent Russian research in Control Theory and Nonlinear Dynamics written by active investigators in these fields.

It is needless to say that the contribution of the scientists of the former Soviet Union to the development of nonlinear dynamics and control was significant and that their scientific schools and research community have highly evolved points of view, accents and depth which complemented, enhanced and sometimes inspired research directions in the West.

With scientific exchange strongly increasing, there is still a considerable number of Eastern publications unknown to the Western community. We have therefore encouraged the authors to produce extended bibliographies in their papers.

The particular emphasis of this volume is on the treatment of uncertain systems in a deterministic setting—a field highly developed in the former Soviet Union and actively investigated in the West.

The topics are concentrated around the three main branches of uncertain dynamics which are the theory of Differential Games, the set-membership approach to Evolution, Estimation and Control and the theory of Robust Stabilization. The application of these techniques to nonlinear systems as well as the global optimization of the latter are also among the issues treated in this volume.

We believe that all students and active researchers in nonlinear systems and control will find this book to be both a source of new ideas and techniques as well as a self-contained introduction to an important, yet heretofore inaccessible body of our literature.

Christopher I. Byrnes
Washington University

Editor's Remarks

The present volume reflects some of the most recent results of a group of active Russian researchers in the field of nonlinear dynamics for control and communicates the flavour of the Russian research environment. The emphasis is on those specific directions that are among the main activities in this community. These are primarily the problems of global optimality, of evolution and control synthesis, of differential games and robust adaptive stabilization for systems with unknown but bounded uncertainties and techniques for the decomposition of nonlinear systems developed in the spirit of other investigations published in this volume.

The techniques used in this volume range from those common in the engineering control communities to those that are specific for investigations carried through in mathematical departments.

We also hope that the extended bibliography of related Russian papers given here in English does perhaps include investigations and results that may not be quite known to the Western reader and will perhaps partly fill a unique niche that exists in Western publications and references.

A. Kurzhanski

The Decomposition of Controlled Dynamic Systems

F.L. Chernousko

ABSTRACT. Nonlinear dynamic systems are considered that are described by the Lagrangian equations and subjected to uncertain disturbances and bounded control forces. Under certain assumptions, feedback control laws are obtained that satisfy the imposed constraints and drive the systems to the prescribed terminal states in finite time. The approaches developed in the paper are based on the decomposition of the given system into subsystems with one degree of freedom each. Methods of optimal control and differential games are then applied to the subsystems. As a result, explicit formulae for the closed-loop control forces and for the time of motion are derived. The obtained feedback controls are time-suboptimal and robust, they can cope with small disturbances and parameter variations. Applications to control of robots are discussed.

KEY WORDS. Feedback control, Optimal control, Nonlinear systems, Uncertain systems.

1 Introduction

The design of feedback controls for nonlinear dynamic systems is often based on using Lyapunov functions. For example, the approach developed in [1–3] makes it possible, under certain assumptions, to obtain controls that ensure the desired behaviour of the system (boundedness and stability) in the presence of uncertainty; the time of motion is infinite.

In this paper, we describe two methods that provide feedback controls to satisfy the imposed constraints and to bring the nonlinear Lagrangian systems to the prescribed terminal states in finite time. Both methods are based upon the decomposition of the original system with many degrees of freedom into subsystems with one degree of freedom each. After that, the approaches of optimal control [4] and differential games [5] are applied. As a result, we obtain explicit formulae representing closed-loop controls for the original nonlinear systems.

We propose two methods that differ in the assumptions made. For the first approach, assumptions must be verified in the $2n$-dimensional phase space, whereas for the second approach the assumptions are to be checked in the n-dimensional coordinate space. Hence, the second method seems to be less restrictive.

The paper consists of 17 Sections. The first approach is described in Sections 3 - 5. Its generalizations are given in Sections 6 and 7. Section 8 is devoted to applications of the method to robotic systems. In Sections 9 - 11 we consider the simplified version of our first method. The second approach is described in Sections 12 - 15. The paper is based on results published in [6–10].

2 Statement of the Problem

Consider a controlled dynamic system governed by the Lagrangian equations

$$\begin{aligned} \frac{d}{dt}\frac{\partial T}{\partial \dot{q}_k} - \frac{\partial T}{\partial q_k} &= Q_k + F_k(q, \dot{q}, t) \\ k &= 1, ..., n \end{aligned} \tag{2.1}$$

Here, t is time, $q = (q_1, ..., q_n)$ is the n-vector of generalized coordinates of the system which has n degrees of freedom. The generalized forces consist of two terms: the control forces Q_k and all the other external and internal forces F_k; the latter also include uncertain perturbations. The kinetic energy of the system T is given by

$$T = \frac{1}{2}\sum_{i,j=1}^{n} A_{ij}(q)\dot{q}_i\dot{q}_j = \frac{1}{2}(A(q)\dot{q}, \dot{q}) \tag{2.2}$$

where A_{ij} are elements of the symmetric positive definite $n \times n$-matrix $A(q)$. Substituting (2.2) into (2.1) we obtain the equations of motion in the form

$$A(q)\ddot{q} = Q + F(q, \dot{q}, t) + S(q, \dot{q}) \tag{2.3}$$

where the following notation is introduced

$$Q = (Q_1, ..., Q_n), \ F = (F_1, ..., F_n), \ S = (S_1, ..., S_n)$$

$$S_k(q, \dot{q}) = \sum_{i,j=1}^{n} \Gamma_{kij}(q)\dot{q}_i\dot{q}_j, \quad k = 1, ..., n$$

$$\Gamma_{kij}(q) = \frac{1}{2}\frac{\partial A_{ij}}{\partial q_k} - \frac{\partial A_{kj}}{\partial q_i} \tag{2.4}$$

The control forces are subjected to the constraints

$$| Q_k | \leq Q_k^0, \quad k = 1, ..., n \tag{2.5}$$

where Q_k^0 are given constants.

We consider the following control problem.

Problem 1. Find a feedback control $Q(q, \dot{q})$ that satisfies the constraints (2.5) and steers the system (2.1) (or (2.3)) from any given initial state

$$q(t_0) = q^0, \quad \dot{q}(t_0) = \dot{q}^0 \tag{2.6}$$

to the prescribed terminal state with zero velocities

$$q(t_1) = q^1, \quad \dot{q}(t_1) = 0 \tag{2.7}$$

in finite time. Here, the instant t_0 and the vectors $q^0, \dot{q}^0$ and q^1 are fixed, whereas $t_1 > t_0$ is free.

We shall propose two approaches to Problem 1 which are both based on the decomposition of system (2.3), but differ in the assumptions made.

3 The First Approach

Let A_1 be a constant symmetric positive definite $n \times n$-matrix. For example, we can set $A_1 = A(q^1)$. Let us multiply both sides of equation (2.3) by $A_1 A^{-1}(q)$. Thus we obtain

$$A_1 \ddot{q} = Q + R(q, \dot{q}, t, Q), \quad R = A_1 A^{-1}(Q + F + S) - Q \tag{3.1}$$

We restrict ourselves with motions lying within some domain W in the $2n$-dimensional $(q, \dot{q})$-space. The initial and the terminal states (2.6) and (2.7) are supposed to belong to the domain W. We assume that

$$| R_k(q, \dot{q}, t, Q) | \leq R_k^0 < Q_k^0, \quad k = 1, ..., n \tag{3.2}$$

for all $t \geq t_0$, all $(q, \dot{q}) \in W$, and all Q satisfying (2.5). Here, R_k^0 are some constants. The conditions (3.2) can be verified by means of the following Lemma.

Lemma. *Suppose the following inequalities*

$$\begin{aligned} &| A_1 z | \geq \mu_1 | z |, \quad | [A(q) - A_1] z | \leq \mu | z | \\ &| F_k + S_k | \leq \nu Q_k^0, \; k = 1, ..., n, \; 0 < \mu < \mu_1, \; \nu > 0 \end{aligned} \tag{3.3}$$

hold for any $z \in R^n$, all $t \geq t_0$, and all $(q, \dot{q}) \in W$, where μ_1, μ, and ν are constants. Then for all $t > t_0$, all $(q, \dot{q}) \in W$ and all Q satisfying (2.5) we have

$$\begin{aligned} \mid R_k \mid &\leq \nu Q_k^0 + \mu(\mu_1 - \mu)^{-1}(1+\nu)Q^0 \\ Q^0 &= [\sum_{k=1}^{n} (Q_k^0)^2]^{1/2}, \quad k = 1, ..., n \end{aligned} \tag{3.4}$$

Remark. The constant μ_1 can be taken equal to (or smaller than) the minimal eigenvalue of the positive definite matrix A_1.

Proof. Denote

$$L = [A(q) - A_1]\, A_1^{-1}, \quad B = A(q)A_1^{-1} \tag{3.5}$$

From the first two inequalities (3.3) we obtain

$$\mid Lz \mid \leq \mu \mid A_1^{-1} \mid \leq \mu\mu_1^{-1} \mid z \mid \tag{3.6}$$

Here, z is an arbitrary n-vector. From (3.5) it follows

$$Bz = z + Lz \tag{3.7}$$

Using (3.6) and (3.7) we obtain

$$\mid Bz \mid \geq \mid z \mid - \mid Lz \mid \geq (1 - \mu\mu_1^{-1}) \mid z \mid \tag{3.8}$$

It follows from (3.3) that $(1 - \mu\mu_1^{-1}) > 0$. By setting $z = B^{-1}z'$ in (3.8), we get

$$\mid B^{-1}z' \mid \leq (1 - \mu\mu_1^{-1})^{-1} \mid z' \mid \tag{3.9}$$

Combining inequalities (3.6) and (3.9), we obtain

$$\mid LB^{-1}z \mid \leq \mu(\mu_1 - \mu)^{-1} \mid z \mid \tag{3.10}$$

Let us substitute $z = B^{-1}z'$ into (3.7)

$$B^{-1}z' = z' - LB^{-1}z' \tag{3.11}$$

Using (3.5) and (3.11), we can rewrite the expression (3.1) for R as follows

$$R = B^{-1}(Q + F + S) - Q = F + S - LB^{-1}(Q + F + S) \tag{3.12}$$

Applying inequalities (3.3) and (3.10) to (3.12), we obtain

$$\begin{aligned} \mid R_k \mid &\leq \mid F_k + S_k \mid + \mid LB^{-1}(Q + F + S) \mid \\ \leq \nu Q_k^0 &+ \mu(\mu_1 - \mu)^{-1} \mid Q + F + S \mid \\ \leq \nu Q_k^0 &+ \mu(\mu_1 - \mu)^{-1}(\mid F + S \mid + \mid Q \mid) \\ \leq \nu Q_k^0 &+ \mu(\mu_1 - \mu)^{-1}(\nu Q^0 + Q^0), \quad k = 1, ..., n \end{aligned}$$

The lemma is proved.

Corollary. *If $\nu < 1$ and μ is sufficiently small, then the conditions (3.2) are satisfied.*

The main assumption (3.2) made above implies that the control forces are greater than all the other terms in the right-hand side of (3.1); these terms can be regarded as "forces" in the transformed system (3.1). The assumption (3.2) seems rather natural and means that the control forces should be great enough to cope with all other forces including the forces of inertia. It follows from Lemma and the Corollary that, in order to ensure the assumption (3.2) for the fixed dynamic system (2.3) or (3.1), we should increase the control bounds Q_k^0 in (2.5) (and thus decrease ν in (3.3)) and diminish the domain W(and thus decrease μ). Of course, we are not always free to fulfil these requirements.

By the change of variables

$$A_1(q - q^1) = y \tag{3.13}$$

we reduce our system (3.1) with the boundary conditions (2.6) and (2.7) to

$$\ddot{y}_k = Q_k + R_k, \quad k = 1, ..., n \tag{3.14}$$

$$\begin{aligned} y_k(t_0) &= \left[A_1(q^0 - q^1)\right]_k, \quad \dot{y}_k(t_0) = (A_1\dot{q}^0)_k \\ y_k(t_1) &= \dot{y}_k(t_1) = 0 \end{aligned} \tag{3.15}$$

We shall regard the terms R_k in equations (3.14) as independent uncertain disturbances subject to the constraints (3.2) and apply the approach of differential games to system (3.14). Hence, we can treat (3.14) as a set of n separate subsystems with one degree of freedom each. Here, Q_k and R_k are the controls of the two players acting on the kth subsystem.

Thus, we come to the following result that can be formalized as a decomposition theorem.

Theorem 1. *Let conditions (3.2) hold and all the considered motions of the system (2.3) lie within the domain W. To solve Problem 1, it is sufficient to find scalar feedback controls $Q_k(y_k, \dot{y}_k)$ that satisfy (2.5) and bring the corresponding subsystems (3.14) from any initial state (3.15) to the zero terminal state in finite time for any admissible disturbances R_k, satisfying (3.2).*

4 Differential Game

By a transposition of variables

$$\begin{array}{ll} y_k = Q_k^0 x, \quad Q_k = Q_k^0 u, & R_k = Q_k^0 v, \rho = R_k^0 / Q_k^0 < 1 \\ \xi = [A_1(q^0 - q^1)]_k \ / \ Q_k^0, & \eta = (A_1 \dot{q}^0)_k \ / Q_k^0, \ k = 1, ..., n \end{array} \tag{4.1}$$

we convert the kth subsystem (3.14) with the constraints and the boundary conditions (3.15) into the following normalized form

$$\begin{array}{ll} \ddot{x} = u + v, \ | \, u \, | \leq 1, & | \, v \, | \leq \rho < 1 \\ x(0) = \xi, \ \dot{x}(0) = \eta, & x(\tau) = \dot{x}(\tau) = 0 \end{array} \tag{4.2}$$

Here, the initial instant of time is taken equal to zero without loss of generality, and the terminal time is denoted by τ. Our problem is reduced to obtaining the feedback control $u(x, \dot{x})$ for system (4.2). Here, v is some unknown but bounded disturbance which can be regarded as the control of another player. This player acts against the control u and tries to prevent the system from reaching the terminal state or, if this is impossible, to increase the time of motion τ. Thus, we come to the differential game described by the equation, the constraints and the boundary conditions given in (4.2). We assume that the performance index of this game is the time τ which the first player *(u)* tries to minimize and the second player *(v)* tries to maximize.

This differential game is a simple linear game with matching constraints imposed on the both players. It was indicated [5] that the solution of this game is reduced to the optimal control problem for the system

$$\ddot{x} = (1 - \rho) u, \quad | \, u \, | \leq 1, \quad \tau \rightarrow \min \tag{4.3}$$

under the same boundary conditions as in (4.2). Equation (4.3) is obtained from (4.2), if the disturbance v is replaced by $-\rho u$. This "worst" disturbance is the optimal reply of the second player to the optimal feedback control of the first player. The optimal feedback control $u(x, \dot{x})$ and the minimal time τ for the game (4.2) are equal to the optimal feedback control and the minimal time, respectively, for the problem (4.3). The control which is sought for is well-known [4] and given by

$$\begin{array}{lll} u(x, \dot{x}) = & \text{sign } \psi_\rho(x, \dot{x}) & \text{if } \ \psi_\rho \neq 0 \\ u(x, \dot{x}) = & \text{sign } x = -\text{sign } \dot{x} & \text{if } \ \psi_\rho = 0 \end{array} \tag{4.4}$$

Here, ψ_ρ is the switching function defined by

$$\psi_\rho(x, \dot{x}) = -x - \dot{x} \, | \, \dot{x} \, | \, [2(1 - \rho)]^{-1} \tag{4.5}$$

The minimal time τ required to reach the terminal state (4.2) is given by

$$\begin{aligned} \tau(\xi,\eta) &= (1-\rho)^{-1}\left\{[2\eta^2 - 4(1-\rho)\xi\gamma]^{1/2} - \eta\gamma\right\} \\ \gamma &= \text{sign } \psi_\rho(\xi,\eta) \end{aligned} \tag{4.6}$$

If $\psi_\rho = 0$, both values $\gamma = \pm 1$ can be taken in (4.6) with the same result.

If we apply the obtained control (4.4) to our system (4.2) and if the disturbance is optimal ($v = -\rho u$), then the system reaches the terminal state in optimal time τ given by (4.6). As a rule, however, the disturbance is not optimal, and the system reaches the terminal state earlier.

Under the optimal control u and the optimal disturbance v, any phase trajectory of our system in the $(x, \dot{x})$ - plane consists of two parabolic arcs. The first arc can be absent, and the second one is a part of the switching curve $\psi_\rho = 0$. If the disturbance is not optimal, the first arc deviates from the optimal trajectory, whereas the second one is still an arc of the switching curve. The motion along this curve is called a sliding regime: the control u switches between $u = -1$ and $u = 1$ infinitely often, so that in the average we have

$$u + v = (1-\rho)\text{sign } x$$

along the switching curve $\psi_\rho = 0$.

5 Feedback Control (the First Approach)

Now we turn to Problem 1 stated in Section 2. According to the changes of variables (4.1) and (3.13) we have

$$\begin{aligned} Q_k(q,\dot{q}) &= Q_k^0 u(x,\dot{x}) \\ x = (Q_k^0)^{-1} y_k &= (Q_k^0)^{-1}\left[A_1(q-q^1)\right]_k \\ \dot{x} = (Q_k^0)^{-1}\dot{y}_k &= (Q_k^0)^{-1}(A_1 q^1)_k, \quad k = 1, ..., n \end{aligned} \tag{5.1}$$

Here, the function $u(x, \dot{x})$ is defined by formulae (4.4) and (4.5). The parameter ρ in these formulae is given in (4.1) for each k. The obtained control is of the bang-bang type: $Q_k = \pm Q_k^0, k = 1, \ldots, n$. Note that motions of different degrees of freedom are coupled through the controls (5.1) and the terms R_k in equations (3.14). The latter terms depend on all components of the vectors q, $\dot{q}$ and Q (see (3.1)). However, if the control is chosen according to (5.1), then for each degree of freedom the motion reaches the prescribed terminal state in the time τ defined in (4.6) and remains there afterwards.

The total time of motion does not exceed the greatest of times calculated for each degree of freedom. According to (4.6) and (4.1), we have

$$\begin{aligned} t_1 \quad - \quad & t_0 \leq \max_{1\leq k\leq n} \tau(\xi_k, \eta_k) \\ \xi_k \quad = \quad & (Q_k^0)^{-1}\left[A_1(q^0 - q^1)\right]_k \\ \eta_k \quad = \quad & (Q_k^0)^{-1}(A_1\dot{q}^0)_k \end{aligned} \tag{5.2}$$

The obtained results are summarized in the following theorem.

Theorem 2. *Let conditions (3.2) hold and all the considered motions of system (2.3) lie within the domain W. Then the feedback control defined by (5.1) gives a solution to Problem 1. Under this control, the system (2.3) reaches the terminal state (2.7) at instant t_1 bounded by the inequality (5.2).*

The obtained control may be called time-suboptimal because it becomes time-optimal for the "worst" (optimal) disturbances.

The approach described above consists of two stages: 1) the decomposition of the system into simple subsystems; 2) the design of a feedback control for the subsystem. Both stages can be modified in various ways.

At the first stage, we can reduce our system to subsystems more complicated than (4.2). In Sections 6 and 7, we consider subsystems with damping [7–9].

At the second stage, the game approach can be replaced by other methods of control theory. The simplest option is to neglect the disturbances altogether.

We shall study this possibility in Sections 9 - 11.

6 System with a Nonlinear Damping

The following nonlinear system is a generalization of (4.2)

$$\begin{aligned} &\ddot{x} + f(\dot{x}) = u + v, \qquad |\, u \,|\leq 1, \ \ |\, v \,|\leq \rho < 1 \\ &x(0) = \xi, \ \dot{x}(0) = \eta, \qquad x(\tau) = \dot{x}(\tau) = 0 \end{aligned} \tag{6.1}$$

Here, we use the same notation as in (4.2). The term $f(\dot{x})$ describes the damping force. The smooth function $f(z)$ has the following properties

$$zf(z) > 0, \ f'(z) > 0 \quad \text{if} \quad z \neq 0, \quad f(0) = 0 \tag{6.2}$$

Consider the differential game described by equations (6.1), where u seeks to minimize, whereas v tries to maximize τ[8]. As in Section 4, we can reduce this game to the optimal control problem for the system

$$\begin{aligned} &\dot{x}_1 = x_2,\ \dot{x}_2 = -f(x_2) + (1-\rho)u, \quad |u| \le 1 \\ &x_1(0) = \xi,\ x_2(o) = \eta, \quad x_1(\tau) = x_2(\tau) = 0 \\ &0 \le \rho < 1, \quad \tau \to \min(\ x_1 = x,\ x_2 = \dot{x}) \end{aligned} \tag{6.3}$$

similar to (4.3). We shall solve the time-optimal control problem (6.3) using Pontryagin's maximum principle [4]. The Hamiltonian H for the autonomous system (6.3) is constant along any optimal trajectory

$$H = p_1 x_2 + p_2\left[(1-\rho)u - f(x_2)\right] = h \ge 0 \tag{6.4}$$

Here, h is a constant. The adjoint variables p_1 and p_2 satisfy the adjoint system

$$\dot{p}_1 = 0, \quad \dot{p}_2 = -p_1 + f'(x_2)p_2 \tag{6.5}$$

It follows from the maximum principle and (6.4) that

$$u = \text{ sign } p_2 \tag{6.6}$$

Suppose an optimal trajectory contains a singular arc where $p_2 = 0$. Then the second equation (6.5) implies that $p_1 = 0$ on this arc. Since $p_1 =$ const according to (6.5), we have $p_1 = 0$ on the whole trajectory. Hence, the second equation (6.5) becomes homogeneous, and since $p_2 = 0$ on the singular arc, p_2 as well as p_1 is identically zero. But according to the maximum principle, the adjoint vector must not be zero on the whole trajectory. The obtained contradiction proves that the optimal trajectories do not contain singular arcs, and p_2 can be equal to zero only at discrete instants of time. Hence, we have $u = \pm 1$ almost everywhere.

Consider first those arcs of optimal trajectories where $p_2 > 0$, $u = 1$. For such arcs we obtain from (6.3)

$$dx_1/dx_2 = x_2[1 - \rho - f(x_2)]^{-1}$$

Therefore, these arcs can be described by the equation

$$x_1 = \varphi_\rho^+(x_2) + c^+ \quad (p_2 > 0) \tag{6.7}$$

Here, c^+ is a constant, and the function φ_ρ^+ is defined by

$$\varphi_\rho^+(y) = \int_0^y \frac{x\,dx}{1 - \rho - f(x)}, \quad 0 \le \rho < 1 \tag{6.8}$$

Let us indicate some properties of the function $\varphi_\rho^+(y)$ that follow from (6.8) and (6.2). As y grows from $-\infty$ to 0, the function $\varphi_\rho^+(y)$ decreases and reaches its zero minimum at $y = 0$.

If the following equation for z^+

$$f(z^+) = 1 - \rho \tag{6.9}$$

has no roots, i.e. $f(z) < 1 - \rho$ for all z, then the function $\varphi_\rho^+(y)$ grows monotonously for all $y \geq 0$. In this case $\varphi_\rho^+(y) > 0$ if $y \neq 0$.

If equation (6.9) has a root, then, due to (6.2), this root z^+ is positive and unique. In this case, the function $\varphi_\rho^+(y)$ increases from *0* to ∞, if $y \in (0, z^+)$, and decreases, if $y > z^+$. A typical curve $x_1 = \varphi_\rho^+(x_2)$ is shown in Fig. 1 for the case where the root $z^+ > 0$ of equation (6.9) exists. The arrows in Fig. 1 show the direction of the time growth.

Arcs of optimal trajectories where $p_2 < 0$ are described by the equation

$$x_1 = \varphi_\rho^-(x_2) + c^- \quad (p_2 < 0) \tag{6.10}$$

where c^- is a constant, and φ_ρ^- is given by

$$\varphi_\rho^-(y) = \int_0^y \frac{x dx}{-1 + \rho - f(x)}, \quad 0 \leq \rho < 1 \tag{6.11}$$

Let us introduce the equation

$$f(z^-) = -(1 - \rho) \tag{6.12}$$

Equations (6.10) - (6.12) are similar to (6.7) - (6.9). The following properties of the function $\varphi_\rho^-(y)$ are also similar to those of $\varphi_\rho^+(y)$.

If equation (6.12) has no solution for z^-, i.e. $f(z) > \rho - 1$ for all z, then $\varphi_\rho^-(y)$ increases for $y < 0$ and decreases for $y > 0$. In this case, $\varphi_\rho^-(y) < 0$ for all $y \neq 0$.

If a root of equation (6.12) exists, then this root z^- is negative and unique. In this case, the function $\varphi_\rho^-(y)$ decreases for $y \in (-\infty, z^-)$, increases for $y \in (z^-, 0)$, and decreases for $y \in (0, \infty)$. The function $\varphi_\rho^-(y)$ tends to $-\infty$, as $y \to z^-$, and reaches its local zero maximum at $y = 0$. We can obtain a typical curve $x_1 = \varphi_\rho^-(x_2)$ by changing the directions of the both axes x_1 and x_2 in Fig. 1.

The curves $x_1 = \varphi_\rho^\pm(x_2)$ describe trajectories passing through the origin of coordinates in the (x_1, x_2)-plane. Other curves containing arcs of optimal trajectories are obtained by translation of the curves $x_1 = \varphi_\rho^\pm(x_2)$ along the x_1-axis, according to (6.7) and (6.10).

If equations (6.9) and (6.12) have a solution, then system (6.3) has the respective solutions

$$x_2 = z^+ \quad (p_2 > 0), \quad x_2 = z^- \quad (p_2 < 0) \tag{6.13}$$

The phase trajectories corresponding to solutions (6.13) are straight lines that are the asymptotes for the respective curves (6.7) and (6.10) (see Fig. 1).

Therefore, the optimal trajectories consist of arcs of curves (6.7) and (6.10) (with different values of c^+ and c^-) and, perhaps, of segments of lines (6.13), in the case when the corresponding equations (6.9) and (6.12) are solvable.

Let us indicate that any optimal trajectory has not more than one switching of the control, i.e., that the function $p_2(t)$ has not more than one zero. Suppose the opposite assertion is true: $p_2(t)$ is equal to zero at two instants t' and t'', and between these instants $p_2(t)$ is non-zero. We assume that

$$p_2(t) > 0, \quad t \in (t', t''), \quad p_2(t') = p_2(t'') = 0 \tag{6.14}$$

Since $p_1 = const$, it follows from (6.4) that

$$p_1 x_2(t') = p_1 x_2(t'') = h \geq 0 \tag{6.15}$$

If $p_1 = 0$, then we obtain from (6.5) a linear homogeneous equation for $p_2(t)$. Since its solution vanishes at two instants t' and t'' (see (6.14)), we have $p_2(t) \equiv 0$. Hence, we come to a contradiction with the maximum principle (the adjoint vector should be non-zero), and, therefore, $p_1 = const \neq 0$. Then we obtain from (6.15) that $x_2(t') = x_2(t'')$. But the analysis of phase trajectories given above shows that $x_2(t)$ is a strictly monotone function of t, with the only exception of the trajectories described by (6.13). Consequently, the equality $x_2(t') = x_2(t'')$ is possible only if our segment of the optimal trajectory for $t \in (t', t'')$ belongs to the line (6.13) with $p_2 > 0$, i.e.

$$x_2(t) = z^+, \quad t \in (t', t'') \tag{6.16}$$

Substituting (6.16) into (6.5), we obtain

$$\dot{p}_2 = -p_1 + cp_2, \; p_1 = \text{ const}, \; c = f'(z^+) > 0 \tag{6.17}$$

Here, the constant c is positive since $z^+ > 0$(see (6.2)). Integrating (6.17), we get

$$p_2(t) = p_1/c + Ce^{ct}, \quad C = \text{ const} \tag{6.18}$$

The obtained solution (6.18) is monotone and does not satisfy conditions (6.14). Thus, we have proved that no optimal trajectory includes the segment defined by (6.14). In a similar way, we show that no optimal trajectory includes a segment with $p_2(t) < 0$ and $p_2(t') = p_2(t'') = 0$. Hence, $p_2(t)$ has not more than one zero for any optimal trajectory, and the optimal control (6.6) has not more than one switching.

The only phase trajectories that reach the origin of coordinates as t grows, are the branch of (6.7) with $c^+ = 0$ lying in the quadrant $x_1 \geq 0,\ x_2 \leq 0$ (see Fig.1) and the branch of (6.10) with $c^- = 0$ lying in the quadrant $x_1 \leq 0,\ x_2 \geq 0$. These branches correspond to controls $u = 1$ and $u = -1$, respectively, and form the switching curve defined by

$$\begin{aligned} x_1 &= \varphi_\rho(x_2), \\ \varphi_\rho(y) &= -\int_0^y \frac{|x|}{1-\rho+|f(x)|}dx = \begin{cases} \varphi_\rho^+(y) & \text{if } y \leq 0 \\ \varphi_\rho^-(y) & \text{if } y \geq 0 \end{cases} \end{aligned} \tag{6.19}$$

formulae (6.19) follow from (6.8), (6.11), and (6.2). Due to the above-mentioned properties of functions φ_ρ^+ and φ_ρ^-, the function $\varphi_\rho(y)$ decreases for all y, and $\varphi_\rho(0) = 0$.

Now we can describe the whole field of optimal phase trajectories. An optimal trajectory that starts at any point (x_1, x_2), consists of an arc of a curve belonging to one of the two families (6.7) or (6.10) (with some non-zero values of c^+ or c^-), and of an arc of the switching curve (6.19). The field of optimal trajectories is shown in Fig. 2 for the case where the roots of both equations (6.9) and (6.12) exist. The switching curve (6.19) is indicated by a thick line, the arrows show the direction of the growth of time.

The feedback optimal control corresponding to this field of trajectories is given by

$$\begin{aligned} u_\rho(x_1, x_2) &= \quad \text{sign}[\varphi_\rho(x_2) - x_1] \quad && \text{if } x_1 \neq \varphi_\rho(x_2) \\ u_\rho(x_1, x_2) &= \text{sign } x_1 = -\text{sign } x_2 && \text{if } x_1 = \varphi_\rho(x_2) \end{aligned} \tag{6.20}$$

Here, φ_ρ is defined by (6.19). The obtained feedback control (6.20) gives the desired solution to the optimal control problem (6.3) and also to the differential game (6.1). If the disturbance v is optimal ($v = -\rho u$), then the motion under the control u from (6.20) occurs along optimal trajectories (see Fig. 2). If $v \neq -\rho u$, then trajectories differ from the optimal ones, and the time of motion becomes smaller than the optimal time. The last part of any trajectory coincides with an arc of the switching curve (6.19); we have here the same sliding regime as described in Section 4.

7 Systems with a Linear Damping

Let us now consider an important particular case where equation (6.1) is linear

$$\ddot{x} + \lambda \dot{x} = u + v, \quad \lambda = \text{ const} > 0 \tag{7.1}$$

All the other conditions and notations are the same as in (6.1). Here we have $f(z) = \lambda z$, and the conditions (6.2) are fulfilled. Calculating the integral in (6.19), we obtain

$$\varphi_\rho(y) = -\lambda^{-1} y + \lambda^{-2}(1-\rho)\ln[1 + \lambda(1-\rho)^{-1} \mid y \mid] \,\text{sign} y \tag{7.2}$$

The feedback control is still given by (6.20). The case of the linear damping was considered in [7] and [9], where the following expression for the optimal time was also derived

$$\begin{aligned} &\tau(\xi,\eta) = 2\lambda^{-1}\ln\{M^{1/2} + [M - 1 + \lambda\eta\gamma(1-\rho)^{-1}]^{1/2}\} \\ &M = \exp[-(\lambda\eta + \lambda^2\xi)\gamma(1-\rho)^{-1}], \quad \gamma = \text{ sign}[\varphi_\rho(\eta) - \xi] \end{aligned} \tag{7.3}$$

As $\lambda \to 0$, we obtain from (7.2)

$$\varphi_\rho(y) = -[2(1-\rho)]^{-1} y \mid y \mid \quad (\lambda = 0) \tag{7.4}$$

Note that formulae (6.20) and (7.4) define the same control law as (4.4) and (4.5), whereas the expression (7.3) for τ reduces to (4.6) as $\lambda \to 0$.

Now we shall describe how the original Lagrangian system (2.3) or (3.1) can be reduced to the set of subsystems (7.1). Suppose the vector R from (3.1) satisfies the following relationships

$$\begin{aligned} &R_k = -\lambda_k (A_1 \dot{q})_k \; + \; R_k', \quad \lambda_k \geq 0 \\ &\mid R_k' \mid \leq R_k^0 < Q_k^0, \qquad R_k^0/Q_k^0 = \rho_k, \; k = 1,...,n \end{aligned} \tag{7.5}$$

for all $t \geq t_0$, all $(q, \dot{q}) \in W$, and all Q satisfying (2.5). Here, λ_k and R_k^0 are constants. The assumption (7.5) is a generalization of (3.2) and implies that linear damping forces in the system are not small and can be separated from other forces and nonlinear terms. Under the conditions (7.5), the transposition of variables (3.13) reduces the system (3.1) to the following equations

$$\ddot{y}_k + \lambda_k \dot{y}_k = Q_k + R_k', \quad k = 1,...,n \tag{7.6}$$

The boundary conditions (2.6) and (2.7) take the form (3.15). As in Section 3, we can regard R_k^0 as independent uncertain disturbances subjected to constraints (7.5). In a procedure similar to the one described in Sections 3-5, we reduce our system to the set of subsystems (7.1) which here replace the subsystems (4.2). Thus, we come to the following theorems that are generalizations of Theorems 1 and 2. As before, we assume that all the motions under consideration lie within the domain W.

Theorem 3. *Under conditions (7.5), Problem 1 is reduced to the specification of feedback controls $Q_k(y_k, \dot{y}_k)$ that satisfy (2.5) and bring the corresponding subsystems (7.6) from any initial state (3.15) to the zero terminal state in finite time for any admissible disturbances R'_k that satisfy (7.5).*

Theorem 4. *Under conditions (7.5), the feedback control (5.1) gives a solution to Problem 1. Here, $u(x, \dot{x})$ is given by formulae (6.20) and (7.2) where $x_1 = x$, $x_2 = \dot{x}$, whereas $\rho = \rho_k$ and $\lambda = \lambda_k$ are defined in (7.5) for each k. Under this control, the system (2.3) reaches the terminal state (2.7) at time t_1 bounded by the inequality (5.2) where the function $\tau(\xi, \eta)$ is defined in (7.3).*

We emphasize that, under appropriate assumptions, these results could well be extended to sets of nonlinear equations (6.1). Here we have restricted ourselves by the linear case because it can be applied to robotic manipulators with electromechanical drives.

8 Applications to Robots

We considered a manipulation robot which has n degrees of freedom and consists of n rigid links connected consecutively by revolute or prismatic joints. Thus, the manipulator forms an open kinematic chain. Angles of relative rotation of links and their relative linear displacements for revolute and prismatic joints, respectively, are denoted through q_k, $k = 1, ..., n$. Equations of motion for the manipulator can be presented in the form (2.1) with the kinetic energy T given by (2.2). The generalized forces here are the torques for revolute joints and the forces for prismatic joints. The terms F_k include all external and internal forces except the controls, namely, the weight, the resistance, the friction, etc. The terms Q_k in (2.1) are the control torques and the forces created by actuators in revolute and prismatic joints, respectively. We assume that each joint is driven by an independent motor, so that the constraints (2.5) are imposed. Thus, the dynamics of the robot is described by equations (2.3) and the constraints (2.5) with the initial and terminal conditions (2.6) and (2.7). Let us consider two versions of Problem 1 for robots.

1) Suppose the control torques and the forces Q_k, $k = 1, ..., n$, can vary arbitrarily within the bounds (2.5), and let the condition (3.2) be

true. This condition can be verified by means of the Lemma of Section 3. It implies that the nonlinear terms, the additional forces, and the perturbations are relatively small compared with the control forces. Then the feedback control can be obtained by means of Theorem 2.

2) A more detailed analysis is required, if the dynamics of the actuators should be taken into account. Let the control torques and the forces Q_k be created by an independent electric DC (direct current) actuators placed at the robot joints. The actuators used in industrial robots usually have reduction gears with high gear ratios N_k. The angular velocity ω_k of the rotor for the kth actuator is given by

$$\omega_k = N_k \dot{q}_k, \quad N_k \gg 1 \tag{8.1}$$

for the revolute joint. The equation of rotation of the rotor for the kth actuator (for $N_k \gg 1$) can be written as follows

$$J_k \dot{\omega}_k = -b_k \omega_k + M_k - N_k^{-1} Q_k, \quad k = 1, \dots, n \tag{8.2}$$

Here, J_k is the moment of inertia of the rotating parts of the kth actuator with respect to the axis of the kth joint, $b_k \omega_k$ is the mechanical resistance, $b_k > 0$ is a constant coefficient, and M_k is the electromagnetic torque created by the actuator. The torque M_k is proportional to the electric current i_k in the kth actuator

$$M_k = c_k i_k, \quad k = 1, \dots, n \tag{8.3}$$

where c_k is a positive constant. The equation of balance of voltages in the circuit of the kth actuator has the form

$$L_k di_k \, /dt + R_k i_k + d_k \omega_k = u_k \tag{8.4}$$

Here, L_k is the inductance, R_k is the electrical resistance, $d_k > 0$ is a constant coefficient, and u_k is the electrical voltage in the circuit of the kth actuator. The first term in (8.4) is usually small in comparison with the other terms and, hence, can be omitted. Then. we obtain from (8.3) and (8.4)

$$M_k = c_k (u_k - d_k \omega_k) R_k^{-1} \tag{8.5}$$

Substituting M_k from (8.5) and ω_k from (8.1) into equation (8.4), we obtain

$$J_k N_k \ddot{q}_k + b_k N_k \dot{q}_k = c_k R_k^{-1} (u_k - d_k N_k \dot{q}_k) - N_k^{-1} Q_k$$

From this equation we determine Q_k:

$$Q_k = -N_k^2 J_k \ddot{q}_k - N_k^2 (b_k + c_k d_k R_k^{-1}) \dot{q}_k + N_k c_k R_k^{-1} u_k, \quad k = 1, ..., n$$

We substitute this expression into Lagrange's equation (2.3). As a result, we obtain the following equations of motion

$$
\begin{aligned}
&[A(q)\ddot{q}]_k + N_k^2 J_k \ddot{q}_k + N_k^2 (b_k + \; c_k d_k R_k^{-1})\dot{q}_k \\
&= F_k + S_k + N_k c_k R_k^{-1} u_k, \qquad k = 1, ..., n
\end{aligned}
\tag{8.6}
$$

Let us introduce a new variable p defined as follows

$$
\begin{aligned}
&p = (p_1, ..., p_n), \quad p_k = N_k q_k, \quad k = 1, ..., n \\
&q = Gp, \quad G = \text{ diag}(N_k^{-1})
\end{aligned}
\tag{8.7}
$$

Here and below, $\text{diag}(a_k)$ denotes the diagonal $n \times n$-matrix with diagonal elements equal to $a_k, k = 1, ..., n$. After the change of variables (8.7), equations (8.6) can be written as follows

$$
\begin{aligned}
&J\ddot{p} + GA(Gp)G\ddot{p} + \Lambda\dot{p} = G(F + S) + Ku \\
&J = \text{ diag}(J_k), \quad \Lambda = \text{ diag}(B_k + c_k d_k R_k^{-1}) \\
&K = \text{ diag}(c_k R_k^{-1}), \quad u = (u_1, ..., u_n)
\end{aligned}
\tag{8.8}
$$

The system (8.8) can be presented in the form (2.3)

$$
\begin{aligned}
&A^*(p)\ddot{p} = Q^* + F^* + S^*, \quad A^* = J + GA(Gp)G \\
&Q^* = Ku, \quad F^* = GF - \Lambda\dot{p}, \quad S^* = GS
\end{aligned}
\tag{8.9}
$$

Here, A^* is a symmetric positive definite matrix.

The voltages u_k of the actuators are usually restricted by the constraints

$$
| u_k | \leq u_k^0, \quad k = 1, ..., n \tag{8.10}
$$

where u_k^0 are constants. Due to (8.8) and (8.9), the constraints (8.10) are transformed into constraints imposed on the components Q_k^* of vector Q^*

$$
| Q_k^* | \leq Q_k^0 = c_k R_k^{-1} u_k, \quad k = 1, ..., n \tag{8.11}
$$

The constraints (8.11) have the same form as in (2.5). Hence, we can consider Problem 1 for the system (8.9) under constraints (8.11). The initial and the boundary conditions for the vector p from (8.7) are similar to (2.6) and (2.7).

To apply the results of Section 7, we should choose the matrix A_1 and verify the conditions (7.5). We set

$$
A_1 = J = \text{ diag}(J_k) \tag{8.12}
$$

and calculate the vector R defined in (3.1) using equations (8.9) and (8.12)

$$\begin{aligned} R = & \ A_1(A^*)^{-1}(Q^* + F^* + S^*) - Q^* \\ = & \ J(J + GAG)^{-1}(Ku + GF + GS - \Lambda\dot{p}) - Ku \end{aligned} \tag{8.13}$$

Representing the first term in (8.13) as $J = (J+GAG)-GAG$, we obtain from (8.13)

$$\begin{aligned} R = -\Lambda\dot{p} + G[F \ & + \ S - AG(J + GAG)^{-1}(Ku + \\ & + \ GF + GS - \Lambda\dot{p})] = -\Lambda\dot{p} + R' \end{aligned} \tag{8.14}$$

Comparing (7.5) and (8.14), we see that the first equation in (7.5) is satisfied, if we take into account the formulae (8.8) for the diagonal matrix Λ and (8.12) for A_1, and set

$$\lambda_k = \Lambda_k J_k^{-1} = (b_k + c_k d_k R_k^{-1}) J_k^{-1}, \quad k = 1, ..., n \tag{8.15}$$

Since $N_k \gg 1$, (see (8.1)), all the elements of the diagonal matrix G defined by (8.7) are small. If follows from (8.14) that $R' \to 0$ as $G \to 0$. Hence, the inequalities (7.5) are satisfied, if $N_k \gg 1$, $k = 1, \ldots, n$. Thus, we have proved that our conditions (7.5) are satisfied for robots driven by electric DC actuators with high gear ratios. Now Theorems 3 and 4 can be applied, and the feedback control voltages can be obtained in the explicit form, as indicated by Theorem 4.

9 Simplified Control: an Auxiliary Problem

As mentioned at the end of Section 5, the game approach to our subsystems can be replaced by other techniques. In Sections 9 - 11, we consider control laws which we call simplified and which are based on the assumption that the unknown disturbances in the subsystems are absent. This approach seems quite natural, if the disturbances are small, and is often used in practical applications. We neglect disturbances only while the control law is designed. After that, we study the dynamics of our subsystems in the presence of disturbances.

In Section 9, we shall consider an auxiliary optimal control problem for the "worst" disturbances. Then we shall analyze the behaviour of our subsystem subjected to the simplified control in the presence of disturbances (Section 10). The discussion of the obtained results is given in Section 11.

Let us again consider the nonlinear system (6.1) and put $v = 0$. Then (6.1) is reduced to the system (6.3) with $\rho = 0$. The optimal control for this system is given by equations (6.20), where we should set $\rho = 0$. Thus, the desired simplified control $u_0(x_1, x_2)$ is defined by equations (6.20) and (6.19) with $\rho = 0$. The corresponding switching curve $x_1 = \varphi_0(x_2)$ for the simplified control is shown in Fig. 3 by a thick line.

Let us now substitute the simplified control $u_0(x_1, x_2)$ into the system (6.1). We obtain

$$\begin{aligned} &\dot{x}_1 = x_2, \qquad \dot{x}_2 = -f(x_2) + u_0(x_1, x_2) + v \\ &| v | \leq \rho < 1, \quad (x_1 = x, \quad x_2 = \dot{x}) \end{aligned} \tag{9.1}$$

To analyze the possible motions of the system (9.1) under bounded disturbances, we consider the following auxiliary optimal control problem.

Problem 2. Find the optimal control $v(x_1, x_2)$ that satisfies the constraint $| v | \leq \rho$ and such that every trajectory of the system (9.1) intersects with the switching curve $x_1 = \varphi_0(x_2)$ as far as possible from the point $x_1 = x_2 = 0$, i.e., at the maximal possible $| x_1 |$ or $| x_2 |$.

Remark. Since the function $x_1 = \varphi_0(x_2)$ is monotone, the maximizations of $| x_1 |$ and $| x_2 |$ are equivalent to each other.

First suppose that the initial point (ξ, η) of our trajectory lies in the domain $x_1 > \varphi_0(x_2)$. Then, according to (6.20), we have $u_0 = -1$ for the whole trajectory. In this case, the trajectory first intersects the part of the switching curve $x_1 = \varphi_0(x_2)$ where $x_1 > 0, x_2 < 0$(see Fig. 3). Hence, Problem 2 can be rewritten as follows

$$\begin{aligned} &\dot{x}_1 = x_2, \quad \dot{x}_2 = -f(x_2) - 1 + v, \quad | v | \leq \rho < 1 \\ &x_1(0) = \xi, \quad x_2(0) = \eta, \quad \xi > \varphi_0(\eta) \\ &x_1(\tau) = \varphi_0(x_2(\tau)), \quad x_1(\tau) > 0, \quad x_2(\tau) < 0, \quad x_1(\tau) \to \max \end{aligned} \tag{9.2}$$

The terminal instant of time τ is not fixed. Due to (9.2), the performance index $x_1(\tau)$ of our problem can be rewritten as follows

$$J = \int_0^\tau (-x_2) dt \to \min \tag{9.3}$$

The Hamiltonian of the optimal control problem defined by (9.2) and (9.3) given by

$$H = p_1 x_2 + p_2 [v - 1 - f(x_2)] + x_2 \tag{9.4}$$

The adjoint variables p_1 and p_2 satisfy the adjoint system

$$\dot{p}_1 = 0, \quad \dot{p}_2 = f'(x_2) p_2 - p_1 - 1 \tag{9.5}$$

and the following transversality conditions corresponding to the boundary conditions given in (9.2)

$$p_1\varphi_0'(x_2) + p_2 = 0, \quad H = 0 \quad (t = \tau) \tag{9.6}$$

From (6.19) under the conditions $\rho = 0$, (6.2), and $x_2(\tau) < 0$ (see (9.2)), we obtain

$$\varphi_0'(x_2) = x_2 \left[1 - f(x_2)\right]^{-1}$$

Inserting this formula into the first condition (9.6), we get

$$p_1 = -p_2 \left[1 - f(x_2)\right] x_2^{-1} \qquad (t = \tau) \tag{9.7}$$

We substitute (9.7) into (9.4) and use the second condition (9.6). We have

$$H = p_2(v - 2) + x_2 = 0 \qquad (t = \tau)$$

Since $x_2(\tau) < 0$ and $|\, v \,| \leq \rho < 1$, we obtain from the previous equation

$$p_2(\tau) < 0 \tag{9.8}$$

Maximizing the Hamiltonian H from (9.4) with respect to v with $|\, v \,| \leq \rho$, we find the optimal control

$$v = \rho \ \text{sign}\ p_2 \tag{9.9}$$

Let us show that the optimal trajectories have no singular arcs. If $p_2 = 0$ on some time interval, then, due to the second equation (9.5), we have $p_1 = -1$. Since $p_1 = \text{const}$, then $p_1 \equiv -1$ on the whole trajectory. Then the second equation (9.5) becomes linear and homogeneous, and its solution $p_2(t)$ under the non-zero initial condition (9.8) cannot vanish at any instant of time. This contradiction proves that singular arcs do not exist.

The optimal control (9.9) switches at the instants when $p_2(t) = 0$. Let us find the switching curve in the (x_1, x_2) - plane. Since our system is autonomous, its Hamiltonian (9.4) is constant along the optimal trajectory and equal to zero, in accord with (9.6). We have

$$H = (p_1 + 1)x_2 + p_2 \left[v - 1 - f(x_2)\right] \equiv 0$$

Therefore, at the instant of switching (when $p_2 = 0$) we have either $p_1 = -1$ or $x_2 = 0$. But, as shown before, the equality $p_1 = -1$ yields that $p_2 \neq 0$ everywhere, and the switching does not exist. This contradiction proves that we have $x_2 = 0$ at the instant of switching. Consequently, the switching curve is a ray $x_2 = 0, \ x_1 > 0$.

To obtain the signs of the control for $x_2 < 0$ and $x_2 > 0$, it is sufficient to find this sign at any point. At the terminal instant τ, we have $x_2(\tau) < 0$ and $p_2(\tau) < 0$. according to (9.2) and (9.8).

Therefore, due to (9.9), we have $v = -\rho$ for $x_2 < 0$. Hence,

$$v(x_1, x_2) = \rho \text{ sign } x_2 \tag{9.10}$$

The optimal feedback control (9.10) is obtained in the domain $x_1 > \varphi_0(x_2)$. To find the control for $x_1 < \varphi_0(x_2)$, we observe the following properties of symmetry.

Let us change the variables in (9.1)

$$x_1 \to -x_1, \quad x_2 \to -x_2, \quad v \to -v, \quad f(z) \to -f(-z)$$

Then, according to (6.19) and (6.20), the function $\varphi_\rho(y)$ will be replaced by $-\varphi_\rho(-y)$, u_0 is replaced by $-u_0$, and the system (9.1) will not change. Therefore, the feedback control and the field of optimal trajectories in the domain $x_1 < \varphi_0(x_2)$ is the same as for $x_1 > \varphi_0(x_2)$, but with $f(z)$ replaced by $-f(-z)$. However, the control (9.10) is independent of the specific function $f(z)$. Hence, the formula (9.10) for the optimal feedback control is true in the whole (x_1, x_2) - plane. Problem 2 is solved. The optimal trajectory corresponding to the control $u_0(x_1, x_2)$ and the disturbance v from (9.10) is shown in Fig. 3 by a thin line.

10 Simplified Control: Dynamics of the Subsystem

Let us now study the motions of the system (9.1) subjected to the control $u_0(x_1, x_2)$ and the "worst" disturbance v from (9.10). Suppose the initial point (ξ, η) lies on the branch of the switching curve $x_1 = \varphi_0(x_2)$ where $x_1 < 0, x_2 > 0$.

Consider the part of the trajectory from the initial point till the intersection with the same branch of the switching curve (with $x_1 < 0$, $x_2 > 0$). This part (shown in Fig. 4 by a thin line) consists of four arcs with the following boundary points and constant controls (the switching curve $x_1 = \varphi_0(x_2)$ is shown in Fig. 4 by a thick line)

$$\begin{array}{lll}
1)(\xi, \eta) \to (x_1^0, 0), & u_0 = -1, & v = \rho \\
2)(x_1^0, 0) \to (\xi', \eta'), & u_0 = -1, & v = -\rho \\
3)(\xi', \eta') \to (x_1^*, 0), & u_0 = 1, & v = -\rho \\
4)(x_1^*, 0) \to (\xi^*, \eta^*), & u_0 = 1, & v = \rho
\end{array} \tag{10.1}$$

The coordinates of the boundary points (10.1) satisfy the following relations reflecting their positions on the switching curve and on the axes of coordinates (see Fig. 4)

$$\begin{array}{llll} \xi = \varphi_0(\eta), & \eta > 0, & \xi < 0, & x_1^0 > 0 \\ \xi' = \varphi_0(\eta'), & \eta' < 0, & \xi' > 0, & x_1^* < 0 \\ \xi^* = \varphi_0(\eta^*), & \eta^* > 0, & \xi^* < 0 & \end{array} \tag{10.2}$$

Substituting u_0 and v from (10.1) into equations (9.1) and integrating them along the four arcs of the trajectory with constant u_0 and v, we obtain

$$\xi' - \xi = \int_{\eta}^{0} \frac{zdz}{-1+\rho-f(z)} + \int_{0}^{\eta'} \frac{zdz}{-1-\rho-f(z)}$$

$$\xi^* - \xi' = \int_{\eta'}^{0} \frac{zdz}{1-\rho-f(z)} + \int_{0}^{\eta^*} \frac{zdz}{1+\rho-f(z)}$$

In these relations, we replace ξ, ξ', and ξ^* by the corresponding expressions following from (10.2) and (6.19) with $\rho = 0$. We have

$$\int_{0}^{\eta'} \frac{zdz}{1-f(z)} - \int_{0}^{\eta} \frac{(-z)dz}{1+f(z)} = \int_{0}^{\eta} \frac{zdz}{1-\rho+f(z)} - \int_{0}^{\eta'} \frac{zdz}{1+\rho+f(z)}$$

$$\tag{10.3}$$

$$\int_{0}^{\eta^*} \frac{(-z)dz}{1+f(z)} - \int_{0}^{\eta'} \frac{zdz}{1-f(z)} = \int_{0}^{\eta'} \frac{(-z)dz}{1-\rho-f(z)} + \int_{0}^{\eta^*} \frac{zdz}{1+\rho-f(z)}$$

Here, $\eta' < 0, \eta > 0$, and $\eta^* > 0$, according to (10.2). Let us set $\eta' = -\eta^0, \eta^0 > 0$, and transform all the integrals in (10.3) into integrals along the segments of the positive semi-axis. After some simplifications, we obtain from (10.3)

$$\Phi_4(\eta^0) = \sigma^2(\rho)\Phi_1(\eta), \quad \Phi_2(\eta^*) = \sigma^2(\rho)\Phi_3(\eta^0) \tag{10.4}$$

Here, the following notation is introduced

$$\Phi_1(y) = \Phi^+(y, f), \quad \Phi_2(y) = \Phi^-(y, f)$$

$$\Phi_3(y) = \Phi^+(y, g), \quad \Phi_4(y) = \Phi^-(y, g)$$

$$\Phi^{\pm}(y, h) = \int_{0}^{y} \frac{zdz}{(1+h)[1 \pm (1 \mp \rho)^{-1} h]} \tag{10.5}$$

$$f = f(z) \geq 0, \quad g = -f(-z) \geq 0$$

$$\sigma(\rho) = [\rho(1 + \rho)]^{1/2}[(1 - \rho)(2 + \rho)]^{-1/2}$$

Consider the transcendental equations (10.4) that define η^0 and η^* for given $\eta > 0$ and $\rho \in (0,1)$. First, we establish some properties of the functions Φ_i, $i = 1,2,3,4$, from (10.5). We shall consider these functions only for such arguments y for which the integrands in (10.5) are positive. Hence, all the derivatives $d\Phi_i/dy$ are positive, and the functions $\Phi_i(y)$ are monotone. Let us recall that, due to (6.2), we have $f(z) > 0$ for $z > 0$ and $f(z) \to 0$ as $z \to 0$.

The denominators of the integrands for the functions Φ_1 and Φ_3 in (10.5) are positive for all $z \geq 0$. Hence, these functions are bounded for all finite $y \geq 0$.

If the equations

$$f(z_2) = 1 + \rho, \quad g(z_4) = -f(-z_4) = 1 + \rho \tag{10.6}$$

have roots z_2 and z_4, then the denominators of the integrands for the corresponding functions Φ_2 and Φ_4 in (10.5) vanish at the respective roots z_2 and z_4. In this case, the functions $\Phi_2(y)$ and $\Phi_4(y)$ increase and tend to infinity as $y \to z_2$ and $y \to z_4$, respectively. If equations (10.6) have no solutions, then the functions Φ_2 and Φ_4 are bounded for all finite $y \geq 0$. In both cases, the denominators of the integrands for the functions Φ_2 and Φ_4 have maximums with respect to $f \geq 0$ and $g \geq 0$, respectively, and both of these maximums are equal to $(2+\rho)^2(1+\rho)^{-1}/4$. Therefore, the following inequalities are true

$$\Phi_2(y) \geq \nu y^2/2, \quad \Phi_4(y) \geq \nu y^2/2, \quad \nu = 4(1+\rho)(2+\rho)^{-2}$$

Thus, in all cases the functions $\Phi_2(y)$ and $\Phi_4(y)$ are positive and increase monotonously, their values varying from *0* to ∞. Consequently, the equations (10.4) have unique positive solutions $\eta^0 > 0$ and $\eta^* > 0$ for all $\eta > 0$ and $\rho \in (0,1)$. These solutions are continuous monotone functions of η.

Let us differentiate equations (10.4) with respect to η. After simple transformations, we obtain

$$\frac{d\eta^*}{d\eta} = \frac{\sigma^2(\rho)\Phi_3'(\eta^0)}{\Phi_2'(\eta^*)}\frac{d\eta^0}{d\eta} = \frac{\sigma^4(\rho)\Phi_3'(\eta^0)\Phi_1'(\eta)}{\Phi_2'(\eta^*)\Phi_4'(\eta^0)} \tag{10.7}$$

Equations (10.5) and properties (6.2) imply the following inequalities

$$\frac{\Phi_1'(y)}{\Phi_2'(y)} < 1, \quad \frac{\Phi_3'(y)}{\Phi_4'(y)} < 1, \quad y > 0$$

Using the second of these inequalities, we obtain from (10.7)

$$\frac{d\eta^*}{d\eta} < \sigma^4(\rho)\frac{\Phi_1'(\eta)}{\Phi'(\eta^*)}, \quad \eta > 0 \tag{10.8}$$

It can be readily verified that the function $\sigma^2(\rho)$ from (10.5) increases monotonously from *0* to ∞ for $\rho \in [0,1]$, and $\sigma = 1$ for the value of ρ equal to the well-known "golden section" ratio

$$\rho^* = (5^{1/2} - 1)/2 = 0.618... \tag{10.9}$$

Let us first assume that $\rho < \rho^*$ and, therefore, $\sigma^2(\rho) < \alpha$, where $\alpha < 1$ is a positive number. Then it follows from (10.8)

$$d\eta^*/d\eta < \alpha^2\Phi_1'(\eta)/\ \Phi_2'(\eta^*), \quad \eta > 0 \tag{10.10}$$

Integrating (10.10), we obtain

$$\Phi_2(\eta^*) < \alpha^2\Phi_1(\eta), \quad \eta > 0 \tag{10.11}$$

Let us prove that $\eta^* < \eta$. Note that (10.5) implies that $\Phi_2(y) > \Phi_1(y)$ for all $y > 0$. Then, since $\Phi_2(y)$ is monotone, we have the following inequalities with $\eta^* \geq \eta$:

$$\Phi_2(\eta^*) \geq \Phi_2(\eta) > \Phi_1(\eta)$$

These inequalities contradict (10.11). Hence, $\eta^* < \eta$.

Now we shall transform the inequality (10.10) by substituting in this relation the expressions for the derivatives Φ_1' and Φ_2' from (10.5) and taking into account that $f(z) > 0$ for $z > 0$. We obtain

$$\frac{d\eta^*}{d\eta} < \frac{\alpha^2\eta\,[1 + f(\eta^*)]\,[1 - (1+\rho)^{-1}f(\eta^*)]}{\eta^*\,[1 + f(\eta)]\,[1 + (1-\rho)^{-1}f(\eta)]} < \frac{\alpha^2\eta\,[1 + f(\eta^*)]}{\eta^*\,[1 + f(\eta)]}$$

$$(\eta > 0,\ \eta^* > 0)$$

To simplify the last inequality, we note that since $f(z)$ is monotone and $\eta^* < \eta$, we have $f(\eta^*) < f(\eta)$. We obtain

$$d\eta^*/d\eta < \alpha^2\eta/\eta^*, \quad \eta > 0$$

Integrating this inequality under the initial condition $\eta^* = 0$ for $\eta = 0$, we obtain $(\eta^*)^2 < \alpha^2\eta^2$, or $\eta^*/\eta < \alpha$.

Thus, if $\rho < \rho^*$, with ρ^* defined by (10.9), we have $\eta^*/\eta < \alpha < 1$,/ so that the phase trajectory approaches the origin. The distance from the phase point to the origin decreases not slower than with the rate of a geometrical progression. Hence, if $\rho < \rho^*$, the system reaches the desired zero state in finite time, though the number of control switches is infinite.

Suppose the system has already reached the small neighbourhood of the zero state, so that η is sufficiently small. Then η^0 and η^* are also small because they are continuous functions of η . Since, according to (6.2), $f(z) \to 0$ as $z \to 0$, we can omit the terms $f(z)$ and $g(z)$ in the integrals (10.5) for small y. Thus, we obtain

$$\Phi_i(y) \sim y^2/2, \; y \to 0, \; i = 1, 2, 3, 4$$

Consequently, for a small η our transcendental equations (10.4) become

$$(\eta^0)^2 = \sigma^2(\rho)\eta^2, \; (\eta^*)^2 = \sigma^2(\rho)(\eta^0)^2$$

From these equations it follows

$$\eta^*/\eta = \sigma^2(\rho) \tag{10.12}$$

Let $\rho > \rho^*$ and, hence, $\sigma^2(\rho) > 1$. Then (10.12) implies $\eta^* > \eta$. Therefore, if the system has arrived at a small neighbourhood of the zero terminal state, the trajectory will not approach this state. In other words, with $\rho > \rho^*$ the system will never reach the prescribed terminal state.

11 Discussion and Particular Cases

The simplified approach applied to system (6.1) provides the control $u_0(x_1, x_2)$ defined by equations (6.20) and (6.19) with $\rho = 0$. The behaviour of the system (6.1) under this control depends essentially on the parameter ρ that is equal to the ratio of the bounds imposed on the disturbances and the control.

If $\rho < \rho^* = 0.618$, then the system reaches the prescribed terminal state in finite time. This statement was proved for the "worst" disturbance and, therefore, it is true also for any admissible disturbance $| v | < \rho$. The number of control switches may be infinite.

If $\rho > \rho^*$, then there exists an admissible disturbance (given by (9.10)) such that the system never reaches the terminal state.

The result presented above for the simplified control of the nonlinear system were obtained in [8]. A more detailed analysis of a system with the linear damping (7.1) was given in [7] and [9]. The case of the linear system (4.2) without damping is especially simple; it was studied in [6] . In this case, we set $f(z) = 0$ in (10.5) and obtain

$$\Phi_i(y) = y^2/2, \; i = 1, 2, 3, 4 \tag{11.1}$$

Substituting (11.1) into (10.4), we have

$$\eta^0 = \sigma(\rho)\eta, \quad \eta^* = \sigma(\rho)\eta^0$$

Hence, we have the equality $\eta^* = \sigma^2(\rho)\eta$ that coincides with the equation (10.12) obtained for the nonlinear system in the neighbourhood of the terminal state. We arrive at the following conclusions for the system (4.2) under the simplified control.

If $\rho < \rho^*$, the system reaches the terminal state in finite time under any admissible disturbance. If $\rho = \rho^*$, then our system under the disturbance (9.10) has a periodic solution: the corresponding phase trajectory is a closed curve. If $\rho > \rho^*$, then there exists a disturbance (given by (9.10)) such that the corresponding phase trajectory never reaches the terminal state and goes to infinity as $t \to \infty$. Typical phase trajectories for the cases $\rho < \rho^*$, $\rho = \rho^*$, and $\rho > \rho^*$ are shown by thin lines in Figs. 5, 6, and 7, respectively. Here, the switching curves are shown by thick lines, and arrows indicate the direction of the time growth.

Hence, the simplified control brings the nonlinear system (6.1) (as well as the linear systems (7.1) and (4.2)) to the terminal state under any admissible disturbance only if $\rho < \rho^*$. Let us recall that the control $u_\rho(x_1, x_2)$ based on the game approach drives our system to the terminal state if $\rho < 1$. The both controls are given by similar formulae (see (6.20) and (6.19)) with the only difference: the game control uses the parameter ρ which is assumed to be known. In other words, for the game approach we should estimate the maximal possible value of the disturbance, and this value must not exceed the maximal possible value of the control. Thus, the game approach is more reliable than the simplified one and is applicable for a broader class of systems. However, with $\rho < \rho^*$, both controls may be used within the framework of our decomposition. In other words, if $\rho < \rho^*$, we can substitute $u_0(x_1, x_2)$ instead of $u_\rho(x_1, x_2)$ into (5.1)(see Theorems 2 and 4).

12 The Second Approach: Main Assumptions

Let us now consider Problem 1 from Section 2 under the following assumptions [10]. We shall consider the motions of system (2.3) that belong to some domain D in the n-dimensional q-space. In particular, D may coincide with R^n. The initial position q^0 from (2.6) and the terminal position q^1 from (2.7) belong to D. Suppose the kinetic energy T defined by (2.2) and the generalized forces F from (2.3) satisfy the following conditions.

Let all the eigenvalues of the matrix $A(q)$ from (2.2) lie within the fixed interval $[m, M]$, i.e.

$$m \mid z \mid^2 \leq (A(q)z, z) \leq M \mid z \mid^2, \; 0 < m < M \tag{12.1}$$

for all $q \in D$ and all $z \in R^n$. Here, m and M are given constants. We also assume that

$$\mid \partial A_{ij}(q)/\partial q_k \mid \leq C, \; q \in D, \quad i, j = 1, ..., n \tag{12.2}$$

where $C > 0$ is a constant.

Let the control forces Q_k be bounded by the constraints (2.5), whereas the forces F_k in (2.3) and (2.4) consist of the dissipative forces G_k and the uncertain disturbances Φ_k

$$F = G + \Phi, \; G = (G_1, ..., G_n), \; \Phi = (\Phi_1, ..., \Phi_n) \tag{12.3}$$

The forces G_k satisfy two conditions.

First, their work is non-positive

$$(G(q, \dot{q}, t), \dot{q}) = \sum_{k=1}^{n} G_k \dot{q}_k \leq 0 \tag{12.4}$$

for all $q \in D$, all $\dot{q}$, and $t \geq t_0$.

Second, there exists a number $\epsilon_0 > 0$ such that, if $\mid \dot{q}_k \mid \leq \beta \leq \epsilon_0$ for all $k = 1, \ldots, n$, then

$$\mid G_k \mid \leq G_k^0(\beta), \quad i = 1, ..., n \tag{12.5}$$

Here, $G_k^0(\beta)$ are some monotone increasing continuous functions defined for $\beta \in [0, \epsilon_0]$ and such that $G_k^0(0) = 0$.

The forces Φ_k in (12.3) are assumed to be bounded by the constraints

$$\mid \Phi_k \mid \leq \Phi_k^0 < Q_k^0, \quad k = 1, ..., n \tag{12.6}$$

for all $q \in D$, all $\dot{q}$, and $t \geq t_0$.

13 The Decomposition

Let us introduce the following two sets in the $(q, \dot{q})$- space

$$\begin{aligned} \Omega_1 = & \; \{(q, \dot{q}) : \; q \in D, \; \exists k, \; \mid \dot{q}_k \mid > \epsilon\} \\ \Omega_2 = & \; \{(q, \dot{q}) : \; q \in D, \; \forall k, \; \mid \dot{q}_k \mid \leq \epsilon\} \end{aligned} \tag{13.1}$$

Here, the number $\epsilon > 0$ will be specified below. We take $\epsilon \leq \epsilon_0$ to ensure the conditions (12.5).

We shall construct the feedback control separately in sets Ω_1 and Ω_2. Due to (12.3) and (12.4), we have

$$dT/dt = (Q + F, \dot{q}) \leq (Q + \Phi, \dot{q}) \tag{13.2}$$

Let us set

$$\begin{array}{lll} Q_k(q, \dot{q}) = & -Q_k^0 \mathrm{sign} \dot{q}_k & \text{if } q_k \neq 0 \\ Q_k(q, \dot{q}) = & 0 & \text{if } \dot{q}_k = 0, \; k = 1, ..., n \end{array} \tag{13.3}$$

in Ω_1. By substituting (13.3) into (13.2) and by taking account of (12.6), we obtain

$$dT/dt \leq -\sum_{k=1}^{n} (Q_k^0 - \Phi_k^0) \mid \dot{q}_k \mid$$

Due to the Cauchy inequality, we have

$$dT/dt \leq -r \mid \dot{q} \mid, \quad r = \left[\sum_{k=1}^{n} (Q_k^0 - \Phi_k^0)^2 \right]^{1/2} \geq 0 \tag{13.4}$$

The upper estimate (12.1) implies that

$$\mid \dot{q} \mid \geq (2T/M)^{1/2}$$

Substituting this inequality into (13.4), we obtain

$$2T^{1/2} dT^{1/2}/dt \leq -r(2T/M)^{1/2}$$

Since $T > 0$ in Ω_1, we have

$$dT^{1/2}/dt \leq -r(2M)^{-1/2}$$

Integrating this inequality, we obtain

$$T^{1/2} - T_0^{1/2} \leq -r(2M)^{-1/2}(t - t_0)$$

$$T_0 = (A(q^0)\dot{q}^0, \dot{q}^0)/2 \tag{13.5}$$

Here, T_0 is the initial value of the kinetic energy (2.2) at the state (2.6). It follows from (13.5) that the kinetic energy may turn to be zero in finite time. Hence, if the initial state (2.6) belongs to Ω_1, then the system reaches the boundary between the sets Ω_1 and Ω_2 from (13.1) at some finite instant of time t_*.

Let us obtain estimates on t_* and the coordinates $q(t_*)$. Due to (12.1) and (13.1), we have the following lower bound on the kinetic energy T_* at instant t_*

$$T_* \geq m(\dot{q}, \dot{q})/2 \geq m\epsilon^2/2 \tag{13.6}$$

The inequalities (13.5) and (13.6) imply the following estimate on t_*

$$t_* - t_0 \leq \tau_1, \; \tau_1 = (2M)^{1/2} r^{-1} \left[T_0^{1/2} - (m/2)^{1/2}\epsilon\right] \tag{13.7}$$

In order to estimate $q(t_*)$, we shall use the following obvious inequalities

$$| q_k(t_*) - q_k^0 | \leq \int_{t_0}^{t_*} | \dot{q}_k | \, dt \leq \int_{t_0}^{t_*} | \dot{q} | \, dt \tag{13.8}$$

The inequalities (12.1) and (13.5) yield the following estimates

$$| \dot{q} | \leq (2T/m)^{1/2} \leq (2/m)^{1/2} \left[T_0^{1/2} - r(2M)^{-1/2}(t - t_0)\right] \tag{13.9}$$

By substituting (13.9) into (13.8) and integrating the result, we obtain

$$| q_k(t_*) - q_k^0 | \leq \varphi(t_* - t_0)$$

$$\varphi(\tau) = (2T_0/m)^{1/2}\tau - r(Mm)^{-1/2}\tau^2/2 \tag{13.10}$$

The function $\varphi(\tau)$ introduced by (13.10) increases in the interval $[0, \tau_1]$ where τ_1 is defined in (13.7). Since $t_* - t_0 \leq \tau_1$, then according to (13.7), we have $\varphi(t_* - t_0) \leq \varphi(\tau_1)$. Hence, we obtain from (13.10) and (13.7)

$$| q_k(t_*) - q_k^0 | \leq \varphi(\tau_1) = (M/m)^{1/2} r^{-1}(T_0 - m\epsilon^2/2) \tag{13.11}$$

Our system reaches the boundary between the sets Ω_1 and Ω_2 from (13.1) at instant t_*. Now we shall design the control within the set Ω_2 in such a way so as to drive the system to the terminal state (2.7) without leaving the set Ω_2. By multiplying both sides of equation (2.3) by A^{-1} and by substituting F from (12.3) into this equation, we obtain

$$\begin{aligned} \ddot{q} &= U + V, \quad U = A^{-1}Q, \\ V &= A^{-1}(G + \Phi + S) \end{aligned} \tag{13.12}$$

Here, U and V can be regarded as a control and a disturbance, respectively. To decouple the equations in the system (13.12), we impose the independent constraints on the components of the vectors U and V. We assume that

$$| U_k | \leq U^0, \; U^0 = r_0 M^{-1} n^{-1/2},$$

$$r_0 = \min_k Q_k^0, \; 1 \leq k \leq n \tag{13.13}$$

Due to (13.12), (12.1), and (13.13), we have

$$Q = AU, \quad | Q_k | \leq | AU | \leq M \mid U \mid \leq M(\sum_{k=1}^{n} U_k^2)^{1/2}$$

$$\leq Mn^{1/2}U^0 = r_0 \leq Q_k^0, \quad k = 1, ..., n \tag{13.14}$$

Hence, the imposed conditions (13.13) guarantee that the original constraints (2.5) are satisfied for all $k = 1, \ldots, n$.

Due to (13.12) and (12.1), we have

$$| V_k | \leq | V | \leq m^{-1}(| G | + | \Phi | + | S |) \tag{13.15}$$

To estimate the components S_k of the vector S from (2.4), we use the inequalities (12.2) and $| q_k | \leq \epsilon$, the latter being true in Ω_2 (see (13.1). Thus, we obtain

$$| S_k | \leq (3/2)Cn^2\epsilon^2, \quad | S | \leq (3/2)Cn^{5/2}\epsilon^2 \tag{13.16}$$

Taking $\epsilon \leq \epsilon_0$ and using the inequalities (12.5), (12.6), and (13.16), we reduce (13.15) to the following inequality

$$| V_k | \leq V^0, \qquad V^0 = m^{-1}\left[G^0(\epsilon) + \Phi^0 + (3/2)Cn^{5/2}\epsilon^2\right]$$

$$G^0(\epsilon) = \left\{\sum_{k=1}^{n}\left[G_k^0(\epsilon)\right]^2\right\}^{1/2}, \quad \Phi^0 = \left[\sum_{k=1}^{n}(\Phi_k^0)^2\right]^{1/2} \tag{13.17}$$

Here, $G^0(\epsilon)$ as well as $G_k^0(\alpha)$ of (12.5), is a monotone increasing continuous function of ϵ for $\epsilon \in [0, \epsilon_0]$, and $G^0(0) = 0$. As a result, our original system (2.3) is reduced to the set (13.12) of subsystems

$$\ddot{q}_k = U_k + V_k, \quad | U_k | \leq U^0,$$

$$| V_k | \leq V^0, \ k = 1, ..., n \tag{13.18}$$

in Ω_2. Here, U_k and V_k could be regarded as controls and disturbances, respectively, which are subjected to independent constraints. The bounds U^0 and V^0 in (13.18) are defined by (13.13) and (13.17), respectively. By the change of variables

$$q_k - q_k^1 = x_1, \ \dot{q}_k = x_2, \ U_k = u, \ V_k = v, \ \rho = V^0/U^0 \tag{13.19}$$

we convert equations (13.18) into

$$\dot{x}_1 = x_2, \ \dot{x}_2 = u + v, \ | u | \leq U^0, \ | v | \leq V^0 = \rho U^0 \tag{13.20}$$

The system (13.20) is similar to (4.2). The only difference is that now we should satisfy also the following state constraint

$$| x_2(t) | \leq \epsilon \tag{13.21}$$

implying that the system does not leave the set Ω_2.

At the instant t_*. our system is on the boundary between the sets Ω_1 and Ω_2. Using (13.19), we obtain the following initial conditions for our system (13.20)

$$x_1(t_*) = q_k(t_*) - q_k^1,\ x_2(t_*) = \dot{q}_k(t_*),\ | x_2(t_*) | \leq \epsilon \tag{13.22}$$

The terminal conditions (2.7) can be rewritten as follows

$$x_1(t_1) = 0,\ x_2(t_1) = 0 \tag{13.23}$$

Thus, we have accomplished the decomposition of our system (2.3) into subsystems (13.20) in the set Ω_2. Our original Problem 1 for a nonlinear system with n degrees of freedom is reduced (in the set Ω_2) to the following problems for subsystems with one degree of freedom each.

Problem 3. Find the feedback control $u(x_1, x_2)$ that satisfies the constraint $| u | \leq U^0$ and the phase constraint (13.21) and brings the system (13.20) from the initial state (13.22) to the zero terminal state (13.23) in finite time under any admissible disturbance v such that $| v | \leq \rho U^0,\ \rho < 1$.

14 The Control of a Subsystem

We can construct the desired control in Problem 3 using the approach of differential games. However, it is simpler to design this control by modifying the law (4.4) in such a way so as to prevent all the phase trajectories from leaving the domain (13.21).

We set

$$\begin{array}{ll} u(x_1, x_2) = U^0 \text{sign}\, [\theta(x_1) - x_2] & \text{if } x_2 \neq \theta(x_1) \\ u(x_1, x_2) = U^0 \text{sign}\ x_1 = -U^0 \text{sign}\ x_2 & \text{if } x_2 = \theta(x_1) \end{array} \tag{14.1}$$

Here, the function $\theta(x_1)$ is defined by

$$\begin{array}{lll} \theta(x_1) = & -\, [2U^0(1-\rho) \mid x_1 \mid]^{1/2} \text{sign}\ x_1 & \text{if } \mid x_1 \mid \leq x^* \\ \theta(x_1) = & -\delta \text{ sign}\ x_1 & \text{if } \mid x_1 \mid > x^* \end{array} \tag{14.2}$$

where δ can be chosen arbitrarily from the interval $(0, \epsilon)$, and x^* is such that the function $\theta(x_1)$ from (14.2) is continuous. The latter condition implies

$$x^* = \delta^2 \left[2U^0(1-\rho)\right]^{-1} \tag{14.3}$$

The switching curve $x_2 = \theta(x_1)$ defined by (14.1) - (14.3) lies in the set (13.21) and is symmetric with respect to the origin. The curve consists of two arcs of parabolas (for $| x_1 | \leq x^*$) and two rays $x_2 = \pm\delta$ (for $| x_1 | > x^*$). The parabolic arcs correspond to the switching curve for the time-optimal control in the absence of a state constraint (13.21) (see (4.4) and (4.5)).

The switching curve is shown in Fig. 8 by a thick line. According to (14.1), this curve divides the set (13.21) into two symmetric parts: X^+, where $x_2 < \theta(x_1)$ and $u = U^0$, and X^-, where $x_2 > \theta(x_1)$ and $u = -U^0$.

Let us prove that the feedback control defined by (14.1) - (14.3) brings the system (13.20) to the terminal state $x_1 = x_2 = 0$ in finite time without violating the constraint (13.21). Equations (13.20) and (14.1) imply that

$$\begin{aligned} \dot{x}_2 &\geq U^0(1-\rho) && \text{if } (x_1, x_2) \in X^+ \\ \dot{x}_2 &\leq -U^0(1-\rho) && \text{if } (x_1, x_2) \in X^- \end{aligned} \tag{14.4}$$

The width of the domains X^+ and X^- along the x_2 - axis does not exceed $\epsilon + \delta$ (see Fig. 8), whereas the velocity along this axis, due to (14.4), is finite and directed towards the switching curve. Hence, the phase trajectory, under the initial condition (13.22), never leaves the set $| x_2 | \leq \epsilon$ and, at some instant t_S, reaches the switching curve $x_2 = \theta(x_1)$.

Suppose the system reaches the "straight" part $x_2 = \pm\rho$ of the switching curve $x_2 = \theta(x_1)$ at $t = t_S$. After that, the system slides along this straight part, because the phase velocities on the both sides of the switching curve are directed towards this curve, in accordance with (14.4). The system moves along these straight parts with a constant speed $| x_1 | = | x_2 | = \delta$ so that $| x_1 |$ decreases. Hence, at some finite time instant $t_p \geq t_S$ the system reaches one of the points $(x^*, -\delta)$ or $(-x^*, \delta)$, where the straight parts of the switching curve $x_2 = \theta(x_1)$ are connected with the parabolic arcs. The parabolic arcs are the phase trajectories of the system (13.20) under the control u given by (14.1) and the disturbance $v = -\rho u$. If $v \neq -\rho u$, then our system follows the parabolic arcs in the sliding regime (see Section 4). Consequently, the system reaches the zero terminal state at some finite instant of time t_1.

Some possible phase trajectories are shown in Fig. 8 by thin lines. The arrows indicate the direction of motion.

Thus, we have obtained a solution to Problem 3. Let us estimate the time of motion $t_1 - t_0$ from the initial state (13.22) to the terminal state

(13.23). The motion from the instant t_* till the instant t_1 consists of three stages; the motion inside the set X^+ or X^-, the motion along the straight lines $x_2 = \pm\delta$, and the motion along the parabolic arcs. Some of these stages may be absent. For instance, the initial state can belong to the switching curve, or the system can reach the parabolic arc directly from the set X^+ or X^-. In all cases, the time of motion $t_1 - t_*$ is finite, and we shall obtain the upper estimate on $t_1 - t_*$ assuming that all the three stages are present.

To evaluate the duration of the first stage (motion inside X^+ or X^-), we shall divide the maximal width $\epsilon + \delta$ of the set X^+ or X^- along the x_2 - axis by the minimal velocity along this axis (see (14.5)). Then we obtain

$$t_s - t_* \leq (\epsilon + \delta)\left[U^0(1-\rho)\right]^{-1} \tag{14.5}$$

Let us estimate the coordinate $x_1(t_s)$ using the constraint (13.21) and the initial conditions (13.22)

$$| x_1(t_s) - x_1(t_*) | \leq \int_{t_*}^{t_s} | x_2 | \, dt \leq \epsilon(t_s - t_*)$$

Taking into account (14.5), we have

$$| x_1(t_s) | \leq | x_1(t_*) | + \epsilon(\epsilon + \delta)\left[U^0(1-\rho)\right]^{-1} \tag{14.6}$$

To estimate the duration of the second stage (motion along the line $x_2 = \pm\delta$), we divide the distance along the x_1-axis by the velocity equal to δ. We obtain

$$t_p - t_s \leq [| x_1(t_s) | - x^*]\,\delta^{-1}$$

Substituting the relations (14.3) and (14.6) into the latter inequality, we have

$$t_p - t_s \leq | x_1(t_*) | \,\delta^{-1} + \epsilon(\epsilon + \\ + \delta)\left[U^0(1-\rho)\delta\right]^{-1} - \delta\left[2U^0(1-\rho)\right]^{-1} \tag{14.7}$$

The duration of the third stage (the motion along the parabolic arc) can be estimated as follows. Due to (13.20), (14.1), and (14.2), along the parabolic arc we have

$$dx_1/dt = x_2 = \theta(x_1) = -\left[2U^0(1-\rho) \,|\, x_1 \,|\right]^{1/2} \operatorname{sign} x_1$$

By integrating this equation under the boundary conditions $x_1(t_p) = x^*$ and $x_1(t_1) = 0$ (see (14.3) and (13.23)), we obtain

$$t_1 - t_p = \int_0^{x^*} \frac{dx}{[2U^0(1-\rho)x]^{1/2}} = \left[\frac{2x^*}{U^0(1-\rho)}\right]^{1/2} = \delta\left[U^0(1-\rho)\right]^{-1} \tag{14.8}$$

The upper estimate on the time $t_1 - t_*$ follows from (14.5), (14.7), and (14.8)

$$t_1 - t_* \leq \mid x_1(t_*) \mid \delta^{-1} + (2\epsilon^2 + 4\epsilon\delta + 3\delta^2)\delta^{-1} \left[2U^0(1-\rho)\right]^{-1}$$

The obtained results are then summarized in the following theorem.

Theorem 5. *The feedback control $u(x_1, x_2)$ defined by (14.1) and (14.2), where x^* is given by (14.3) and $\delta \in (0, \epsilon)$, gives a solution to Problem 3, i.e., satisfies the constraints $\mid u \mid \leq U^0$ and (13.21), and brings the system (13.20) from the initial state (13.22) to the terminal state (13.23) under any admissible disturbance $\mid v \mid \leq \rho$ in finite time bounded by the estimate (14.9).*

15 Feedback Control (the Second Approach)

Let us now design the feedback control for the original system (2.3) using the results of Sections 13 and 14. In the set Ω_1, the desired control is defined by equations (13.3). In Ω_2, we have, according to (13.14) and (13.19)

$$Q(q, \dot{q}) = A(q)U(q, \dot{q}), \; U_k = (q_k - q_k^1, \dot{q}_k), \; k = 1, ..., n \qquad (15.1)$$

Here the function $u(x_1, x_2)$ is defined by equations (14.1) - (14.3). The feedback control $u(x_1, x_2)$ was obtained under the condition $\rho < 1$, where ρ is defined in (13.19). By substituting U^0 from (13.13) and V^0 from (13.17) into the inequality $\rho = V^0/U^0 < 1$, we obtain

$$\rho = (M/m)\left[G^0(\epsilon) + \Phi^0 + (3/2)Cn^{5/2}\epsilon^2\right] r_0^{-1} n^{1/2} < 1 \qquad (15.2)$$

From the properties of the function $G^0(\epsilon)$ introduced in (13.17), it follows that the left-hand side of (15.2) is a monotone increasing function of ϵ, and $G^0(\epsilon) \to 0$ as $\epsilon \to 0$. Hence, we can choose $\epsilon \in (0, \epsilon_0]$ is so as to satisfy the condition (15.2), whenever

$$(M/m)\Phi^0 r_0^{-1} n^{1/2} < 1$$

Using (13.13), we can rewrite this inequality as follows

$$\Phi^0 < (m/M) r_0 n^{-1/2}, \quad r_0 = \min_k Q_k, \; k = 1, ..., n \qquad (15.3)$$

If the condition (15.3) holds, we can choose an ϵ that satisfies (15.2). Taking any δ such that $0 < \delta < \epsilon$ and substituting the parameters $\epsilon, \delta,$

and ρ from (15.2) into (14.1) - (14.3) and (15.1), we obtain the desired solution of Problem 1.

Let us now estimate the total time of motion $t_1 - t_0$ as the sum of the times for the sets Ω_1 and Ω_2 (see (13.7) and (14.9)).

Applying the inequality (14.9), we should replace $| x_1(t_*) |$ by the maximal (with respect to k) expression $| q_k(t_*) - q_k^1 |$ (see (13.19)), because the system reaches the terminal state only when all its coordinates and velocities reach the respective terminal values. Using (13.11), we have

$$| x_1(t_*) | = \max_k | q_k(t_*) - q_k^1 | \leq \max_k \left[| q_k(t_*) - q_k^0 | + | q_k^0 - q_k^1 | \right]$$

$$\leq \max_k | q_k^0 - q_k^1 | + (M/m)^{1/2} r^{-1} (T_0 - m\epsilon^2/2), \ k = 1, ..., n$$

Substituting this inequality into (14.9), we obtain from (13.7) and (14.9)

$$t_1 - t_0 \leq \delta^{-1} \max_k | q_k^0 - q_k^1 | + (2M)^{1/2} r^{-1} \left[T_0^{1/2} - (m/2)^{1/2} \epsilon \right]$$
$$+ (M/m)^{1/2} r^{-1} \delta^{-1} (T_0 - m\epsilon^2/2) \qquad (15.4)$$
$$+ (2\epsilon^2 + 4\epsilon\delta + 3\delta^2) \delta^{-1} \left[2U^0 (1 - \rho) \right]^{-1}, \ k = 1, ..., n$$

Here, the parameters U^0 and ρ are given by equations (13.13) and (15.2), respectively. The obtained results are summed up in the following theorem [10].

Theorem 6. *Under the conditions (12.1) - (12.6) and (15.3), the feedback control solving Problem 1 can be described by equations (13.3) and (15.1) in the respective sets* Ω_1 *and* Ω_2 *defined in (13.1). The total time of motion satisfies the inequality (15.4). In these formulae, the parameter* ϵ *should satisfy the conditions* $\epsilon \in (0, \epsilon_0]$ *and (15.2),* δ *should be chosen from the interval* $(0, \epsilon)$*, while the parameters* $U^0, r_0, \rho, r, T_0, \Phi^0$*, and the function* $G^0(\epsilon)$ *are defined by the corresponding formulae (13.13), (15.2), (13.4), (13.5), and (13.17).*

Remark. The parameter δ can be chosen arbitrarily from the interval $(0, \epsilon)$. The closer is δ to ϵ, the smaller is the time of motion.

Corollary. *Let the uncertain disturbances be absent, namely let* $\Phi = 0$ *in (12.3), and* $\Phi^0 = 0$ *in (13.17). Then the condition (15.3) is always satisfied, and, therefore, the Problem 1 is solvable under the conditions* $\Phi = 0, (12.1), (12.2), (12.4)$*, and (12.5).*

In other words, under the conditions mentioned above, our nonlinear Lagrangian system can be driven to the prescribed terminal state in finite time by control forces as small as desired (the bounds Q_k^0 in (2.5) may be very small).

16 Conclusions

We have presented two methods for obtaining feedback controls in nonlinear dynamic systems described by Lagrangian equations. Under corresponding conditions, both methods can bring our systems to the prescribed terminal states with zero velocities in finite time. The first approach is based on the assumptions imposed in the $2n$-dimensional phase space, whereas the second one deals with assumptions in the n-dimensional coordinate space. Hence, the assumptions of the second approach are less restrictive. On the other hand, the second method includes motions with low velocities (in the set Ω_2), and, hence, the total time of motion may be greater for the second method.

Both approaches are based on the decomposition of the Lagrangian system into simple subsystems. Methods of optimal control and differential games are used to obtain explicit formulae for feedback control. Since both approaches use time-optimal controls, they can be called time-suboptimal.

The closed–loop control laws are obtained under certain conditions which can be regarded as sufficient conditions of controllability for our nonlinear Lagrangian systems.

Both approaches do not presume that the external forces are known. These forces may be uncertain, with only the bounds on them being essential. The feedback controls obtained are robust, i.e., they can cope with additional small disturbances and parameter variations. To ensure the robustness, we should increase the parameter $\rho < 1$, creating thus a sufficient margin in the control possibilities.

Since the Lagrangian equations can serve as a model for robots, our approach can be, in principle, applied in robotics. Preliminary results of computer simulation for robots are rather satisfactory. However, in order to obtain feedback control laws suitable for practical applications, it seems that our technique should be combined with other methods, and a considerable computer simulation should be accomplished.

REFERENCES

[1]. Leitmann, G.,*Deterministic control of uncertain systems.* Acta Astronautica, 7, 1457 - 1461, 1980.

[2]. Corless, M., and Leitmann, G.,*Adaptive control of systems containing uncertain functions and unknown functions with uncertain bounds.* J. Optimization Theory Appl., 41, 155 - 168, 1983.

[3]. Corless, M., and Leitmann, G.,*Adaptive controllers for a class of uncertain systems.* Annales Fond. de Broglie,. 9, 65 - 95, 1984.

[4]. Pontryagin, L.S., Boltyansky, V.G., Gamkrelidze, R.V., and Mishchenko, E.F.,*Mathematical Theory of Optimal Processes*, Wiley-Interscience, 1972.

[5]. Krasovskii, N.N.,*Game Problems of the Encounter of Motions*, Nauka, Moscow, 1970 (in Russian).

[6]. Chernousko F.L., *Decomposition and suboptimal control in dynamical systems*, J.Appl. Maths Mechs (PMM), 54, 6, 727 - 734, 1990.

[7]. Chernousko, F.L.,*Decomposition and synthesis of control in dynamical systems*, Soviet J. Computer and Systems Sciences, 29, 5, 126 - 144, 1990.

[8]. Chernousko, F.L.,*Control synthesis in a system with a non-linear damping*, J. Appl. Maths Mechs (PMM), 55, 6, 1991.

[9]. Chernousko, F.L.,*Decomposition and suboptimal control in dynamic systems*, Optimal Control Appl. and Methods, 1993, to appear.

[10].Chernousko, F.L.,*Control synthesis in a non-linear dynamical system*, J. Appl. Maths Mechs (PMM), 56, 2, 157 - 166, 1992.

Institute for Problems in Mechanics,
Russian Academy of Sciences,
pr. Vernadskogo 101, Moscow 117526, Russia

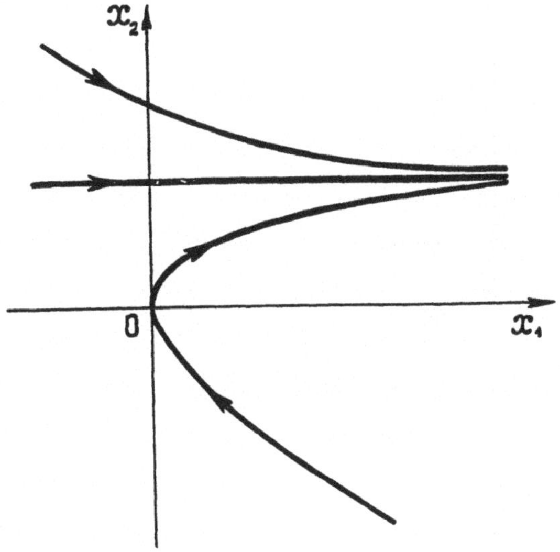

Figure 1.

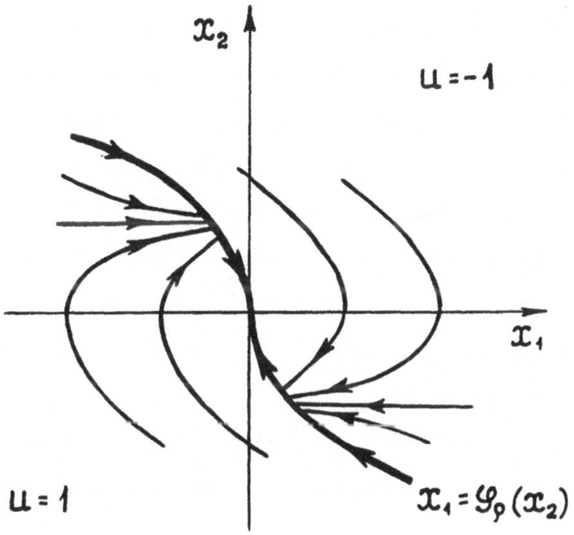

Figure 2.

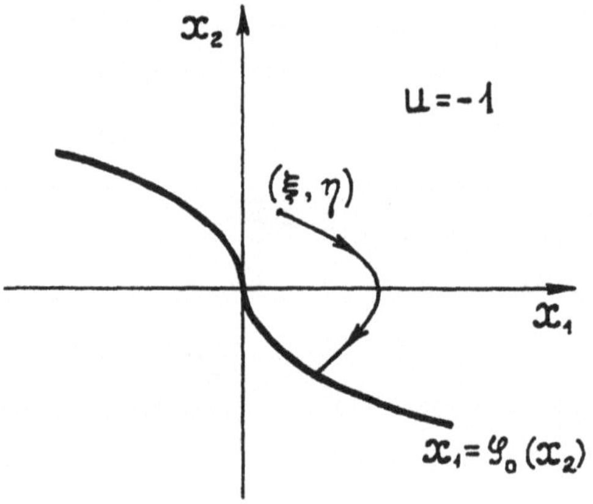

Figure 3.

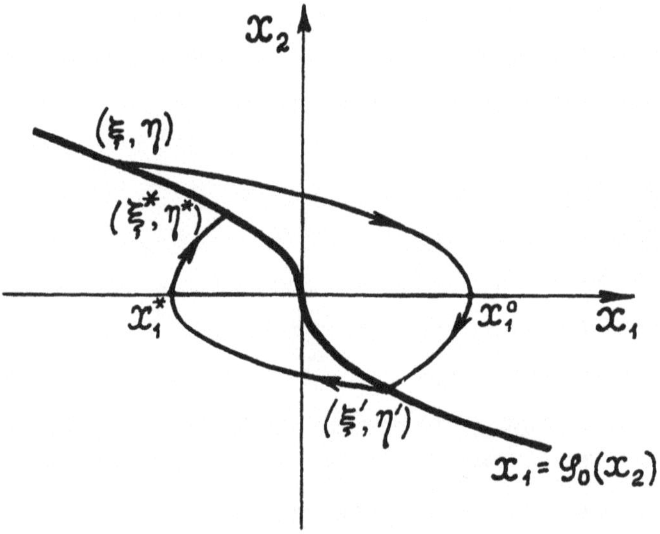

Figure 4.

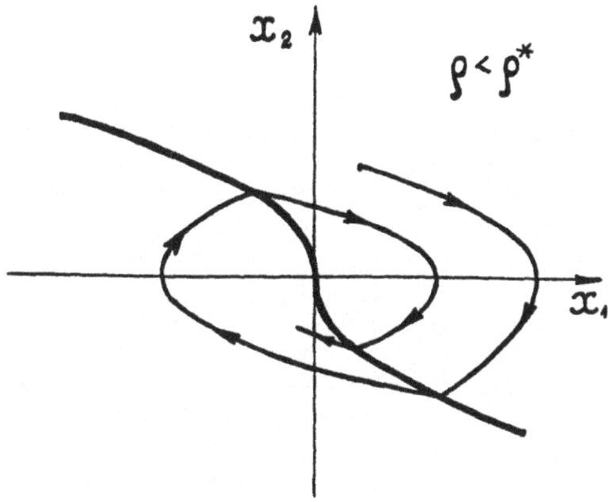

Figure 5.

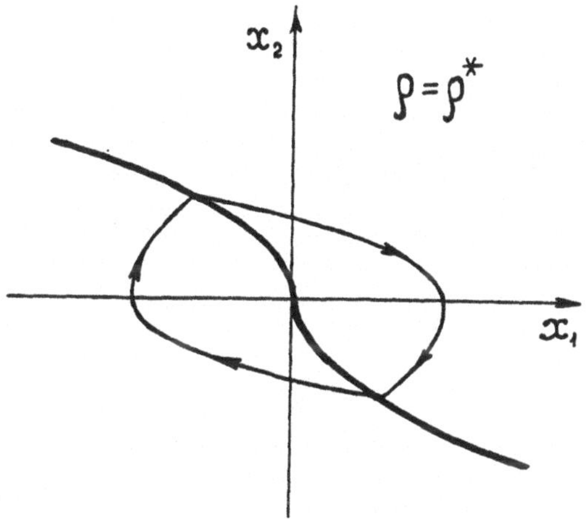

Figure 6.

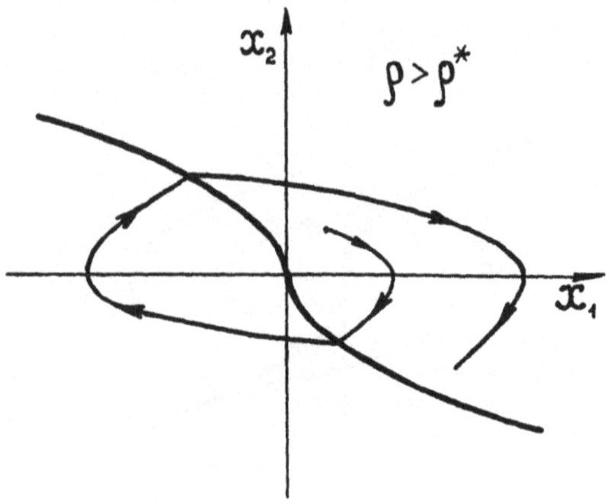

Figure 7.

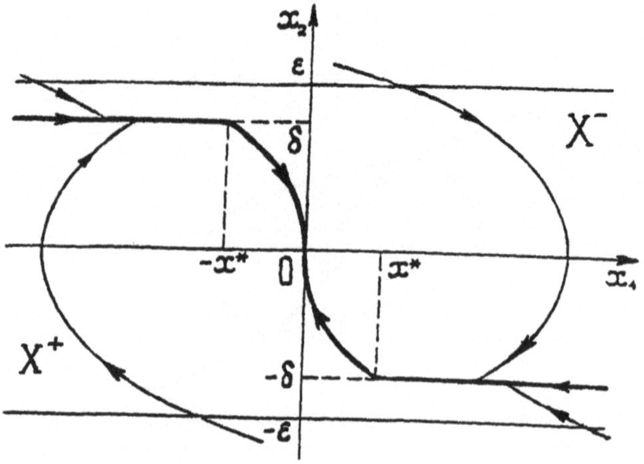

Figure 8.

A Differential Game for the Minimax of a Positional Functional

N.N. Krasovskii and A.N. Krasovskii

In this paper a feedback control problem for a dynamical system under a lack of information is considered. The optimal control problem for the minimax of a given positional functional can be formalized within the framework of differential games. The present paper is devoted to a formalization developed in Ekaterinburg [1-11].

The quality index is positional [8,10] and has the form of a functional which is the maximum of a seminorm of the phase vector which is estimated at every time moment. A special case of this quality index is the terminal norm of the phase vector.

The problem will be solved within the framework of pure positional strategies. The main result is a new effective calculation procedure for a value of the game and a new construction of optimal strategies. The idea of this construction was suggested in [9]. It is connected with the method of stochastic program synthesis [2]. This construction uses recurrence procedures of calculating the upper convex hulls of some auxiliary functions. The theory is illustrated by examples of computer simulation of the control process for a model system.

1 The Equation of Motion

Assume that the motion of the controlled dynamical system is described by the ordinary vector differential equation

$$\dot{x} = f(t, x, u, v), \quad t_0 \leq t \leq \vartheta \tag{1.1}$$

where x is an n-dimensional phase vector, t is the time, t_0 and ϑ are fixed instants, u is an r-dimensional vector which characterizes the control, v is an s-dimensional vector of the disturbance. For a fixed t the function $f(\cdot)$ is continuous in x, u, v, for fixed x, u, v it is semicontinuous in t and the points of discontinuity of $f(\cdot)$ are independent of x, u, v and at these points the function $f(t,\cdot)$ is continuous from the right. Assume that $f(\cdot)$ satisfies the condition

$$| f(t,x,u,v) | \leq \chi \, (1+ | x |), \quad \chi = \text{const} \tag{1.2}$$

Here $| x |$ denotes the Euclidean norm of x, i.e.

$$| x | = (x_1^2 + \dots x_n^2)^{1/2} \tag{1.3}$$

Also, we shall assume that in each bounded region G the function $f(\cdot)$ satisfies a Lipschitz condition in x with a constant L_G , i.e.

$$| f(t,x^{(1)},u,v) - f(t,x^{(2)},u,v) | \leq L_G \cdot | x^{(1)} - x^{(2)} | \tag{1.4}$$

where $x^{(i)} \in G$, $i = 1,2$; $t_0 \leq t \leq \vartheta$.

It is assumed that the admissible values of u and v are restricted by the inclusions

$$u \in P, \tag{1.5}$$

$$v \in Q \tag{1.6}$$

Here P and Q are compact sets in Euclidean spaces R^r and R^s respectively. The sets P and Q describe the abilities of the regulator U and the device V that generates the disturbance.

As usual, the pair $\{t,x\}$, $t \in [t_0,\vartheta]$, $x \in R^n$ will be called the position of the controlled system (1.1) at time t. Let some time interval $[t_*,t^*] \subset [t_0,\vartheta]$ and initial position t_*, x_* be chosen. Suppose that some control action

$$u[t_*[\cdot]t^*) = \{u[t] \in P, \ t_* \leq t < t^*\} \tag{1.7}$$

and some disturbance action

$$v[t_*[\cdot]t^*) = \{v[t] \in Q, \ t_* \leq t < t^*\} \tag{1.8}$$

are realized. The actions $u[\cdot]$ (1.7) and $v[\cdot]$ (1.8) may be arbitrary Borel-measurable [2] functions in $t \in [t_*,t^*]$.

The function

$$x[t_*[\cdot]t^*] = \{x[t], \ t_* \leq t \leq t^*\} \tag{1.9}$$

is called the motion of the system (1.1) generated from the initial position $\{t_*,x_*\}$ by the actions $u[\cdot]$ (1.7) and $v[\cdot]$ (1.8) if it is determined at $t_* \leq t \leq t^*$ as the solution of the differential equation

$$\dot{x}[t] = f(t,x[t],u[t],v[t]), \tag{1.10}$$

$$t_* \leq t \leq t^*, \quad x[t_*] = x_*$$

where $u[t]$ and $v[t]$, $t_* \leq t < t^*$ are the functions (1.7) and (1.8) respectively.

The function $x[\cdot]$ (1.9) satisfies the Lipschitz condition in t. Consequently the function $x[\cdot]$ (1.9), (1.10) is absolutely continuous.

We shall assume that the initial position $\{t_*, x_*\}$ satisfies the condition

$$\{t_*, x_*\} \in G. \tag{1.11}$$

Here G is a compact set in the space $\{t, x\}$, $t \in [t_0, \vartheta]$, $x \in R^n$ which satisfies the following condition. It contains all the possible positions $\{t, x[t]\}$, $t \in [t_*, \vartheta]$ of the system (1.1), (1.2) that can be realized on the basis of all possible actions $u[\cdot]$ (1.7) and $v[\cdot]$ (1.8) from a given initial position $\{t_*, x_*\}$ (1.11). Such a set G can be defined by conditions

$$G = \{\ \{t, x\} :\ t \in [t_0, \vartheta],\ \ | x | \leq R[t]\ \} \tag{1.12}$$

where

$$R[t] = (1 + R_0) \cdot \exp\{\chi \cdot (t - t_0)\} - 1, \quad R_0 > 0 - \ \text{const} \tag{1.13}$$

and χ is the constant in (1.2).

2 The Quality Index

We consider the problems of minimax-maximin over u and v for the functional (the quality index)

$$\gamma = \varphi(x[t_*[\cdot]\vartheta]) \tag{2.1}$$

that evaluates the motion

$$x[t_*[\cdot]\vartheta] = \{x[t],\ \ t_* \leq t \leq \vartheta\},\ \ t_* \in [t_0, \vartheta] \tag{2.2}$$

of the system (1.1), (1.2).

The functionals (2.1) are called positional [8] if they can be represented in the form

$$\begin{aligned} &\gamma(x[t_*[\cdot]\vartheta]) = \beta(x[t_*[\cdot]t^*), \alpha), \\ &\alpha = \gamma(x[t^*[\cdot]\vartheta]),\ \ t_* \leq t^* \leq \vartheta \end{aligned} \tag{2.3}$$

where the functional $\beta(x[t_*[\cdot]t^*), \alpha)$ (for a fixed history $x[t_*[\cdot]t^*]$) is continuous and nondecreasing in α.

In particular the following functionals are positional

$$\gamma = \sigma(x[\vartheta]),\ \ \gamma = \int_{t_*}^{\vartheta} \omega(t, x[t])dt + \sigma(x[\vartheta]),$$

$$\gamma = \sum_{i=1}^{N} \mid x[t^{[i]}] - c^{[i]} \mid,$$

$$\gamma = \max_{i=1,\dots,N} \mid x[t^{[i]}] - c^{[i]} \mid \tag{2.4}$$

where ω and σ are given continuous functions that are Lipschitz in x, and $t^{[i]}$ are given instants, $t^{[i]} \in [t_*, \vartheta]$, $c^{[i]}$ - given points, $c^{[i]} \in R^n$.

In this paper we consider only the case of a positional functional which is defined as the supremum of some semi-norm-function of the phase vector $x[t]$ on the time interval $[t_*, \vartheta] \subset [t_0, \vartheta]$.

Namely, on the interval $[t_0, \vartheta]$ we fix the moments $t_*^{[i]}$, $i = 1, \dots, N_*$, $t^{[N_*]} = \vartheta$, the semi-norms $\mu_*^{[i]}(x)$ and the semi-norm-function $\mu_*(t, x)$ that are piecewise continuous in t. Let the collection of integers $\nu^{[i]} = \nu[t^{[i]}] \in [1, n]$, $i = 1, \dots, N_*$ and the collection of constant $(\nu^{[i]} \times n)$-matrix $D_*^{[i]}$, $i = 1, \dots, N_*$ be given. The semi-norm $\mu_*^{[i]}(x[t_*^{[i]}])$ is defined as some norm $\mu^{[i]}(D_*^{[i]} x[t_*^{[i]}])$ and the semi-norm-function $\mu_*(t, x)$ - as the norm-function $\mu(t, D_*(t)x)$ piecewise continuous in t. Here $D_*(t)$ is a piecewise constant $(\nu[t] \times n)$-matrix-function, $\nu[t] \in [1, n]$, $t_0 \leq t \leq \vartheta$. All the functions are continuous from the right.

We consider the quality index

$$\gamma = \max[\sup_{t_* \leq t \leq \vartheta} \mu(t, D_*(t)x[t]), \max_{i=g_*,\dots,N} \mu^{[i]}(D_*^{[i]} x[t_*^{[i]}])] \tag{2.5}$$

where $t_* \in [t_0, \vartheta]$ is the starting instant of time for the controlled process, $g_* = \min_{i: t^{[i]} \geq t_*} i$.

3 The Statement of the Problem

At first we shall explain some symbols. The symbol $< a, b >$ denotes a scalar product of two m-dimensional vectors a and b, i.e.

$$< a, b >= a_1 b_1 + \dots + a_m b_m = a^T b \tag{3.1}$$

Vectors in the Euclidean space are understood as column vectors. The upper index T denotes transposition. Thus the column vector c can be written as $(c_1, \dots, c_m)^T$.

We shall also assume that the function $f(\cdot)$ (1.1), (1.2) satisfies the following condition:

For any $\{t, x\} \in G$ and $l \in R^n$ the equality

$$\min_{u \in P} \max_{v \in Q} < l, f(t, x, u, v) >= \max_{v \in Q} \min_{u \in P} < l, f(t, x, u, v) > \tag{3.2}$$

is true.

This condition is called the saddle point condition in a minor game [1,2,5].

The informal description of our problem indicates that we consider a feedback control minimax-maximin problems with respect to the controls u and v , i.e. - the problems of finding the control u that minimizes the quality index γ (2.5) and of finding the disturbance v that maximizes γ.

Within the framework of differential games this problem can be formalized as follows [1-11].

In the case (3.2) the problem can be formalized in pure positional strategies [1,2].

Let us call a function

$$u(\cdot) = \{u(t, x, \epsilon_u) \in P,\ t \in [t_0, \vartheta],\ x \in R^n,\ \epsilon_u > 0\} \qquad (3.3)$$

a pure positional strategy of the first player. Here $\{t, x\}$ is the position of the system (1.1), (1.2), (3.2); $\epsilon_u > 0$ is a parameter [2,8]. The parameter $\epsilon_u > 0$ is not an informational argument of the strategy $u(\cdot)$ (3.3). It is a parameter of accuracy. The meaning of this parameter ϵ_u and its role in the solution of the problem will become evident in subsequent paragraphs.

The strategy $v(\cdot)$ of the second player is defined as a function

$$v(\cdot) = \{v(t, x, \epsilon_v) \in Q,\ t \in [t_0, \vartheta],\ x \in R^n,\ \epsilon_v > 0\} \qquad (3.4)$$

where $\epsilon_v > 0$ is the parameter of accuracy of the second player.

The motion

$$x_u[t_*[\cdot]\vartheta] = x_u[t],\quad t_* \leq t \leq \vartheta \qquad (3.5)$$

generated by the strategy $u(\cdot)$ (3.3) is defined as follows. Let the strategy $u(\cdot)$ (3.3) be chosen and $\epsilon_u > 0$ be fixed. Let an initial position $t_*, x_* \in G$ (1.12) be given. Suppose that a partition

$$\Delta_u\{t_i\} = \{t_* = t_1^{(u)},\ t_i^{(u)} < t_{i+1}^{(u)},\ i = 1, \ldots, k_u,\ t_{k_u+1} = \vartheta\} \qquad (3.6)$$

of the time interval $[t_*, \vartheta] \subset [t_0, \vartheta]$ is selected. For the position t_*, x_* and the parameter ϵ_u the strategy $u(\cdot)$ (3.3) assigns a control action

$$u[t_*[\cdot]t_2^{(u)}) = \{u[t] = u(t_*, x_*, \epsilon_u) \in P,\ t_* \leq t < t_2^{(u)}\} \qquad (3.7)$$

where $t_* = t_1^{(u)} \in \Delta_u\{t_i\}$, $t_2^{(u)} \in \Delta_u\{t_i\}$.

The disturbance or the control action of the second player $v[t_*[\cdot]t_2^{(u)})$ may be an arbitrary function (1.8) where $t^* = t_2^{(u)}$. On the first step

$t_* \le t \le t_2^{(u)}$ this pair $u[\cdot]$ and $v[\cdot]$ generates a motion $x[t_*[\cdot]t_2^{(u)}]$ which is a solution of (1.10) where $t^* = t_2^{(u)}$. Thus a new position $\{t_2^{(u)},\ x[t_2^{(u)}]\} = \{t_2^{(u)},\ x[t_*[t_2^{(u)}]t_2^{(u)}]\} \in G$ is obtained. And so on, step by step. Thus the motion $x_u[t_*[\cdot]\vartheta]$ (3.5) is determined at $t_* \le t \le \vartheta$ as the step-by-step solution to the differential equation

$$\dot{x}_u[t] = f(t, x_u[t], u(t_i^{(u)},\ x[t_i^{(u)}], \epsilon_u), v[t]),$$

$$t_i^{(u)} \le t \le t_{i+1}^{(u)}, \qquad i = 1, \dots, k_u \tag{3.8}$$

with the initial conditions $x[t_1] = x_*$ and $x[t_i] = x[t_{i-1}[t_i]t_i]$, $i = 2, \dots, k$.

Here the realization

$$v[t_*[\cdot]\vartheta) = \{v[t_i^{(u)}[\cdot]t_{i+1}^{(u)}),\ \ i = 1, \dots, k_u\} \tag{3.9}$$

may be an arbitrary admissible function (1.8) where $t^* = \vartheta$.

The motion $x_v[t_*[\cdot]\vartheta]$ generated by the strategy $v(\cdot)$ (3.4) is defined similarly. Namely the motion $x_v[t_*[\cdot]\vartheta]$ is determined as a solution of the stepwise equation

$$\dot{x}_v[t] = f(t, x_v[t], u[t], v(t_i^{(v)}, x[t_i^{(v)}], \epsilon_v)),$$

$$t_i^{(v)} \le t \le t_{i+1}^{(v)}, \ \ i = 1, \dots, k_v \tag{3.10}$$

where $x[t_1^{(v)}] = x_*$, $x[t_i^{(v)}] = x[t_{i-1}^{(v)}[t_i^{(v)}]t_i^{(v)}]$. Here $u[t_*[\cdot]\vartheta) = \{u[t_i^{(v)}[\cdot]t_{i+1}^{(v)})$, $i = 1, \dots, k_v\}$ is a realization of control action (1.7) where $t^* = \vartheta$.

For a strategy $u(\cdot)$ (3.3) and for an initial position $\{t_*, x_*\} \in G$ (1.12) we say that the quantity

$$\rho(u(\cdot), t_*, x) = \tag{3.11}$$

$$= \overline{\lim_{\epsilon_u \to 0}}\ \lim_{\delta_u \to 0}\ \sup_{\Delta_{u_\delta}}\ \sup_{v[t_*[\cdot]\vartheta)} \gamma(x_u[t_*[\cdot]\vartheta])$$

is a guaranteed result. Here symbol Δ_{u_δ}, where $\delta_u > 0$, denotes the partition $\Delta_u\{t_i\}$ (3.6) that satisfies the condition

$$\max_{i=1,\dots,k_u} (t_{i+1}^{(u)} - t_i^{(u)}) \le \delta_u \tag{3.12}$$

The guaranteed result for a strategy $v(\cdot)$ (3.4) and a position $\{t_*, x_*\} \in G$ is defined by the equality

$$\rho(v(\cdot), t_*, x_*) = \tag{3.13}$$

$$= \underline{\lim_{\epsilon_v \to 0}}\ \lim_{\delta_v \to 0}\ \inf_{\Delta_{v_\delta}}\ \inf_{u[t_*[\cdot]\vartheta)} \gamma(x_v[t_*[\cdot]\vartheta])$$

The strategies $u^o(\cdot) = u^o(t, x, \epsilon)$ and $v^o(\cdot) = v^o(t, x, \epsilon)$ are optimal if

$$\rho(u^o(\cdot), t_*, x_*) = \min_{u(\cdot)} \rho(u(\cdot), t_*, x_*) = \rho_u^o(t_*, x_*) \tag{3.14}$$

and

$$\rho(v^o(\cdot), t_*, x_*) = \max_{v(\cdot)} \rho(v(\cdot), t_*, x_*) = \rho_v^o(t_*, x_*) \tag{3.15}$$

for every initial position $\{t_*, x_*\} \in G$.

For an initial position $\{t_*, x_*\} \in G$ we shall say that quantities $\rho_u^o(t_*, x_*)$ (3.14) and $\rho_v^o(t_*, x_*)$ (3.15) are the optimal guaranteed results for the first and the second players respectively.

The pair $\{u^o(\cdot), v^o(\cdot)\}$ forms a positional saddle point and gives the value $\rho^o(t, x)$ of the differential game if the condition

$$\rho^o(t, x) = \rho_u^o(t, x) = \rho_v^o(t, x) \tag{3.16}$$

holds for all $\{t, x\} \in G$ (1.12).

The definitions (3.3)-(3.16) mean the following. For an arbitrary $\eta > 0$ the inequality

$$\gamma(x_{u^o}[t_*[\cdot]\vartheta]) \leq \rho^o(t_*, x_*) + \eta \tag{3.17}$$

holds for $\epsilon_u \leq \epsilon_u(\eta, t_*, x_*)$ and $\delta_u \leq \delta_u(\eta, \epsilon, t_*, x_*)$, for any motion $x_u[\cdot]$ (3.5) generated by the strategy $u^o(\cdot)$ (3.14). Similarly,

$$\gamma(x_{v^o}[t_*[\cdot]\vartheta]) \geq \rho^o(t_*, x_*) - \eta \tag{3.18}$$

for $\epsilon_v \leq \epsilon_v(\eta, t_*, x_*)$ and $\delta_v \leq \delta_v(\eta, \epsilon, t_*, x_*)$ for any motion $x_{v^o}[\cdot]$ generated by the strategy $v^o(\cdot)$ (3.15).

The strategies $u^o(\cdot)$ and $v^o(\cdot)$ will be called uniformly optimal in G if $\epsilon_u(\eta)$, $\epsilon_v(\eta)$, $\delta_u(\eta, \epsilon)$, $\delta_v(\eta, \epsilon)$ can be chosen independently of $\{t_*, x_*\} \in$ G.

Therefore the informal description of the saddle point $\{u^o(\cdot), v^o(\cdot)\}$ and the value of the game $\rho^o(t_*, x_*)$ means, that for any motion $x_{u^o,v^o}[t_*[\cdot]\vartheta]$ generated from any initial position $\{t_*, x_*\} \in G$ by the pair of optimal strategies $\{u^o(\cdot), v^o(\cdot)\}$ we have the result

$$\gamma(x_{u^o,v^o}[t_*[\cdot]\vartheta]) \approx \rho^o(t_*, x_*) \tag{3.19}$$

The problem consists in constructing a saddle point $\{u^o(\cdot), v^o(\cdot)\}$.

We have the following result [8].

Theorem 3.1. *The differential game for the system (1.1), (1.2), (3.2) with a positional functional γ (2.3) has a (uniform) saddle point $\{u^o(\cdot) = u^o(t, x, \epsilon),\ v^o(\cdot) = v^o(t, x, \epsilon)\}$ and the value of the game $\rho^o(t, x)$.*

The optimal strategies $u^o(t, x, \epsilon)$ and $v^o(t, x, \epsilon)$ are constructed as the extremal ones with respect to the function $\rho^o(t, x)$ [2, 8].

Description of these constructions will be given in the following paragraphs.

4 The Extremal Strategy $u_e(\cdot)$

Let us consider a motion of any abstract w-model together with the real motion of the x-system (1.1). The current state of this model at time $t \in [t_0, \vartheta]$ is determined by its n-dimensional phase vector $w[t]$. It is assumed that the motion of the phase vector $w[t]$ is described by a differential equation similar to (1.1), i.e.

$$\dot{w} = f(t, w, u_*, v_*), \quad t_0 \leq t \leq \vartheta \tag{4.1}$$

It is assumed that the function $f(\cdot)$ coincides with the function $f(\cdot)$ from (1.1), (1.2), (3.2) and the auxiliary control actions u_* and v_* satisfy the restrictions (1.5) and (1.6), i.e.

$$u_* \in P, \quad v_* \in Q \tag{4.2}$$

We shall assume that each motion

$$w[t_*[\cdot]\vartheta] = \{w[t], \quad t_* \leq t \leq \vartheta\} \tag{4.3}$$

satisfies the conditions

$$\{t, w[t]\} \in G^*, \quad t_* \leq t \leq \vartheta \tag{4.4}$$

where G^* is a region in the space $\{t, w\}$, $t \in [t_0, \vartheta], w \in R^n$. We shall suppose that $G_\lambda \subset G^*$, $\lambda > 0$ and the set G^* satisfy all the properties imposed on G (1.12). Here G_λ is composed of the λ-neighbourhoods of sections G_t of G.

Let us now assume that from the position $\{t, w\} \in G^*$ we can construct any function $\rho(t, w)$ that satisfies the following conditions.

1. For each fixed $t \in [t_0, \vartheta]$ the function $\rho(t, w)$ satisfies the Lipschitz condition in w, i.e.

$$| \rho(t, w^{(1)}) - \rho(t, w^{(2)}) | \leq L^* \cdot | w^{(1)} - w^{(2)} | \tag{4.5}$$

2. The equality

$$\gamma(w[\vartheta[\vartheta]\vartheta]) = \rho(\vartheta, w[\vartheta]) \tag{4.6}$$

holds for each $\vartheta, w[\vartheta] \in G^*$.

3u. Suppose that given are a position $\{\tau_*, w_*\} \in G^*$, a number $\epsilon_* > 0$, a moment $\tau^* > \tau_*$ and a control action $v_*[\tau_*[\cdot]\tau^*)$. Then there exists an $\epsilon^* > 0$ and a control action $u_*[\tau_*[\cdot]\tau^*)$ which in combination with $v_*[\tau_*[\cdot]\tau^*)$ generates from the initial position $\{\tau_*, w_*\}$ a motion $w[\tau_*[\cdot]\tau^*]$ of the model (4.1) such that

$$\beta(w[\tau_*[\cdot]\tau^*), \rho(\tau^*, w[\tau^*]) + 2\epsilon^*) \leq \rho(\tau_*, w_*) + 2\epsilon_* \qquad (4.7)$$

where $\beta(\cdot)$ is the function in (2.3), $w[\tau^*] = w[\tau_*[\tau^*]\tau^*]$.

The last condition 3u is called the property of u-stability for the function $\rho(t, w)$.

Let $u_e(\cdot) = u_e(t, x, \epsilon)$ be the extremal strategy with respect to the function $\rho(t, w)$ (4.5)-(4.7). This strategy is constructed in the following way [8].

We consider the function

$$\lambda(t, x, w) = | x - w |^2 \exp\{-2L \cdot (t - t_0)\} \qquad (4.8)$$

where L is the constant in (1.4).

We select a sufficiently small parameter $\epsilon > 0$ in accordance with λ from $G_\lambda \subset G^*$. Suppose that a partition $\Delta\{t_i\}$ (3.6) is selected and a position $\{t_i, x[t_i]\}$ has realized. By $K(\epsilon, t_i)$ we denote the set of values w that satisfy the condition $K(\epsilon, t_i) \subset G^*_{t_i}$ where $K(\epsilon, t_i)$ is composed of all the vectors w that satisfy the inequality

$$\lambda(t_i, x[t_i], w) \leq \epsilon + \epsilon \cdot (t_i - t_0) \qquad (4.9)$$

An accompanying point

$$w_u^o[t_i] \in K(\epsilon, t_i) \qquad (4.10)$$

is a point that satisfies the condition

$$\rho(t_i, w_u^o[t_i]) = \min_{w \in K(\epsilon, t_i)} \rho(t_i, w) \qquad (4.11)$$

This point can be non-unique for a fixed $t_i, x[t_i]$, ϵ. We select one of these for every given t_i, $x[t_i]$, ϵ.

An extremal strategy $u_e(\cdot) = u_e(t, x, \epsilon)$ is a function $u^o(t_i, x[t_i], \epsilon) \in P$ which satisfies the condition

$$\max_{v \in Q} < s_u^o[t_i], f(t_i, x[t_i], u^o(t_i, x[t_i], \epsilon), v) > =$$

$$= \min_{u \in P} \max_{v \in Q} < s_u^o[t_i], f(t_i, x[t_i], u, v) > \tag{4.12}$$

where

$$s_u^o[t_i] = x[t_i] - w_u^o[t_i] \tag{4.13}$$

Suppose $\epsilon > 0$ and a partition $\Delta\{t_i\}$ (3.6) $t_1 = t_*$, $t_{k+1} = \vartheta$, are chosen so that $\max_i \mid t_{i+1} - t_i \mid \leq \delta$. The strategy $u_e(\cdot)$ and these data generate a motion $x_{u_e}[t_*[\cdot]\vartheta]$ (3.5) due to the scheme (3.8). Here the positions $\{t_i, x[t_i]\} \in G$ and the accompanying points $w_u^o[t_i]$ are realized at instants of time t_i. By selecting for $x_{u_e}[t_*[\cdot]\vartheta]$ some suitable functions $w[t_*[\cdot]\vartheta]$, where $w[t_i[t_i]\vartheta] = w_u^o[t_i]$ and by employing the property of u-stability 3u of the function $\rho(t, w)$, we obtain

$$\gamma(w[t_*[\cdot]\vartheta]) \leq \rho(t_*, w_*) + 2\epsilon \tag{4.14}$$

According to [2] the motion $x_{u_e}[t_*[\cdot]\vartheta]$ is close to the function $w[t_*[\cdot]\vartheta]$ provided the value $\delta(\epsilon, \eta) > 0$ is sufficiently small. Therefore we obtain

$$\gamma(x_{u_e}[t_*[\cdot]\vartheta]) \leq \rho(t_*, x_*) + \eta(\epsilon), \tag{4.15}$$

Here $\lim_{\epsilon \to 0} \eta(\epsilon) = 0$.

The next result given below follows from (4.15).

Lemma 4.1. *Suppose we have the function $\rho(t, w)$ (4.5)-(4.7) and that the position $\{t_*, x_*\} \in G$ has realized. If starting from time t_*, an extremal strategy $u_e(\cdot)$ is utilized then for an arbitrary small $\eta > 0$ it is possible to find an $\epsilon(\eta) > 0$ and a $\delta(\epsilon, \eta) > 0$ so that for any motion $x_{u_e}[t_*[\cdot]\vartheta]$ the inequality*

$$\gamma(x_{u_e}[t_*[\cdot]\vartheta) \leq \rho(t_*, x_*) + \eta \tag{4.16}$$

is true for $\epsilon \leq \epsilon(\eta)$ and $\delta \leq \delta(\epsilon, \eta)$.

5 The Extremal Strategy $v_e(\cdot)$

Suppose that the function $\rho(t, w)$ satisfies the conditions (4.5), (4.6) and also the following property of v-stability.

4v. Suppose that given are a position $\{\tau_*, w_*\} \in G^*$, a number $\epsilon_* > 0$, a moment $\tau^* > \tau_*$ and a control $u_*[\tau_*[\cdot]\tau^*)$. Then there exists an $\epsilon^* > 0$ and a control $v_*[\tau_*[\cdot]\tau^*)$ that together with $u_*[\tau_*[\cdot]\tau^*)$ generate a motion $w[\tau_*[\cdot]\tau^*]$ of a w-model (4.1) such that

$$\beta(w[\tau_*[\cdot]\tau^*), \rho(\tau^*, w[\tau^*]) - 2\epsilon^*) \geq \rho(\tau_*, w_*) - 2\epsilon_* \tag{5.1}$$

where $\beta(\cdot)$ is the function in (2.3), $w[\tau^*] = w[\tau_*[\tau^*]\tau^*]$.

An extremal strategy $v_e(\cdot) = v_e(t, x, \epsilon)$ is a function $v^o(t_i, x[t_i], \epsilon) \in Q$ that satisfies the condition

$$\min_{u \in P} < s_v^o[t_i], f(t_i, x[t_i], u, v^o(t_i, x[t_i], \epsilon)) >=$$

$$= \max_{v \in Q} \min_{u \in P} < s_v^o[t_i], f(t_i, x[t_i], u, v) > \tag{5.2}$$

where

$$s_v^o[t_i] = w_v^o[t_i] - x[t_i], \tag{5.3}$$

and $w_v^o[t_i]$ is an accompanying point determined now by the condition

$$\rho(t_i, w_v^o[t_i]) = \max_{w \in K(\epsilon, t_i)} \rho(t_i, w) \tag{5.4}$$

It can be shown that in this case the following result holds.

Lemma 5.1. *Suppose that we have the function $\rho(t, w)$ (4.5), (4.6), (5.1) and that the position $\{t_*, x_*\} \in G$ is realized. If beginning from the time instant t_* an extremal strategy $v_e(\cdot)$ is used then for any arbitrary small $\eta > 0$ there are a sufficiently small $\epsilon(\eta) > 0$ and a $\delta(\eta, \epsilon) > 0$ such that for any motion $x_v[t_*[\cdot]\vartheta]$ the inequality*

$$\gamma(x_{v_e}[t_*[\cdot]\vartheta]) \geq \rho(t_*, x_*) - \eta \tag{5.5}$$

is valid for $\epsilon \leq \epsilon(\eta)$ and $\delta \leq \delta(\eta, \epsilon)$.

Thus, if both players follow to their own extremal strategies $u_e(\cdot)$ and $v_e(\cdot)$ then for any number $\eta > 0$ each of the two players can assign numbers $\epsilon_u(\eta) > 0$, $\delta_u(\epsilon, \eta) > 0, \epsilon_v(\eta) > 0$ and $\delta_v(\epsilon, \eta) > 0$ such that for the motion $x_{u_e, v_e}[t_*[\cdot]\vartheta]$ the inequalities

$$\gamma(x_{u_e, v_e}[t_*[\cdot]\vartheta]) \leq \rho(t_*, x_*) + \eta \tag{5.6}$$

$$\gamma(x_{u_e, v_e}[t_*[\cdot]\vartheta]) \geq \rho(t_*, x_*) - \eta \tag{5.7}$$

are valid for any initial position $\{t_*, x_*\} \in G$ if the parameter ϵ and the partition step δ satisfy the respective inequalities

$$\epsilon \leq \epsilon_u(\eta), \qquad \delta \leq \delta_u(\epsilon, \eta) \tag{5.8}$$

$$\epsilon \leq \epsilon_v(\eta), \qquad \delta \leq \delta_v(\epsilon, \eta) \tag{5.9}$$

Thus we have the following result.

Theorem 5.1. *Let the function $\rho(t, w)$ satisfy the conditions 1, 2, 3u (4.5)-(4.7) and 4v (5.1). Then the extremal strategies $u_e(\cdot)$ and $v_e(\cdot)$ are* *optimal strategies* $u^o(\cdot) = u^o(t, x, \epsilon)$ *and* $v^o(\cdot) = v^o(t, x, \epsilon)$ *which form a (uniform) positional saddle point* $\{u^o(\cdot),\ v^o(\cdot)\}$. *The value of the game* $\rho^o(t, x)$ *is equal to* $\rho(t, x)$.

The given method of constructing the strategies $u^o(\cdot)$ and $v^o(\cdot)$ in the form of extremal strategies with respect to the function $\rho^o(t, x)$ is known as the method of an extremal shift to the accompanying points [2].

6 An Approximating Functional

We shall consider the functional γ (2.5). According to §2 the functional γ (2.5) should be positional, i.e. one shall be able to present it in the form (2.3). In fact we have

$$\gamma(x[t_*[\cdot]\vartheta]) = \beta(x[t_*[\cdot]t^*), \alpha) = \max[\sup_{t_* \le t < t^*} \mu(t, D_*(t)x[t]),$$

$$\max_{i=g_*,\dots,g^*} \mu^{[i]}(D_*^{[i]}x[t_*^{[i]}]), \alpha], \tag{6.1}$$

$$\alpha = \gamma(x[t^*[\cdot]\vartheta]) = \max\,[\sup_{t^* \le t \le \vartheta} \mu(t, D_*(t)x[t]),$$

$$\max_{i=g^*+1,\dots,N} \mu^{[i]}(D_*^{[i]}x[t_*^{[i]}])] \tag{6.2}$$

where the numbers g_* and g^* are defined by

$$g_* = \min_{i:t_* \le t_*^{[i]}} i \tag{6.3}$$

$$g^* = \max_{i:t_{[i]} < t^*} i. \tag{6.4}$$

For the positional functional γ (2.5) Theorem 5.1 holds, i.e. the differential game for the system (1.1), (1.2), (3.2) has a saddle point $\{u^0(\cdot) = u^0(t, x, \epsilon),\ v^0(\cdot) = v^0(t, x, \epsilon)\}$ and the value $\rho^0(t, x)$.

For the functional γ (2.5) we now consider an approximating functional γ_* which is constructed in the following way. Let

$$\Delta\{\tau_*^{[h]}\} = \{t_0 = \tau_*^{[1]},\ \tau_*^{[h]} < \tau_*^{[h+1]},\ h = 1, \dots, M, \tau_*^{[M]} = \vartheta\} \tag{6.5}$$

be a partition for the time interval $[t_0, \vartheta]$. This partition includes all the instants $t_j^0 \in [t_0, \vartheta]$ divide the intervals where the matrix-function $D_*(t)$ (2.5) is constant. We also assume that $\Delta\{\tau_*^{[h]}\}$ includes all the instants $t_j^* \in [t_0, \vartheta]$ that are the dividing points for the intervals of continuity of the piecewise continuous norm-function $\mu(t, D_*(t)x)$ from (2.5).

We suppose

$$D[\tau_*^{[h]}] = D_*(t) \tag{6.6}$$

if $t = \tau_*^{[h]}$. We select only those $\tau_*^{[h]}$ that are such that $\mu(\tau_*^{[h]}, \cdot) \not\equiv 0$. Let us combine the points $t_*^{[i]}$ in (2.5) with the points $\tau_*^{[h]}$ of (6.5) and enumerate them anew. We denote these instants by $t^{[1]} \leq t^{[2]} \leq \ldots \leq t^{[N]} = \vartheta$. Assume

$$\mu^{[i]}(D^{[i]}x) = \mu(\tau_*^{[h]}, D_*[\tau_*^{[h]}]x) \tag{6.7}$$

if $t^{[i]} = \tau_*^{[h]} \neq t_*^{[i]}$ and

$$\mu^{[i]}(D^{[i]}x) = \mu^{[j]}(D_*^{[j]}x) \tag{6.8}$$

if $t^{[i]} = t_*^{[j]}$. Also we assume that all the points $\tau_*^{[h]}$ and $t_*^{[j]}$ are different. Otherwise the changes are obvious.

Thus we obtain the functional

$$\gamma_* = \max_{i=g,\ldots,N} \mu^{[i]}(D^{[i]}x[t^{[i]}]) \tag{6.9}$$

where

$$g = \min_{i:\ t_* \leq t^{[i]}} i \tag{6.10}$$

We shall say that γ_* is an approximating functional for γ (2.5).

The functional γ_* (6.9) is positional, i.e. it is presented in the form (2.3). Therefore Theorem 5.1 is true for the functional γ_* of (6.9). In other words, the differential game for the system (1.1), (1.2), (3.2) with the functional γ_*(6.9) has a saddle point $\{u_*^o(\cdot) = u_*^o(t, x, \epsilon), v_*^o(\cdot) = v_*^o(t, x, \epsilon)\}$ and a value $\rho_*^o(t, x)$. Here the lower index * denotes that the corresponding functional is approximating (6.9).

The following assertion is true.

Lemma 6.1. *For each $\zeta > 0$ there exists a number $\delta(\zeta) > 0$ such that*

$$| \rho_*^o(t_*, x_*) - \rho^o(t_*, x_*) | \leq \zeta \tag{6.11}$$

for all the positions $\{t_*, x_*\} \in G$ if the partition step for (6.5) satisfies the following condition

$$\max_h(\tau_*^{[h+1]} - \tau_*^{[h]}) \leq \delta(\zeta) \tag{6.12}$$

In formula (6.11) it is $\rho^o(t, x)$ that is the value of game for the functional γ (2.5) and $\rho_*^o(t, x)$ is the value of game for γ_* of (6.9).

According to Lemma 6.1 the value $\rho_*^o(t, x)$ and the optimal strategies $u_*^o(\cdot)$ and $v_*^o(\cdot)$ solve the problem from §3 for the initial functional γ (2.5).

7 The Maximum over Stochastic Programs

The main result of this paper is a new effective construction of the value $\rho_*^o(t,x)$ of the game for the cost functional γ_* (6.9). Here we shall describe a construction which is based on the idea of a method known as the stochastic program synthesis [2]. This construction was proposed in the paper [9] and was used in [7,10,11].

According to §§4, 5 we shall construct the function $\rho_*(t,x)$ that satisfies the conditions 1, 2, 3u and 4v.

Here we shall consider the construction of the value $\rho_*^o(t,x)$ for a controlled system described by a differential equation

$$\dot{x} = A(t)x + f(t,u,v), \quad t_0 \le t \le \vartheta \tag{7.1}$$

under restrictions (1.5) and (1.6). The matrix-valued function $A(t)$ and the vector-valued function $f(t,u,v)$ are semicontinuous in t for $t \in [t_0, \vartheta]$ and $f(t,u,v)$ is continuous in u and v within the sets (1.5) and (1.6).

We confine ourselves to the case when the vector $f(t,u,v)$ fills a convex set when u runs over P and the function $f(\cdot)$ (7.1) also satisfies a condition similar to (3.2), i.e.

$$\min_{u\in P} \max_{v\in Q} < l, f(t,u,v) >= \max_{v\in Q} \min_{u\in P} < l, f(t,u,v) > \tag{7.2}$$

With the given controlled x-system (7.1), (7.2) we associate a stochastic w-model. The current state of this model is described by the vector $w[t,\omega]$ whose evolution is subjected to the differential equation

$$\dot{w} = A(t)w + f(t,u,v), \quad t_o \le t \le \vartheta \tag{7.3}$$

where $w \in R^n$, $u \in P$, $v \in Q$, $A(t)$ and $f(t,u,v)$ are the same as in (7.1).

Let $\{\tau_*, w_*\}$ be the initial position for w-model (7.3). Let τ be a conceptual time variable for the model. This variable τ can vary in the interval $[t_o, \vartheta]$. For the interval $[\tau_*, \vartheta] \subset [t_*, \vartheta]$ we choose a partition

$$\Delta\{\tau_j\} = \{\tau_* = \tau_1, \ \tau_j < \tau_{j+1}, \ j = 1, \dots, k, \tau_{k+1} = \vartheta\} \tag{7.4}$$

and a probability space $\{\Omega, B, P\}$ [14]. This space is generated by independent random variables ξ_j, $j = 1, \dots, k$, each uniformly distributed on the semi-open interval $0 \le \xi_j < 1$. We can interpret that in the following way. A source Ξ of random events gives a number $\xi_j \in [0,1)$ at the time instant τ_j. Further, all the values of ξ_j are regarded as equally probable in advance (for $t < \tau_j$). Each collection $\{\xi_1, \ldots, \xi_k\}$ of numbers $\xi_j \in [0,1)$ is considered as an elementary event ω, i.e. $\omega = \{\xi_1, \ldots, \xi_k\} \in \Omega$. Hence Ω is the unit cube in the k-dimensional space $\{\xi_1, \ldots, \xi_k\}$, $B = B_\Omega$ is the Borel σ-algebra for the cube, and $P(B)$ is Lebesgue measure on the cube, $B \in \mathcal{B}$.

Let

$$u[\cdot] = \{u[\tau, \omega] \in P, \quad \tau_* \leq \tau < \vartheta, \quad \omega \in \Omega\} \tag{7.5}$$

and

$$v[\cdot] = \{v[\tau, \omega] \in Q, \quad \tau_* \leq \tau < \vartheta, \quad \omega \in \Omega\} \tag{7.6}$$

be the stochastic programs that are nonanticipating with respect to $\xi(\tau_j) = \xi_j$, i.e.

$$u[\tau, \omega] = u[\tau, \xi_1, \ldots, \xi_j], \quad \tau_j \leq \tau < \tau_{j+1}, \quad j = 1, \ldots, k \tag{7.7}$$

$$v[\tau, \omega] = v[\tau, \xi_1, \ldots, \xi_j], \quad \tau_j \leq \tau < \tau_{j+1}, \quad j = 1, \ldots, k \tag{7.8}$$

where the functions $u[\tau, \xi_1, \ldots, \xi_j]$ and $v[\tau, \xi_1, \ldots, \xi_j]$ are jointly measurable in the variables τ, $\xi_1, \ldots, \xi_j$.

Suppose some stochastic programs $u[\cdot]$ (7.5) and $v[\cdot]$ (7.6) are chosen. With position $\{\tau_*, w_*\}$ and partition $\Delta\{\tau_j\}$ given the chosen programs $u[\cdot]$ and $v[\cdot]$ determine a random motion of the model (7.3) or namely a solution

$$w[\tau_*[\cdot]\vartheta, \cdot] = \{w[\tau, \omega], \quad \tau_* \leq \tau \leq \vartheta, \quad \omega \in \Omega\} \tag{7.9}$$

of the stochastic differential equation

$$\dot{w}[\tau, \omega] = A(\tau) w[\tau, \omega] + f(\tau, u[\tau, \omega], v[\tau, \omega]), \tag{7.10}$$

with the initial condition

$$w[\tau_*, \omega] = w_*, \qquad \omega \in \Omega \tag{7.11}$$

We call the quantity

$$\rho_*(\tau_*, w_*, \Delta) = \max_{v[\cdot]} \min_{u[\cdot]} M\{ \max_{g \leq i \leq N} \mu^{[i]}(D^{[i]} w[t^{[i]}, \omega]) \} \tag{7.12}$$

the stochastic programed maximin. Here $M\{\cdots\}$ denotes the expectation [14]. In (7.12) we have $w[t^{[i]}, \omega] = w[t, \omega]$ at $t = t^{[i]}$, $i = g, \ldots, N$ where $w[\tau_*[\cdot]\vartheta, \cdot]$ is the motion (7.9).

Taking the dual vector random variable $l(\cdot)$ we can calculate $\rho_*(\cdot)$ (7.12) as an open loop extremum $e_*(\tau_*, w_*, \Delta)$. This extremum $e_*(\cdot)$ is defined in the following way.

Let $l^{[i]}(\omega)$ be an $\nu^{[i]}$-dimensional random vector variable defined over the probability space $\{\Omega, B, P\}$. Here $\nu^{[i]} \in [1, n]$ is defined by the dimension of the matrix $D^{[i]}$.

Let $l(\cdot)$ be a variable with components $l^{[i]}(\omega)$ and let

$$\| \, l(\cdot) \, \|^{*} = \operatorname{vrai}\max_{\omega} \, [\sum_{i=g}^{N} \mu^{[i]*}(\, l^{[i]}(\omega) \,) \,] \tag{7.13}$$

denote the norm of $l(\cdot)$. Here $\mu^{[i]*}(l)$ is the conjugate norm for $\mu^{[i]}(l)$. Denote by $X(t, \tau)$ the fundamental matrix of the differential equation $dx/dt = A(t)x$ and by $M\{\cdots \mid \cdots\}$ the conditional mathematical expectation.

The value of the extremum $e_*(\cdot)$ is defined by the equality

$$e_*(\tau_*, w_*, \Delta) = \max_{\|l(\cdot)\|^* \le 1} [\sum_{i=g}^{N} M\{l^{[i]T}(\omega)\} \cdot D^{[i]} X(t^{[i]}, \tau_*) w_* +$$

$$+ M\{ \sum_{j=1}^{k} (\tau_{j+1} - \tau_j) \max_{v \in Q} \min_{u \in P} M\{ \sum_{i=d(j)}^{N} l^{[i]T}(\omega) \times$$

$$\times D^{[i]} X(t^{[i]}, \tau_j) f(\tau_j, u, v) \mid \xi_1, \ldots, \xi_j \, \} \, \} \,] \tag{7.14}$$

where $d(j) = \min_{t^{[i]} > \tau_j} i$.

The equalities

$$\rho_*(\tau_*, w_*, \Delta_\delta) = e_*(\tau_*, w_*, \Delta_\delta) + O(\delta) \tag{7.15}$$

$$\rho^o(\tau_*, w_*) = \lim_{\gamma_*, \gamma, \delta \to 0} e_*(\tau_*, w_*, \Delta_\delta) \tag{7.16}$$

are true. Here $\rho^o(\cdot)$ is the value of the game for the functional $\gamma(2.5)$ and $\rho_*(\cdot)$ is the maximin over the stochastic programs (7.12), Δ_δ is a partition (7.4)that satisfies the condition $\max_j \, (\tau_{j+1} - \tau_j) \le \delta$.

These constructions are the concretization of the idea of the stochastic program method [2,9] for the problem considered here.

Denote

$$\tilde{w}_* = X(\vartheta, \tau_*) w_*, \quad \tilde{f}(\tau, \cdot) = X(\vartheta, \tau) f(\tau, \cdot) \tag{7.17}$$

$$m^{(i)} = X^T(t^{[i]}, \vartheta) \; M \; \{ \; D^{[i]T} l^{[i]}(\omega) \; \} \tag{7.18}$$

$$m_j^{(i)T} = M \; \{ \; l^{[i]T}(\omega) D^{[i]} X(t^{[i]}, \vartheta) \mid \; \xi_1, \ldots, \xi_j \; \} \tag{7.19}$$

According to (7.13), (7.17)-(7.19) we obtain that the quantity $e_*(\cdot)$ can be represented in the form

$$e_*(\tau_*, w_*, \Delta) = \max_{m^*, m^*, \nu} [\, m^{*T} \tilde{w}_* + \kappa(\tau_*, m_*, \Delta, \nu) \,] \tag{7.20}$$

where

$$\kappa(\tau_*, m_*, \Delta, \nu) = \max_{\|l(\cdot)\|^* \leq 1} [\, M\{ \sum_{j=1}^{k} (\tau_{j+1} - \tau_j) \times$$

$$\times \max_{v \in Q} \min_{u \in P} (\sum_{i=d(j)}^{N} m_j^{(i)T}) \tilde{f}(u, v) \} \,]. \tag{7.21}$$

Here we must consider two cases. In the first case which is $\tau_* < t^{[g]}$ we have

$$m^* = m_* = \sum_{i=h}^{N} m^{(i)}, \quad h = g, \quad \sum_{i=g}^{N} \mu^{[i]*}(l^{[i]}) = \nu = 1 \tag{7.22}$$

In the second case ($\tau_* = t^{[g]}$) we have

$$m^* = \sum_{i=g}^{N} m^{(i)}, \quad \sum_{i=g}^{N} \mu^{[i]*}(l^{[i]}) = \nu = 1 \tag{7.23}$$

$$m_* = \sum_{i=h}^{N} m^{(i)}, \quad \sum_{i=h}^{N} \mu^{[i]*}(l^{[i]}) = \nu \in [0, 1] \tag{7.24}$$

where $h = g + 1$, $\mu^{[i]*}(l^{[g]}) = 1 - \nu$.

The main result is the following procedure for the construction of the values $\kappa(\cdot)$ of (7.21) and $e_*(\cdot)$ of (7.20) and for the proof of the required properties of these quantities.

8 Recurrence Procedures for the Program Extremum $e_*(\cdot)$

We offer a construction for $e_*(\cdot)$ (7.20) which uses a recurrence procedure for determining the upper convex hulls $\varphi_j(m) = \bar{\psi}_j(m)$ for some functions $\psi_j(m)$ whose argument m is a conditional mathematical expectation $m \in G$. Here G is a convex set. We define the upper convex hull $\bar{\psi}(m)$, $m \in G$ of a function $\psi(m)$ as a function $\varphi(m)$ that satisfies the following conditions.

1. The function $\varphi(m)$ is concave in $m \in G$, i.e. for any given $m^{(1)} \in G$, $m^{(2)} \in G$ and for any number $\lambda \in [0, 1]$ the inequality

$$\varphi(\lambda m^{(1)} + (1 - \lambda) m^{(2)}) \geq \lambda \varphi(m^{(1)}) + (1 - \lambda) \varphi(m^{(2)}) \tag{8.1}$$

is true.

2. The inequality

$$\psi(m) \leq \varphi(m) \tag{8.2}$$

holds for any $m \in G$.

3.The upper convex hull $\varphi(m)$ is the minimal concave function that satisfies conditions 1 and 2, i.e. for any concave function $\xi(m)$ that majorizes $\psi(m),\ m \in G$ the inequality

$$\varphi(m) \leq \xi(m), m \in G \tag{8.3}$$

is true.

We use the following property for the considered lower concave function.

4. Let $r = r_1, \ldots, r_n$ be an n-dimensional vector. For any $m,\ m \in G$ there exists a probability measure $\eta(R \mid m)$ on the set $r \in G$ such that the equalities

$$\varphi(m) = \int_{r \in G} \psi(r) \cdot \eta(dr \mid m) \tag{8.4}$$

$$m = \int_{r \in G} r \cdot \eta(dr \mid m) \tag{8.5}$$

are true.

In reality due to the theorem of Caratheodory one can select a measure $\eta(R \mid m)$ concentrated on a finite collection of points $r^{[s]}(m)$. This simplifies the theoretical proofs and practical constructions.

The calculation of the value $\kappa(\cdot)$ (7.21) is reduced to the following procedure.

According to (7.12), (7.13) we have the following restrictions

$$\sum_{i=g}^{N} \mu^{[i]*}(l^{[i]}(\omega)) \leq 1 \tag{8.6}$$

for the random variable $l^{[i]}(\omega)$.

Let $\nu \in [0, 1]$ be a parameter that evaluates an admissible sum of the norms $\mu^{[i]*}(l^{[i]})$ for the time interval $[\tau_j, \tau_{k+1}]$. Therefore we shall consider functions $\Delta\psi(\cdot)$, $\psi(\cdot)$ and $\varphi(\cdot)$ of the arguments τ, m and ν.

Let us assume that for the interval $[\tau_*, \vartheta]$ with a fixed τ_* the following equalities

$$\varphi_{k+1}(\tau_*, m, \nu) = 0, \quad \nu \in [0, 1] \tag{8.7}$$

$$\Delta\psi_k(\tau_*, m, \nu) = J(\tau_k, \tau_{k+1}, m, \nu) \tag{8.8}$$

are satisfied where the symbol $J(\tau_j, \tau_{j+1}, m, \nu)$ denotes the quantity

$$J(\tau_j, \tau_{j+1}, m, \nu) =$$

$$= (\tau_{j+1} - \tau_j) \max_{v \in Q} \min_{u \in P} \Big(\sum_{i=d(j)}^{N} m_j^{(i)T} \Big) \tilde{f}(\tau_j, u, v) \tag{8.9}$$

if $j \leq k$. Here $\tau_j \in \Delta\{\tau_j\}$, $m^{(i)}$, $\tilde{f}(\cdot)$ are the quantities in (7.21).

Let

$$\psi_k(\tau_*, m, \nu) = \Delta\psi_k(\tau_*, m, \nu) \tag{8.10}$$

and

$$\varphi_k(\tau_*, m, \nu) = \bar{\psi}_k(\tau_*, m, \nu) \tag{8.11}$$

Here $\bar{\psi}_k(m)$ denotes the upper convex hull of function $\psi_k(m)$, $m \in G_{k,\nu}(\tau_*)$, where

$$G_{k,\nu}(\tau_*) = \{\, m : m = D^{[N]T} l : \ \mu^{[N]*}(l) \leq \nu \,\} \tag{8.12}$$

Let us form a recurrent sequence of functions $\varphi_j(\cdot)$. Assume that a function $\varphi_{j+1}(\tau_*,\ m,\ \nu)$ and a set $G_{j+1,\nu}(\tau_*)$ are already constructed for $\nu \in [0, 1]$ and $m \in G_{j+1,\nu}(\tau_*)$ where

$$G_{j+1,\nu}(\tau_*) = \{\, m : m = \sum_{p=i+1}^{N} X^T(t^{[p]}, \vartheta) D^{[p]T} l^{[p]},$$

$$\sum_{p=i+1}^{N} \mu^{[p]*}(l^{[p]}) \leq \nu \,\} \tag{8.13}$$

In addition we have $i + 1 > g$ and $\tau_{j+1} < t^{[i+1]}$. Let us consider two cases. In the first case we have $\tau_{j+1} > t^{[i]}$. Then we assume that

$$G_{j,\nu}(\tau_*) = G_{j+1,\nu}(\tau_*), \tag{8.14}$$

$$\Delta\psi_j(\tau_*, m, \nu) = J(\tau_j, \tau_{j+1}, m, \nu) \tag{8.15}$$

$$\psi_j(\tau_*, m, \nu) = \varphi_{j+1}(\tau, m, \nu) + \Delta\psi(\tau_j, m_*, \nu), \tag{8.16}$$

$$\varphi(\tau_j, m_*, \nu) = \bar{\psi}(\tau_j, m_*, \nu) \tag{8.17}$$

for $m \in G_{j,\nu}(\tau_*)$. In (8.15) the term $J(\cdot)$ is the quantity (8.9).

In the second case we have $\tau_{j+1} = t^{[i]}$. Then we assume that

$$\psi_j(\tau_*, m, \nu) = \max_{m_*, \nu_*} [\Delta\psi_j(\tau_*, m^*, \nu) + \varphi_{j+1}(\tau_*, m_*, \nu)] \tag{8.18}$$

for

$$m^* = m_* + m, \qquad m_* \in G_{j+1,\nu_*}(\tau_*); \qquad \nu_* \leq \nu,$$

$$m = X^T(t^{[i]}, \vartheta) D^{[i]T} l, \qquad \mu^{[i]*}(l) \leq \nu - \nu_* \tag{8.19}$$

Let $G_{j,\nu}(\tau_*)$ be the set of all m^*(8.18) and let

$$\varphi_j(\tau_*, m, \nu) = \bar{\psi}_j(\tau_*, m, \nu), \quad m \in G_{j,\nu}(\tau_*) \tag{8.20}$$

Using an induction in j from $j = k$ to $j = 1$ we come to the value $\varphi_1(\tau_*, m, \nu)$. The equalities

$$\kappa(\tau_*, m_*, \Delta, \nu) = \varphi_1(\tau_*, m_*, \nu),$$
$$m_* \in G_{1,\nu}(\tau_*), \quad \nu = 1 \tag{8.21}$$

if $\tau_* < t^{[g]}$ and

$$\kappa(\tau_*, m_*, \Delta, \nu) = \varphi_1(\tau_*, m_*, \nu),$$
$$m_* \in G_{1,\nu_*}(\tau_*), \quad \nu_* = 1 - \mu^{[g]*}(l^{[g]}) \tag{8.22}$$

if $\tau_* = t^{[g]}$ holds for $\kappa(\cdot)$ (7.21).

These equalities (8.20), (8.21) can be proved through the properties 1-4 for the function $\varphi(m)$, $m \in G$.

From (7.20), (8.20), (8.21) we obtain the equality

$$e_*(\tau_*, w_*, \Delta) = \max_{m^*, m_*, \nu} [\, m^{*T} \tilde{w}_* + \varphi_1(\tau_*, m_*, \nu) \,] \tag{8.23}$$

where in the case $\tau_* < t^{[g]}$ we have the condition

$$m^* = m_* \in G_{1,\nu}(\tau_*), \quad \nu = 1 \tag{8.24}$$

and in the case $\tau_* = t^{[g]}$ we have

$$m^* = m_* + m, \qquad m_* \in G_{1,\nu_*},$$
$$m = X^T(\tau_*, \vartheta) D^{[g]T} l, \qquad \mu^{[g]*}(l) = 1 - \nu_*. \tag{8.25}$$

This concludes the construction of the value $e_*(\cdot)$ (7.13)-(7.15), (7.20).

We would like to emphasize that with all the functions $\psi_j(m), m \in G$, $j = 1, \ldots, k$ being concave we are within the regular case [2] and so that the random vector $l(\cdot)$ turns to be a deterministic vector $l(\cdot)$. In this case we can use the deterministic program synthesis [1,2]. One should note that once we follow the procedure for calculating the value $e_*(\tau_*, w_*, \Delta)$ we deal with functions of the argument m which is an n-dimensional vector. We also ought to note that our construction is based but on the dual variable l rather than the phase vector w of the system (7.3) or actually on the expectation m. Here the value m performs the same role as the modernized Lagrange multipliers in Pontryagin's maximum principle. We must also mention the important fact, that the construction offered here is based on recurrent sequence of functions $\varphi_j(\cdot)$ and directly permits to check the required properties of u-stability and v-stability for the value $e_*(\cdot)$ through the properties of function $\varphi(\cdot)$. The description of these properties will be given in the following section.

9 The u-Stability and v-Stability Conditions for the Program Extremum $e_*(\cdot)$

The functional γ_* (6.9) is positional, i.e. it can be represented in the form (2.3). It follows from Theorem 5.1 that any function $\rho_*(t,w)$ is the value $\rho_*^O(t,x)$ if it satisfies conditions 1, 2, 3u and 4v from §§4, 5 for the functional γ_*. It is obvious that the function $e_*(\tau_*,w_*,\Delta)$ constructed in §8 satisfies conditions 1 and 2.

Let us show that the function $e_*(\tau_*,w_*,\Delta)$ (8.22) also satisfies the following condition of u-stability.

3*u. Suppose we know a position $\{\tau_*,w_*\}\in G^*$, a parameter $\epsilon>0$, the instants of time $\tau_*=\tau_1\in\Delta_\delta$ and $\tau^*=\tau_2\in\Delta_\delta$ ($\Delta_\delta=\Delta\{\tau_j\}$ (7.4): $\tau_{j+1}-\tau_j\le\delta$, where $\delta>0$ is sufficiently small) and a control $v[\tau_*[\cdot]\tau^*)=\{v[\tau]=v[\tau_*]\in Q,\ \tau_*\le\tau<\tau^*\}$. Then there exists a control $u[\tau_*[\cdot]\tau^*)=\{u[\tau]=u[\tau_*]\in P,\ \tau_*\le\tau<\tau^*\}$ which in combination with $v[\tau_*[\cdot]\tau^*)$ generates from the initial position $\{\tau_*,w_*\}$ deterministic motion $w[\tau_*[\cdot]\tau^*]=\{w[\tau],\ \tau_*\le\tau\le\tau^*\}$ of the w-model (7.3) that comes to the state $w[\tau_*[\tau^*]\tau^*]=w[\tau^*]=w^*$ such that

$$e_*(\tau^*,w^*,\Delta^*)\le e_*(\tau_*,w_*,\Delta_*)+\epsilon(\tau^*-\tau_*) \tag{9.1}$$

whenever

$$\tau_*<\tau^*\le t^{[g]} \tag{9.2}$$

and that whenever

$$\tau_*=t^{[g]} \tag{9.3}$$

the inequality

$$\max[\mu^{[g]}(D^{[g]}w[t^{[g]}]),e_*(\tau^*,w^*,\Delta^*)]\le$$
$$\le e_*(\tau_*,w_*,\Delta_*)+\epsilon(\tau^*-\tau_*) \tag{9.4}$$

is true. Here Δ_* and Δ^* are the following partitions

$$\Delta_*=\Delta_*\{\tau_j\}=\{\ \tau_*=\tau_1,\ \tau^*=\tau_2,\ \tau_j<\tau_{j+1},\ \tau_{k+1}=\vartheta\ \} \tag{9.5}$$

$$\Delta^*=\Delta^*\{\tau_j\}=\{\ \tau^*=\tau_2,\ \ \tau_j<\tau_{j+1},\ \ \tau_{k+1}=\vartheta\ \} \tag{9.6}$$

It should be noted that the condition 3*u is a convenient variant of the general condition 3u from §4 for the particular case considered here.

Let us prove condition 3*u. At first we consider the case (9.2) and particularly the case $\tau^*<t^{[g]}$. Here g is a number (6.10). We then have

to prove that the inequality (9.1) is true. According to (8.22), (8.23) we take the difference

$$\Delta e_* = e_*(\tau^*, w^*, \Delta^*) - e_*(\tau_*, w_*, \Delta_*) = \max_{m^*_{\tau^*}}[m^{*T}_{\tau^*}\tilde{w}_{\tau^*}+$$

$$+\varphi_1(\tau^*, m^*_{\tau^*}, \nu_{\tau^*})] - \max_{m^*_{\tau^*}}[m^{*T}_{\tau_*}\tilde{w}_{\tau_*} + \varphi_1(\tau_*, m^*_{\tau_*}, \nu_{\tau_*})] =$$

$$= m^{*oT}_{\tau^*}\tilde{w}_{\tau^*} + \varphi_1(\tau^*, m^{*o}_{\tau^*}, \nu_{\tau^*}) - m^{*oT}_{\tau_*}\tilde{w}_{\tau_*} - \varphi_1(\tau_*, m^{*o}_{\tau_*}, \nu_{\tau_*}) \qquad (9.7)$$

Here in the case $\tau^* < t^{[g]}$ according to (8.23) we have $m^{*o}_{\tau^*} \in G_{1,\nu^o_{\tau^*}}(\tau^*)$, $m^{*o}_{\tau_*} \in G_{1,\nu^o_{\tau^*}}(\tau^*)$ and $\nu^o_{\tau^*} = \nu^o_{\tau_*} = 1$. The symbols τ_* and τ^* emphasize that the extremal values $e_*(\tau_*, \cdot)$, $e_*(\tau^*, \cdot)$ and the corresponding quantities m^*, $\tilde{w}$ and ν be defined in §§7, 8 are calculated for the initial times τ_* and τ^*.

If we substitute in (9.7) the values $m^{*o}_{\tau_*}$ and $\nu^o_{\tau_*}$ that maximize $e_*(\tau_*, \cdot)$ by $m^{*o}_{\tau^*}$ and $\nu^o_{\tau^*}$ for $e_*(\tau^*, \cdot)$, then the following estimate will be true

$$\Delta e_* \leq m^{*oT}_{\tau^*}(\tilde{w}_{\tau^*} - \tilde{w}_{\tau_*})+$$

$$+\varphi_1(\tau^*, m^{*o}_{\tau^*}, \nu^o_{\tau^*}) - \varphi_1(\tau_*, m^{*o}_{\tau^*}, \nu^o_{\tau^*}) \qquad (9.8)$$

According to (7.17) and the formula of Cauchy the equality

$$\tilde{w}_{\tau^*} = \tilde{w}_{\tau_*} + \int_{\tau_*}^{\tau^*} \tilde{f}(\tau, u[\tau], v[\tau])d\tau \qquad (9.9)$$

is valid. Hence

$$m^{*oT}_{\tau^*}(\tilde{w}_{\tau^*} - \tilde{w}_{\tau_*}) = m^{*oT}_{\tau^*}\int_{\tau_*}^{\tau^*} \tilde{f}(\tau, u[\tau], v[\tau])d\tau \qquad (9.10)$$

For any $\epsilon > 0$ there exists a number $\delta_\Delta(\epsilon) > 0$ such that

$$m^{*oT}_{\tau^*}(\tilde{w}_{\tau^*} - \tilde{w}_{\tau_*}) \leq m^{*oT}_{\tau^*} \cdot \tilde{f}(\tau_*, u[\tau_*], v[\tau_*])\times$$

$$\times (\tau^* - \tau_*) + \epsilon(\tau^* - \tau_*), \qquad (9.11)$$

provided $\tau^* - \tau_* \leq \delta_\Delta(\epsilon)$.

Then according to (9.8), (9.11) the estimate

$$\Delta e_* \leq m^{*oT}_{\tau^*} \cdot \tilde{f}(\tau_*, u[\tau_*], v[\tau_*])(\tau^* - \tau_*)+$$

$$+\varphi_1(\tau^*, m^{*o}_{\tau^*}, \nu^o_{\tau^*}) - \varphi_1(\tau_*, m^{*o}_{\tau^*}, \nu^o_{\tau^*}) + \epsilon(\tau^* - \tau_*) \qquad (9.12)$$

does hold.

With regards to the calculations of the quantity $\varphi_1(\cdot)$ in §8 the inequality

$$\varphi_1(\tau_*, m^{*o}_{\tau^*}, \nu^o_{\tau^*}) \geq \varphi_1(\tau^*, m^{*o}_{\tau^*}, \nu^o_{\tau^*})+$$

$$+ (\tau^* - \tau_*) \max_{v \in Q} \min_{u \in P} m_{\tau^*}^{*oT} \tilde{f}(\tau_*, u, v) \tag{9.13}$$

is true. Besides, we also have the trivial inequality

$$\max_{v \in Q} \min_{u \in P} m_{\tau^*}^{*oT} \tilde{f}(\tau_*, u, v) \geq \min_{u \in P} m_{\tau^*}^{*oT} \tilde{f}(\tau_*, u, v[\tau_*]) \tag{9.14}$$

for each $v[\tau_*] \in Q$.

According to (9.12) - (9.14) we then get the following estimate

$$\Delta e_* \leq (\tau^* - \tau_*)[\ m_{\tau^*}^{*o} \cdot \tilde{f}(\tau_*, u[\tau_*], v[\tau_*]) -$$

$$- \min_{u \in P} m_{\tau^*}^{*o}\ \tilde{f}(\tau_*, u, v[\tau_*])\ ] + \epsilon(\tau^* - \tau_*) \tag{9.15}$$

The set $m_{\tau^*}^{*o}$ and the set of vectors $\tilde{f}(\cdot, u)$ where $u \in P$ are convex and compact. Similarly to [2] we now employ a fixed point theorem (see, for example [13], p.495). According to this Theorem we establish that there exists a pair $\{m_{\tau^*}^{*o},\ u^o[\tau_*]\}$, $m_{\tau^*}^{*o} \in G_{1,\nu_{\tau^*}^o}(\tau^*)$, $u^o[\tau_*] \in P$ that satisfies the following condition

$$\min_{u \in P} m_{\tau^*}^{*oT} \tilde{f}(\tau_*, u, v[\tau_*]) = m_{\tau^*}^{*oT} \tilde{f}(\tau_*, u^o[\tau_*], v[\tau_*]) \tag{9.16}$$

At the same time the pair of controls $v[\tau_*]$, $\tau_* \leq \tau < \tau^*$ and $u^o[\tau_*]$, $\tau_* \leq \tau < \tau^*$ generate from the initial position $\{\tau_*, w_*\}$ a state $\tilde{w}_{\tau^*} = X(\vartheta, \tau^*)w[\tau^*]$. Here the vector $m_{\tau^*}^{*o}$ of the pair $\{m_{\tau^*}^{*o},\ u^o[\tau_*]\}$ maximizes $e_*(\tau^*, w^*, \Delta^*)$, where $w^* = w[\tau^*]$. According to (9.15), (9.16) the condition (9.1) is true in the case of (9.2), once $\tau^* < t^{[g]}$.

In the other case, namely when $\tau^* = t^{[g]}$ in (9.2), we prove condition (9.1) within a similar scheme but with some variations. These variations take into account a transformation of the constructions for the sets $G_{j,\nu}(\tau^*)$ and $G_{j,\nu}(\tau_*)$. Namely with $\tau^* = t^{[g]}$ we have the following equalities due to (8.22), (8.24)

$$e_*(\tau^*, w^*, \Delta) = \max_{m_{\tau^*}^*, m_{*,\tau^*}, \nu_{\tau^*}} [\ m_{\tau^*}^{*T} \tilde{w}_{\tau^*} + \varphi_1(\tau^*, m_{*\tau}*, \nu_\tau *)\ ] =$$

$$= m_{\tau^*}^{*oT} \tilde{w}_{\tau^*} + \varphi_1(\tau^*, m_{*\tau}^o*, \nu_\tau^o*) \tag{9.17}$$

Here with regards to (8.24) the quantities $m_{\tau^*}^*$, $m_{*\tau}*$ and $\nu_\tau*$ satisfy the following conditions

$$m_{\tau^*}^* = m_{*\tau} * + m, \qquad m_{*\tau}* \in G_{2,\nu_\tau *}(\tau^*) \tag{9.18}$$

$$m = X^T(\tau^*, \vartheta) D^{[g]T} l, \qquad \mu^{[g]*}(l) = 1 - \nu_\tau * \tag{9.19}$$

$$m_\tau^* * \in G_{1,\nu}(\tau^*), \qquad \nu = 1 \tag{9.20}$$

In this case, namely, when $\tau^* = t^{[g]}$ we have $\tau_* < t^{[g]}$ according to (8.22), (8.23) the quantity $e_*(\tau_*, w_*, \Delta_*)$ is defined by the equality

$$e_*(\tau_*, w_*, \Delta_*) = \max_{m^*_{\tau_*}, \nu^{\tau_*}} [\, m^{*T}_{\tau_*} \tilde{w}_{\tau_*} + \varphi_1(\tau_*, m^*_{\tau_*}, \nu_{\tau_*})\,] =$$

$$= m^{*oT}_{\tau_*} \tilde{w}_{\tau_*} + \varphi_1(\tau_*, m^{*o}_{\tau_*}, \nu^o_{\tau_*}) \tag{9.21}$$

where $m^*_{\tau_*} \in G_{1,\nu_{\tau_*}}(\tau_*)$, $\nu_{\tau_*} = \nu^o_{\tau_*} = 1$.

Taking into account the constructions in §8 we have the following equality

$$\varphi_1(\tau_*, m^*_{\tau_*}, 1) = \bar{\psi}_1(\tau_*, m^*_{\tau_*}, 1), \tag{9.22}$$

where

$$\psi_1(\tau_*, m^*_{\tau_*}, 1) = J(\tau_1, \tau_2, m^*_{\tau_*}) + \max_{m_{*\tau_*}, \nu} \varphi_2(\tau_*, m_{*\tau_*}, \nu) =$$

$$= J(\tau_1, \tau_2, m^*_{\tau_*}) + \max_{m_{*\tau_*}, \nu} \varphi_1(\tau^*, m_{*\tau_*}, \nu) \tag{9.23}$$

Here $J(\cdot)$ is the quantity (8.9).

Hence the inequalities

$$\varphi_1(\tau_*, m^{*o}_{\tau^*}, 1) \geq J(\tau_1, \tau_2, m^{*o}_{\tau^*}, 1) + \varphi_1(\tau^*, m^o_{*\tau}*, \nu^o_{\tau^*}) \tag{9.24}$$

and

$$e_*(\tau_*, w_*, \Delta_*) \geq m^{*oT}_{\tau^*} \tilde{w}_{\tau_*} + \varphi_1(\tau^*, m^o_{*\tau}*, \nu^o_{\tau^*}) \tag{9.25}$$

are true. After that we follow (9.7) - (9.16). This proves condition (9.1) for case (9.2) when $\tau^* = t^{[g]}$.

We have thus proved that condition (9.1) of u-stability 3*u is valid for case (9.2).

Consider now the case (9.3) with condition (9.4). Let partition Δ_* (9.5) have a sufficiently small step so that

$$\tau_* = t^{[g]}, \qquad \tau^* < t^{[g+1]} \tag{9.26}$$

In this case according to (8.22), (8.23), (8.24) we arrive at the following equalities

$$e_*(\tau^*, w^*, \Delta^*) = m^{*oT}_{\tau^*} \tilde{w}_\tau * + \varphi_1(\tau^*, m^{*o}_{\tau^*}, \nu^o_{\tau^*}) \tag{9.27}$$

$$e_*(\tau_*, w_*, \Delta_*) = \max_{m^*_{\tau_*}, m_{*\tau_*}, \nu_{\tau_*}} [\, m^{*T}_{\tau_*} \tilde{w}_{\tau_*} + \varphi_1(\tau_*, m_{*\tau_*}, \nu_{\tau_*})\,] \tag{9.28}$$

where $m^*_{\tau_*} \in G_{1,\nu_\tau}(\tau_*)$, $\nu_{\tau_*} = 1$, $m_{*\tau_*} \in G_{2,\nu}(\tau_*)$ and

$$m^*_{\tau_*} = m_{*\tau_*} + m, \quad m = X^T(\tau_*, \vartheta) D^{[g]T} l, \quad \mu^{[g]*}(l) \leq 1 - \nu \tag{9.29}$$

According to (9.28), (9.29) the following equality

$$e_*(\tau_*, w_*, \Delta_*) = \max_{m_{*\tau_*},\nu}[(1-\nu)\mu^{[g]}(D^{[g]}w[\tau_*]) + m^T_{*\tau_*}\tilde{w}_{\tau_*} +$$

$$+\varphi_1(\tau_*, m_{*\tau_*}, \nu)] = (1-\nu^o_{\tau_*})\mu^{[g]}(D^{[g]}w_*) + m^{oT}_{*\tau_*}\tilde{w}_{\tau_*} +$$

$$+ \varphi_1(\tau_*, m^o_{*\tau_*}, \nu^o_{\tau_*}) \tag{9.30}$$

is also true.

Consider the difference

$$\Delta^o = \max\ [\ \mu^{[g]}(D^{[g]}w_*), e_*(\tau^*, w^*, \Delta^*)\ ] - e_*(\tau_*, w_*, \Delta_*) \tag{9.31}$$

which will be estimated in order that the condition (9.4) could be proved.

With regards to (9.27), (9.30) we have the following estimate

$$\Delta e_* = e_*(\tau^*, w^*, \Delta) - e_*(\tau_*, w_*, \Delta) \le m^{*oT}_{\tau^*}\tilde{w}_{\tau^*} +$$

$$+\varphi_1(\tau^*, m^{*o}_{\tau^*}, \nu^o_{\tau^*}) - m^{*oT}_{\tau^*}\tilde{w}_{\tau_*} - \varphi_1(\tau_*, m^{*o}_{\tau^*}, \nu^o_{\tau^*}) -$$

$$- (1-\nu^o_{\tau^*})\mu^{[g]}(D^{[g]}w_*) \tag{9.32}$$

Again, as in the first case (9.2), by selecting for the given $v[\tau_*]$ a suitable pair $\{m^{*o}_{\tau^*}, u^o[\tau_*]\}$ we come to the inequality

$$m^{*0T}_{\tau^*}\cdot\tilde{w}_{\tau^*} + \varphi_1(\tau^*, m^{*0}_{\tau^*}, \nu^0_{\tau^*}) \le m^{*0T}_{\tau^*}\cdot\tilde{w}_{\tau} +$$

$$+ \varphi_1(\tau_*, m^{*0}_{\tau^*}, \nu^0_{\tau^*}) + \epsilon\cdot(\tau^* - \tau_*) \tag{9.33}$$

where the state $\tilde{w}_{\tau^*} = X(\vartheta, \tau^*)w[\tau^*]$ is generated from the initial position $\{\tau_*,\ w_*\}$ by the pair of controls $u^o[\tau_*]$, $v[\tau_*]$, $\tau_* \le \tau < \tau^*$.

By using (9.31), (9.32) we derive the estimate

$$\Delta^0 \le \max\ [\ \mu^{[g]}(D^{[g]}w_*), m^{*0T}_{\tau^*}\cdot\tilde{w}_{\tau^*} + \varphi_1(\tau^*, m^{*0}_{\tau^*}, \nu^0_{\tau^*})\ ] -$$

$$-\ [\ m^{*0T}_{\tau^*}\cdot\tilde{w}_{\tau_*} + \varphi_1(\tau_*, m^{*0}_{\tau^*}, \nu^0_{\tau^*}) + (1-\nu^0_{\tau^*})\cdot\mu^{[g]}(D^{[g]}w_*)\ ] \tag{9.34}$$

Suppose at first that condition

$$m^{*0T}_{\tau^*}\cdot\tilde{w}_{\tau^*} + \varphi_1(\tau^*, m^{*0}_{\tau^*}, \nu^0_{\tau^*}) = e_*(\tau^*, w^*, \Delta^*) > \mu^{[g]}(D^{[g]}w_*) \tag{9.35}$$

be true. Then according to (9.33)-(9.35) we get the inequality

$$\max\ [\ \mu^{[g]}(D^{[g]}w_*), e_*(\tau^*, w^*, \Delta^*)\ ] - e_*(\tau_*, w_*, \Delta_*) \le$$

$$\le \epsilon\cdot(\tau^* - \tau_*) - (1-\nu^0_{\tau^*})\cdot\mu^{[g]}(D^{[g]}w_*) \le \epsilon\cdot(\tau^* - \tau_*) \tag{9.36}$$

that coincides with condition (9.4) for 3*u.

Now assume that condition

$$\mu^{[g]}(D^{[g]}w_*) \geq m_{\tau^*}^{*0T} \cdot \tilde{w}_{\tau^*} + \varphi_1(\tau^*, m_{\tau^*}^{*0}, \nu_{\tau^*}^0) = e_*(\tau^*, w^*, \Delta^*) \qquad (9.37)$$

is true. Then in accordance with the definition of the value $e_*(\tau_*, w_*, \Delta_*)$ and regarding the fact that (8.22) is also maximized over $\nu \in [0, 1]$ we deduce the following inequality

$$e_*(\tau_*, w_*, \Delta_*) = \max_{m_*, \nu} [(1-\nu) \cdot \mu^{[g]}(D^{[g]}w_*), m_*^T \cdot \tilde{w}_{\tau_*} +$$

$$+ \varphi_1(\tau_*, m_*, \nu)] \geq \mu^{[g]}(D^{[g]}w_*) \qquad (9.38)$$

and, therefore, the inequality

$$\mu^{[g]}(D^{[g]}w_*) \leq e_*(\tau_*, w_*, \Delta_*) + \epsilon \cdot (\tau^* - \tau_*) \qquad (9.39)$$

holds.

With the aid of (9.37), (9.39) we then come to the required condition (9.4). The property of u-stability 3*u is thus completely proved.

Similarly to the property 3*u we are now able to formulate a particular case of the v-stability property for the function $e_*(\tau_*, w_*, \Delta_*)$ (8.22). The formulation of this property can be obtained if we interchange the symbols u and v, and substitute the symbols "$\leq$" and "+" by the symbols "$\geq$" and "- " respectively.

Namely the property of v-stability for program extremum $e_*(\cdot)$ is formulated in the following form.

4*v. Suppose that given are a position $\{\tau_*, w_*\} \in G^*$, a parameter $\epsilon > 0$, the times $\tau_* = \tau_1$ and $\tau^* = \tau_2$ and a control $u[\tau_*[\cdot]\tau^*) = \{ u[\tau] = u[\tau_*] \in P, \tau_* \leq \tau < \tau^* \}$. Then there exists a control $v[\tau_*[\cdot]\tau^*) = \{ v[\tau] = v[\tau_*] \in Q, \ \tau_* \leq \tau < \tau^* \}$ which in combination with $u[\tau_*[\cdot]\tau^*)$ generates from the initial position $\{\tau_*, w_*\}$ the deterministic motion $w[\tau_*[\cdot]\tau^*]$ of the w-model (7.3) that reaches to the state $w[\tau^*] = w^*$ such that

$$e_*(\tau^*, w^*, \Delta^*) \geq e_*(\tau_*, w_*, \Delta_*) - \epsilon \cdot (\tau^* - \tau_*) \qquad (9.40)$$

when $\tau_* \leq \tau^* = t^{[g]}$, or that

$$\max\ [\mu^{[g]}(D^{[g]}w[t^{[g]}]), e_*(\tau^*, w^*, \Delta^*)] \geq$$

$$\geq e_*(\tau_*, w_*, \Delta_*) - \epsilon \cdot (\tau^* - \tau_*) \qquad (9.41)$$

when $\tau_* = t^{[g]}$.

We can prove the property 4*v similarly to 3*u with some changes. These changes are the following. At first for the estimation of Δe_* which is done similarly to (9.7) we now use the quantities $m_{\tau_*}^{*0}$ and $\nu_{\tau_*}^0$ instead of the maximizers $m_{\tau^*}^{*0}$ and $\nu_{\tau^*}^0$. Besides that we now use the 3-rd property (8.3) for the upper convex hull $\varphi(\cdot)$. This will allow to estimate $\varphi_1(\tau^*, m_{\tau_*}^{*0}, \nu_{\tau_*}^0)$ $-\varphi_1(\tau_*, m_{\tau_*}^{*0}, \nu_{\tau_*}^0)$ and therefore to obtain the proof of the property 4*v.

10 The Construction of the Optimal Strategies

According to the procedures of §§4,5 the desired optimal strategies $u^0(\cdot) = u^0(t, x, \epsilon)$ and $v^0(\cdot) = v^0(t, x, \epsilon)$ can be formed as extremal control forces $u^0(t_i, x[t_i], \epsilon)$ and disturbance forces $v^0(t_i, x[t_i], \epsilon)$ given by conditions (4.12), (4,13) and (5.2), (5.3) respectively. Here the accompanying points $w_u^0[t_i]$ and $w_v^0[t_i]$ satisfy by conditions (4.8) - (4.11) and (5.4) with respect to the function $\rho(t, w) = \rho^0(t, w)$, which is u-stable and v-stable. As it was shown in §§7-9 we can approximate a value $\rho^0(t, x)$ by the function $e_*(t, x, \Delta)$ (7.20), (8.22) which is also $u-$ and v-stable. Here we transform the conditions that describe the points $w_u^0[t_i]$ and $w_v^0[t_i]$ to the following form

$$e_*(t_i, w_u^0[t_i], \Delta) = \min_{w \in K(\epsilon, t_i)} e_*(t_i, w, \Delta) =$$

$$= \min_{w \in K(\epsilon, t_i)} \max_{m^*, m_*, \nu} [< m^*, X(\vartheta, t_i)w > + \varphi_1(t_i, m_*, \nu)] \tag{10.1}$$

$$e_*(t_i, w_v^0[t_i], \Delta) = \max_{w \in K(\epsilon, t_i)} e_*(t_i, w, \Delta) =$$

$$= \max_{w \in K(\epsilon, t_i)} \min_{m^*, m_*, \nu} [< m^*, X(\vartheta, t_i)w > + \varphi_1(t_i, m_*, \nu)] \tag{10.2}$$

where m^*, m_* and ν satisfy the conditions (8.23), (8.24) at $\tau_* = t_i$.

According to (4.8), (4.9),(4.13) and (5.3) we can transform the condition $w \in K(\epsilon, t_i)$ into conditions

$$| s_u[t_i] | = | x[t_i] - w | \leq \tilde{\lambda}(\epsilon, t_i) \tag{10.3}$$

$$| s_v[t_i] | = | w - x[t_i] | \leq \tilde{\lambda}(\epsilon, t_i) \tag{10.4}$$

where

$$\tilde{\lambda}^2(\epsilon, t_i) = (\epsilon + \epsilon \cdot (t_i - t_0)) \exp 2M \cdot (t_i - t_0) \tag{10.5}$$

Here for the system (7.1) we have $M = \max_{t_0 \leq t \leq \vartheta} | A(t) |$, where $| A(t) | = \max_{|x| \leq 1} | A(t)x |$.

Besides that, the concavity of the function $\varphi_1(\cdot, m_*)$ in m_* makes it possible to transpose the minimum and the maximum operations in (10.1). We therefore have the following conditions

$$e_*(t_i, w_u^0[t_i], \Delta) = \max_{m^*, m_*, \nu} \min_{|s_u[t_i]| \leq \tilde{\lambda}(\epsilon, t_i)} [< m^*, X(\vartheta, t_i)x[t_i] > -$$

$$- < m^*, X(\vartheta, t_i)s_u[t_i] > + \varphi_1(t_i, m_*, \nu)] \tag{10.6}$$

$$e_*(t_i, w_v^0[t_i], \Delta) = \max_{m^*, m_*, \nu} \max_{|s_v[t_i]| \leq \tilde{\lambda}(\epsilon, t_i)} [< m^*, X(\vartheta, t_i)x[t_i] > +$$

$$+ < m^*, X(\vartheta, t_i)s_v[t_i] > + \varphi_1(t_i, m_*, \nu)] \tag{10.7}$$

In (10.6) we use the equality (4.13), i.e. $w = x[t_i] - s_u[t_i]$, in (10.7) - the equality (5.3), i.e. $w = x[t_i] + s_v[t_i]$.

It follows from (10.3)-(10.7) that vectors $s_u^0[t_i]$ and $s_v^0[t_i]$ are given by the equalities

$$s_u^0[t_i] = \tilde{\lambda}(\epsilon, t_i) \cdot \frac{X^T(\vartheta, t_i) m_u^{*0}}{| X^T(\vartheta, t_i) m_u^{*0} |} \tag{10.8}$$

$$s_v^0[t_i] = \tilde{\lambda}(\epsilon, t_i) \cdot \frac{X^T(\vartheta, t_i) m_v^{*0}}{| X^T(\vartheta, t_i) m_v^{*0} |} \tag{10.9}$$

where the vectors m_u^{*0} and m_v^{*0} are the solutions of the following problems

$$< m_u^{*0}, X(\vartheta, t_i)x[t_i] > + \varphi_1(t_i, m_{*u}^0, \nu) - \tilde{\lambda}(\epsilon, t_i) \mid X^T(\vartheta, t_i) m_u^{*0} \mid =$$

$$= \max_{m^*, m_*, \nu} [< m^*, X(\vartheta, t_i)x[t_i] > + \varphi_1(t_i, m_*, \nu) -$$

$$- \tilde{\lambda}(\epsilon, t_i) \mid X^T(\vartheta, t_i) m^* \mid] \tag{10.10}$$

$$< m_v^{*0}, X(\vartheta, t_i)x[t_i] > + \varphi_1(t_i, m_{*v}^0, \nu) + \tilde{\lambda}(\epsilon, t_i) \mid X^T(\vartheta, t_i) m_v^{*0} \mid =$$

$$= \max_{m^*, m_*, \nu} [< m^*, X(\vartheta, t_i)x[t_i] > + \varphi_1(t_i, m_*, \nu) +$$

$$+ \tilde{\lambda}(\epsilon, t_i) \mid X^T(\vartheta, t_i) m^* \mid] \tag{10.11}$$

under the conditions (8.23), (8.24) at $\tau_* = t_i$.

The optimal control $u^0(t_i, x[t_i], \epsilon)$ and the optimal disturbance $v^0(t_i, x[t_i], \epsilon)$ are determined by the following conditions

$$\max_{v \in Q} [m_u^{*0T} X(\vartheta, t_i) f(t_i, u^0(t_i, x[t_i], \epsilon), v)] =$$

$$= \min_{u \in P} \max_{v \in Q} [m_u^{*0T} X(\vartheta, t_i) f(t_i, u, v)] \tag{10.12}$$

$$\min_{u \in P} [m_v^{*0T} X(\vartheta, t_i) f(t_i, u, v^0(t_i, x[t_i], \epsilon)] =$$

$$= \max_{v \in Q} \min_{u \in P} [m_v^{*0T} X(\vartheta, t_i) f(t_i, u, v)] \tag{10.13}$$

where $f(\cdot)$ is a function of (7.1).

Thus, if we admit that the control loop contains a high-speed computer, we can organize the following control process. During a very short time interval $t_i^{(u)} \leq t < t_i^{(u)} + \alpha$ following a time instant $t_i^{(u)} \in \Delta t_i^{(u)}$ (3.6), (3.12) the problem (10.7), (10.9) is solved for a known value $x[t_i^{(u)}]$ and the control force $u^0[t_i^{(u)}] = u^0(t_i^{(u)}, x[t_i^{(u)}], \epsilon_u) \in P$ is calculated with the help of condition (10.11). This control is applied to the controlled system (7.1) during the time interval $t_i^{(u)} + \alpha \leq t < t_{i+1}^{(u)} + \alpha$. In some cases it is possible to simulate this a control process using available computers. Similar considerations are valid for the second player who forms the optimal disturbance $v^0[t_i^{(u)}]$ within the time step $t_i^{(v)} \leq t < t_{i+1}^{(v)}$.

11 Examples

In accordance with the procedure described above in this paper we shall consider some simple model examples where the value $\rho^o(t, x)$ and the optimal strategies $u^o(t, x, \epsilon)$ and $v^o(t, x, \epsilon)$ are constructed for particular cases.

Let the motion of a controlled system be described by the differential equation in the Newtonian form

$$\ddot{h} = f(t)u + g(t)v, \quad t_o \leq t \leq \vartheta \tag{11.1}$$

Here h is scalar, $f(t)$ and $g(t)$ are given continuous functions of time t and the controls u and v of the players satisfy the constraints

$$| u | \leq 1, \quad | v | \leq 1 \tag{11.2}$$

The quality index in the differential game considered above in §§1-3 is defined here by the equality

$$\gamma = \max \left[\, | \, \dot{h}[t^{[1]}] \, |, \, | \, h[\vartheta] \, | \, \right] \tag{11.3}$$

where $t^{[1]} \in (t_o, \vartheta)$ is a given instant of time and ϑ is a fixed termination time for the process.

Introduce the phase vector

$$x = \begin{bmatrix} x_1 \\ x_2 \end{bmatrix} = \begin{bmatrix} h \\ \dot{h} \end{bmatrix} \tag{11.4}$$

Vector x satisfies the differential equation

$$\dot{x} = Ax + F(t)u + G(t)v \tag{11.5}$$

where

$$A = \begin{pmatrix} 0 & 1 \\ 0 & 0 \end{pmatrix}, \quad F(t) = \begin{bmatrix} 0 \\ f(t) \end{bmatrix}, \quad G(t) = \begin{bmatrix} 0 \\ g(t) \end{bmatrix} \tag{11.6}$$

The quality index γ has the form

$$\gamma = \max\ [\ |\ D^{[1]}x[t^{[1]}]\ |,\ |\ D^{[2]}x[t^{[2]}]\ |\ ] \tag{11.7}$$

where $t^{[2]} = \vartheta$,

$$D^{[1]} = [0\ 1],\quad D^{[2]} = [1\ 0] \tag{11.8}$$

The equation (11.5) is a special case of the general equation (7.1) and the index γ (11.7) is a special case of γ (2.5). In the case of equation (11.5) the condition (7.2) is satisfied.

The considered problem for system (11.5), (11.6) with the quality index γ was solved according to §§7 - 10. Thus, we construct the value of the game $\rho^o(t, x)$ and the optimal strategies $u^o(\cdot) = u^o(t, x, \epsilon)$ and $v^o(\cdot) = v^o(t, x, \epsilon)$.

The control process for the considered system was simulated by a computer IBM PC/AT for the following parameters

$$t_* = t_0 = 0,\ \vartheta = 4,\ t^{[1]} = 1,\ x_* = \{4.0, 2.0\},\ f(t) = \vartheta - t,\ g(t) \equiv 1 \tag{11.9}$$

For the selected initial position $\{t_*, x_*\}$ the value of the game was $\rho^o(t_*,\ x_*) = 0.163$. In Figure 1 we depict in the plane $\{x_1 = h,\ x_2 = \dot{h}\}$ the corresponding realization $x_{u^o,v^o}[t] = \{x_{1_{u^o,v^o}}[t], x_{2_{u^o,v^o}}[t]\}$, $t_* \le t \le \vartheta$, generated by the strategies $u^o(\cdot)$ and $v^o(\cdot)$ with a time step $\delta = t_{i+1} - t_i = 0.005$ and a parameter $\epsilon = 0.01$. In this case the simulation produce the value γ close to $\rho^o(t_*, x_*)$, as expected :

$$\gamma = \max\ [\ |\ x_{2_{u^o,v^o}}[t^{[1]}]\ |,\ |\ x_{1_{u^o,v^o}}[\vartheta]\ |\ ] =$$

$$= 0.167 \approx \rho^o(t_*, x_*) = 0.163 \tag{11.10}$$

For the system (11.5), (11.6) we also consider the quality index

$$\gamma = \max\ [\ |\ x_1[t^{[1]}] - c^{[1]}\ |,\ |\ x_1[\vartheta] - c^{[2]}\ |\ ] \tag{11.11}$$

where $c^{[1]}$ and $c^{[2]}$ are given points.

This problem was solved for the parameters (11.9) and for

$$c^{[1]} = 9.0,\quad c^{[2]} = -6.0 \tag{11.12}$$

For the selected initial position $\{t_*,\ x_*\}$ (11.9) for γ (11.11) the value of the game was $\rho^o(t_*,\ x_*) = 4.546$. The motion $x_{u^o,v^o}[t] = \{x_{1u^o,v^o}[t],$ $x_{2u^o,v^o}[t]\}$, $t_* \le t \le \vartheta$ for this case is shown in Figure 2. Here

$$\gamma = \max\ [\ |\ x_{1u^o,v^o}[t^{[1]}] - c^{[1]}\ |,\ |\ x_{1u^o,v^o}[\vartheta] - c^{[2]}\ |\ ] =$$

$$= 4.623 \approx \rho^o(t_*, x_*) = 4.546 \tag{11.13}$$

We note that in both cases {(11.5), (11.7)} and {(11.5), (11.11)} the value $\rho^0(t_*, x_*)$ was calculated in advance, i.e. at the initial instant t_* or before and for the initial parameters (11.9), (11.12). But the values γ (11.10) and (11.13) were found only at the termination instant ϑ for the corresponding realization of the motion $x_{u^o,v^o}[t_*[\cdot]\vartheta]$.

References

[1] Krasovskii, N.N. and Subbotin, A.I. (1988). *Game-Theoretical Control Problems*, Springer-Verlag, New-York.

[2] Krasovskii, N.N. (1985). *Controlling of a Dynamical System*, Nauka, Moscow (in Russian).

[3] Kurzhanskii A.B. (1977). *Control and Observation under Conditions of Uncertainty*, Nauka, Moscow (in Russian).

[4] Osipov, Ju.S. (1975). *An Informational Game Problem.* Optimization Techniques IFIP Technical Conference. Springer- Verlag, Berlin, Heidelberg, New-York.

[5] Subbotin, A.I. and Chentsov, A.G. (1981). *Optimization of a Guarantee in Control Problems*, Nauka, Moscow (in Russian).

[6] Krasovskii, N.N. and Tret'jakov, V.E. (1983). *Stochastic program synthesis of an ensuring control.* Probl. Cont. Inf. Theory. 12, 2, p. 70 - 95.

[7] Krasovskii, N.N.and Reshetova, T.N. (1988). *On the program synthesis of a guaranteed control.* Probl. Cont. Inf. Theory. 17, 6 p. 333 - 343.

[8] Krasovskii, A.N.(1980). *A differential game for the positional functional.* Dokl. Akad. Nauk SSSR. 253, 6, p. 1303 - 1307; English transl. in Soviet Math. Dokl., 22, 1, p. 251 - 255.

[9] Krasovskii, A.N.(1987). *Construction of mixed strategies on the basis of stochastic programs.* Prikl. Mat. Mekh. 51, 2, p. 186 - 192; English transl. in J. Appl. Math. Mech. 51.

[10] Krasovskii, A.N.(1991). *Control under minimax of an integral functional.* Dokl. Akad. Nauk SSSR, 320, 4, p. 785 - 788; English transl. in Soviet Math. Dokl. 44, 2, p. 525 - 528.

[11] Lokshin, M.D.(1990). *On the optimal control of a linear system under the condition of the integral disturbance constraint.* Probl. Cont. Inf. Theory, 19, 2, p. 111 - 127.

[12] Arkin, V.I. and Levin, V.L.(1972). *Convexity of the values of vector integrals, theorems of measurable selection and variational problems.* Uspekhi matem. nauk, 12, 3; English transl. in Russian Math. Surveys, 27, 3, p. 21 - 86.

[13] Bohnenblust, H.F. and Karlin, S.(1950). *On a Theorem of Ville, Contributions to the theory of games, I,* Ann. Math. Studies, Princeton, p. 155 -160.

[14] Liptser, R.S. and Shirjaev, A.N.(1977). *Statistics of Random Processes,* Nauka, Moscow (in Russian); English transl. vols. I, II, Springer-Verlag.

N.N.Krasovskii
Institute of Mathematics and Mechanics
Ural Branch, Russian Academy of Sciences
S.Kovalevsky st. 16, Ekaterinburg 620219, Russia

A.N.Krasovskii
University of Ural
Leninski Prospect 52, Ekaterinburg 620151, Russia

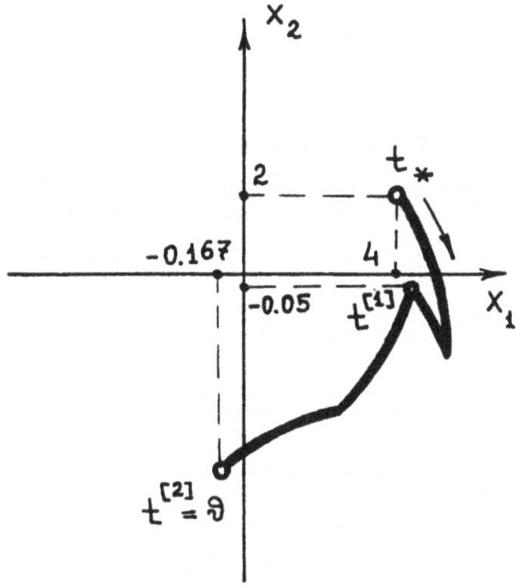

Figure 1

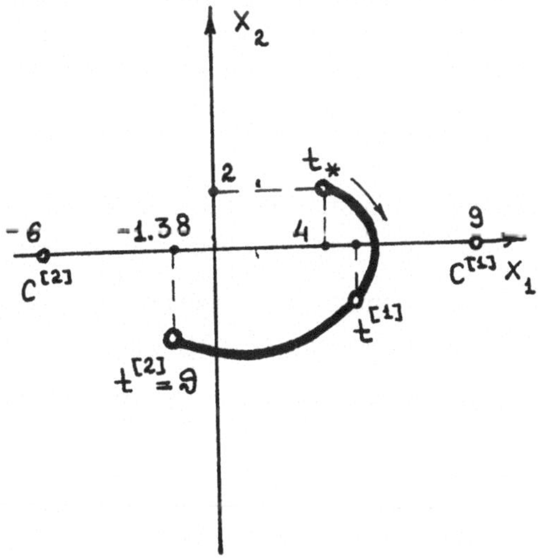

Figure 2

Global Methods in Optimal Control Theory

V.F. Krotov

The paper contains a survey of theoretical and practical results connected with sufficient conditions for global optimality of controlled dynamic processes. Both discrete and continuous time systems and systems with distributed parameters are discussed.

0 Introduction

Consider a controlled dynamic system whose state at any instant of time t is described by the vector x. The term "controlled" means that the system equations are sub-definite. Let this sub-definiteness with any fixed t, x be represented by some non-fixed element u of the set U. To control the system means to close its equations by the assigned function $u(t)$ or $u(t, x)$. The first method of closing is referred to as the one of programmed control while the second as the one of feedback control. The aims of control are:

to bring the system to the assigned conditions which are reflected in the constraints of the state x at the end of the control period $t = T$;

to provide process admissibility which is connected with the fulfillment of current constraints imposed on the state $x(t)$;

to fulfill some qualitative dynamic requirements on the system such as stability.

These problems require to be solved in an optimal way with a minimum expenditure of the costly resources: energy, time, etc. At the same time the control possibilities are limited and there can also be other complicated factors that affect the process, for example: incomplete information on the system.

The functional optimization problems are the object of variational calculus. However the above problems of optimal control for dynamic systems essentially differ from the canonical statements of variational calculus. In the process of formalizing and analyzing these problems some new mathematical ideas were suggested. And in the fifties and sixties a new branch of applied mathematics, namely the theory of optimal control has been developed as a result of a synthesis of these ideas with classical variational calculus.

Variational calculus provides us with two methods to investigate optimal processes: the Euler-Lagrange variation method and the Hamilton-Jacobi method. The first one is connected with a generalization of conditions for a local minimum of the real variable functions $f'(x) = 0$, $f''(x) \geq 0$ for the problems of functional analysis. The second one is connected with embedding of the optimal process construction problem in the family of these problems for various initial conditions. Since the second one is designed to solve a more general problem, it is, in general, more complex. Therefore the main tool of the variational calculus is the method of variations . Among the new mathematical ideas of the theory of optimal control the following two should be noted: Pontryagin's maximum principle and Bellman's method of dynamic programming. The first one develops a variation method for modern problems and the second one the Hamilton-Jacobi method. The first method happened to be adequate for the problem of programmed control synthesis while the second one is aimed at to the problem of feedback control synthesis.

Among the new ideas of optimal control theory lies the approach based on sufficient conditions for global optimality of control processes and the mathematical techniques of global bounds. This approach was proposed in the beginning of the sixties. In the course of the recent 30 years a great deal of investigations were carried out to master this method, to substantiate and develop it. A significant experience in its application had also been acquired.

In the paper we describe the basics of this approach and give a survey of related results . The first of these were given in [1-5], see also [6,7]. A development of this approach went in the following directions:

(i) The elaboration of new analytical tools for particular classes of problems, of mainly singular nature, and those that have no solution in the admissible set. The specification of new classes of solutions and their describing equations.
(ii) The elaboration of optimization algorithms of a broad destination that use non-local operations for improving controls or lower bounds.
(iii) The extension of sufficient conditions of optimality to necessary con-

ditions.

(iv) The adjustment of the approach to new classes of optimal control problems (with a multidimensional argument, a delayed argument, with incomplete information etc.) and to other similar problems like global bounds for functionals in the feasible regions or the synthesis of invariant controls.

(v) The application of the elaborated methodology to practical problems of automatic control, mechanics, applied physics, and mathematical economics which led to an extensive experience.

1 Problem Statement

Consider a triplet of variables $t \in A$, $x \in X$, $u \in U$, where the sets A, X, U are given. These variables will be called the *argument*, the *state* and the *control*, respectively. We also introduce a pair of functions $v = (x(t), u(t))$, $v : A \times X \Rightarrow U$. The latter altogether will be called *an admissible controlled process* (the set of all those processes is denoted by D) if the following conditions are satisfied:

1. The state and control restrictions

$$(x(t), u(t)) \in V(t), \quad \forall t \in A, \tag{1.1}$$

where $V(t)$ is a given family of subsets of the direct product $X \times U$. From the relation (1.1) it is possible to distinguish a separate state and a control restriction:

$$x(t) \in V_x(t); \qquad u(t) \in V^x[t, x(t)]; \tag{1.2}$$

where $V_x(t)$ and $V^x(t, x)$ are the projections of the set $V(t)$ on the set X and of the intersection of the set $V(t)$ with the set defined by a given constant value x, respectively. We also further admit the following notation: the superscript will denote the corresponding intersection and the subscript will denote the projection.

2. A process equation (an equation of motion) is satisfied that defines a process class. Three kinds of these equations are then considered:

a) A multistage (discrete argument) process. Here the set A is an integer sequence $\{\tau,\ \tau + 1,\ \tau + 2,\ \dots, T\}$ and the process equation has the form:

$$x(t+1) = f[t, x(t), u(t)], \quad t = \tau, \tau + 1, ..., T - 1, \tag{1.3}$$

where $f(t, x, u)$ is a given function, $f : A \times X \times U \Rightarrow X$

b) An ordinary process. Here the set A is a segment $[\tau, T]$ of the real line; X, U are real vector spaces R^n and R^r, respectively, and the process equation has the form:

$$\dot{x}(t) = f[t, x(t), u(t)], \quad t \in (\tau, T), \tag{1.4}$$

where $f(t, x, u)$ is a given continuous vector-function: $A \times X \times U \Rightarrow X$.

c) A multiargument process (a process with distributed parameters). The argument t is a real vector, $t = (t^1, t^2, ..., t^m) \in R^m$.

The set A is the region: $0 \le t^j \le T^j$, $j = 1, 2, ..., m$. The sets X, U are real vector spaces R^n and R^r, respectively, and the process equation has the form:

$$dx(t)/dt = f[t, x(t), u(t)], \tag{1.5}$$

where $dx/dt = \{\partial x^i / \partial t^j\}$, $f = \{f_j^i\}$; $(i = 1, 2, ..., n), j = 1, 2, \ldots m$, and the functions $f_j^i(t, x, u)$ are defined and continuous on the direct product $A \times X \times U$.

3. It is necessary to impose some function-theoretical restricts on the admissible process $v = (x(t), u(t))$ in the cases b) and c) in order to satisfy the equations (1.4), (1.5). In the case b) this is usually the piecewise continuity of the function $u(t)$ and the continuity and piecewise differentiability of the function $x(t)$ or the measurability of $u(t)$ and the absolute continuity of $x(t)$. In the case c) this is the continuity of the function $u(t)$ and the differentiability of the function $x(t)$.

The functions $x(t)$ and $u(t)$ are called *the trajectory* and *the control program*, respectively. Introduce also a set D_u of admissible control programs $u(t)$ which is the projection of the set D on the set of all functions $u(t), A \Rightarrow U$. Similarly introduce a set D_x of admissible trajectories $x(t)$. The process argument t will be also named as *time* although it is not always that the variable t has this physical meaning. When state set X is a vector space we shall name the coordinates of vector x as *phase coordinates*.

Let us define an optimality criterion for "ordinary processes" by

$$I(v) = \int_A f^0[t, x(t), u(t)]dt + F[x(T)] + F_0[x(0)] \tag{1.6}$$

where the functions $f^0(t, x, u)$, $F(x)$, $F_0(x)$ are defined and continuous over the direct product $A \times X \times U$ and the space X, respectively. The integral form $(F(x) = F_0(x) = 0)$ of the functional and the terminal form $(f^0 = 0)$ are also used. These three forms are equivalent and may be mutually reduced. We come to the criterion for multistage processes by substituting the integral (1.6) by the sum from τ to $T - 1$ and to the criterion for a multiargument process if we substitute the term $F[x(T)]$ for a functional $F[x(\vartheta)]$, $\vartheta \in S$, where S is the bound on the region A.

The object of our analysis is the problem of optimizing a dynamic process. Let us consider some basic statements of the mathematical problems related to optimization.

Problem 1. Find an element $\bar{v} \in D$ named *optimal* such that

$$I(\bar{v}) = \min I(v), \quad v \in D. \tag{1.7}$$

This problem appears to be most laconic and clear. But it is not always solvable. The set D considered here may not always contain an element $\bar{v}$ that satisfies the exact equality (1.1). It is therefore necessary to prove some special existence theorems to apply this statement. We can add that the absence of a minimum is a typical property of many applied variational problems.

Problem 2. Find a sequence $\{v_s\} \subset D$ such that:

$$I(v_s) \to d = \inf I(v), \quad v \in D, \quad s \to \infty. \tag{1.8}$$

This sequence is known as a *minimizing sequence.* It always exists due to the definition of the exact lower bound $\inf I(v)$, $v \in D$, if the set D is not empty.

A control process $v \in D$ (an admissible process) is a solution of an independent complicated problem, namely.

Problem 3. Let us assume that there is a set E such that $D \subset E$ and that there exists a sufficiently simple algorithm for constructing the element $v \in E$. From these elements it is required to find an admissible one: $v \in D$. In more details : let a functional $\rho(v)$, $v \in N$ be defined which satisfies the conditions:

$$\rho(v) = 0 \ \text{ for } v \in D, \quad \rho(v) > 0 \text{ for } v \in E \backslash D.$$

This functional is called the distance between the element $v \in E$ and the set D. It is required to find a sequence $\{v_s\} \subset E$ such that $\rho(v_s) \to 0$.

In fact the Problems 1(2) and 3 are solved simultaneously:

Problem 4. Introduce a functional $J(v)$, $v \in E$, such that $J(v) = I(v)$ for $v \in D$. It is required to find a sequence $\{v_s\} \subset E$ that satisfies the condition

$$I(v_s) \to d = \inf I(v), \quad v \in D; \quad \rho(v_s) \to 0, \quad s \to \propto . \tag{1.9}$$

The approximate variants of these problems will also be considered. Let us introduce a set $D_\epsilon(\rho)$ which satisfies the following conditions: $v \in N$, $\rho(v) < \epsilon$, $\epsilon > 0$. We shall call $D_\epsilon(\rho)$ an ϵ-extension of the set D in the metric ρ.

The approximate variant of the Problem 3 is : to find an ϵ - extended solution $v \in D_\epsilon(\rho)$. In the approximate variant of the Problem 4 it is required that the solution is also η-optimal on $D_\epsilon(\rho)$: $I(v) - d_\epsilon(\rho) < \eta$, $\eta > 0$, $d_\epsilon(\rho) = \inf_{D_\epsilon(\rho)} I(v)$. Two other variants can also be considered independently, namely, when $\epsilon = 0$ i.e an η-optimality strictly on D, and $\eta = 0$ i.e. a strong optimality on the ϵ-extension of D. In connection with the definitions of the above there arises a question of well posedness of the problem (correctness), which is whether we will have $d_\epsilon(\rho) \to d$ if $\epsilon \to 0$.

In this paper we define the set E for ordinary processes as follows. The pair of functions $v = (x(t), u(t)) \in E$ satisfies restrictions (1.1) but not necessarily (1.4). Namely, we assume also that both functions are piecewise continuous and $x(t)$ is piecewise differentiable. The distance ρ is defined as:

$$\rho(v) = \int_A \mid z(t) \mid dt + \sum_{t \in \beta} \mid x(t) \mid \, |_{t-0}^{t+0}; \tag{1.10}$$

$$z(t) = \dot{x}(t) - f[t, x(t), u(t)];$$

where β is the set of arguments in which the function $x(t)$ allows jumps. The second element in the sum (1.10) is taken into account only in the continuous case.

We call a solution to the described problem an *optimal open loop (program) control* according to the nature of its control i.e. the control $u(t)$. An important role in control theory is played by solutions given in the feedback form. Let there be given a function $u(t,x)$, $T \times X \Rightarrow U$, such that equation (1.4) when "closed" by function $u(t) = u(t, x(t))$ with the initial condition $x(\tau) = y$ does have an admissible solution $x(t) \in D_x$

We will say that the trajectory $x(t)$ and the process $v = (x(t), u(t))$ *correspond* to the control (or feedback policy) function $u(t,x)$.

Let us consider a family $A(\tau, y)$ of optimization problems with assigned initial conditions $(\tau,\ y = x(\tau))$. Let there be a set B of initial conditions (τ, y) and let the rest of the conditions for the problem be fixed. We shall include the dependence on the initial conditions into the notation like $D(\tau, y)$, $d(\tau, y)$, $v(\tau, y)$ etc. Let there exist a unique process $v(\tau, y) \in D(\tau, y)$ that corresponds to $u(t,x)$ for every $(\tau, y) \in B$. We will then call the function $u(t,x)$ a *synthesizing control function* or simply a *control synthesis* for set B of the initial conditions.

Let a synthesis $\bar{u}(t,x)$ satisfy the following condition:

$$I(\bar{v}(\tau,y)) - d(\tau,y) \leq \epsilon, \quad \epsilon > 0, \quad \forall \tau, y \in B.$$

The function $\bar{u}(t,x)$ will be called an *ϵ-optimal control synthesis* or an *optimal synthesis* if $\epsilon = 0$. This means that the construction of an optimal synthesizing function is equivalent to the solution of a family of open loop optimal control problems with initial conditions $(\tau, y) \in B$.

2 Equivalent Representations

2.1 The ordinary processes

Let us introduce a class Π of real differentiable functions $\varphi(t,x) : A \times X \Rightarrow R^1$ and the following constructions:

$$R(t,x,u) = \partial\varphi/\partial x \; f(t,x,u) - f^0(t,x,u) + \partial\varphi/\partial t \; ; \qquad (2.1)$$

$$G_0(x) = F_0(x) + \varphi(\tau,x); \quad G(T,x) = F(T,x) + \varphi(T,x); \qquad (2.2)$$

$$L(v,\varphi) = G(T,x(T)) - \int_A R(t,x(t),u(t))dt - G_0(x(\tau)); \qquad (2.3)$$

where A is a segment $[\tau, T]$, $f(t,x,u) = \{f^i(t,x,u)\}$, $i = \overline{1,n}$, is the right-hand side of the process equation (1.4), $f^0(t,x,u)$ is an integrand (1.6), $F_0(x)$, $F(T,x)$ are the initial and the terminal terms in the optimality criterion (1.6), $\partial\varphi/\partial x \; f(t,x,u) = \partial\varphi/\partial x^i \; f^i(t,x,u)$ is a Euclidean product. Here and below repeated indices will indicate a summation of the respected terms.

The functionals $L(v,\varphi)$ and $I(v)$ coincide for any admissible process $v \in D$ and functions $\varphi(t,x) \in \Pi$:

$$L(v,\varphi) = I(v), \quad \forall v \in D, \quad \varphi \in \Pi; \qquad (2.4)$$

Indeed, using (2.1) and (1.4), we get $R[t,x(t),u(t)] = d\varphi[t,x(t)]/dt - f^0$, $(x(t), u(t)) \in D$. Substituting the latter into (2.3) and remembering that functions $x(t)$ are continuous, we obtain $L = I$.

Equation (2.3) defines a family of representations of the functional $I(v)$ on the set D with a parameter-function $\varphi(t,x)$ that corresponds to a family of equivalent optimal control problems (L, D). The representations are used to define the above mentioned extended family $(L(v,\varphi), E)$ and the respective optimality conditions. They are also useful for other purposes. For example, we shall use this to solve the improvement problem for a process $v_0 \in D$ by choosing $L(v,\varphi)$ in such a way that it is obvious how to choose an admissible process $v \in D$ such that $L(v, \varphi) = I(v) < I(v_0) = L(v_0, \varphi)$.

In general the equation (2.4) is not valid for nonadmissible processes, when $v \notin D$. Let $v \in E$, where E is the class of piecewise-continuous functions pairs $(x(t), u(t))$ described above which does not satisfy equation (1.4). It is easy to obtain:

$$L = I(v) + \int_A \varphi_x(t, x(t))z(t)dt + \sum_{t\in\beta} \varphi(t, x(t)) \,|_{t-0}^{t+0}, \tag{2.5}$$

where β is the set of points of discontinuity of the function $x(t)$, and $z(t)$ is defined by (1.10).

The functional $L(v, \varphi)$ has an obvious property: $L(v, \varphi_1) = L(v, \varphi_2) = I(v)$ for any $v \in D$, $\varphi_1, \varphi_2 \in \Pi$. But in general: $L(v, \varphi_1) \not\equiv L(v, \varphi_2)$, if $v \notin D$. The functional $L(v, \varphi)$ is invariant relative to special transformation:

$$L(v, \varphi_1) = L(v, \varphi_2), \quad \varphi_1 = \varphi_2 + c(t), \quad \forall\, v \in E \tag{2.6}$$

where $c(t)$ is any differentiable function of the argument t.

Let us single out the representations $L(v, \varphi)$ where the parameter φ is linear with respect to the state function: $\varphi(t, x) = \psi(t)x$. These are the usual Lagrange's functionals and the coefficient ψ is the Lagrange multiplier for the constraint (1.4).

2.2 Multistage processes

Let us introduce a class Π of functionals $\varphi(t, x)$: $A \times X \Rightarrow R^1$, where A is a sequence of integers $\{\tau, 1, \ldots, T\}$ and X is the set of states. Let us define the following functionals:

$$R(t, x, u) = \varphi[t+1, f(t, x, u)] - \varphi(t, x) - f^0(t, x, u); \tag{2.7}$$

$$L(v, \varphi) = \sum_{t=\tau}^{T-1} R[t, x(t), u(t)] \;-\; G_0[x(\tau)] \;+\; G[T, x(T)]; \tag{2.8}$$

where $G_0[x]$, $G[T, x]$ are defined by (2.2); f, f^0 are the terms mentioned above in the Problem Statement. It is not difficult to prove properties (2.4) and (2.6) for these constructions and also the following representation:

$$L(v, \varphi) = I(v) + \sum_{t=\tau+1}^{T} \{\varphi[t, x(t)] - \varphi[t, x(t) - z(t-1)]\},$$

$$z(t) = x(t+1) - f[t, x(t), u(t)].$$

2.3 Multiargument processes

Let us introduce a class Π of vector-functions $\varphi(t,x) = \{\varphi^k\}$, $k = 1, ..., m$: $A \times X \Rightarrow R^m$ defined and continuously-differentiable for all t, x. We form the functions

$$R(t,x,u) = \varphi^j_{x^i}\, f^i_j(t,x,u) + \varphi^j_{x^j}(t,x) - f^0(t,x,u), \tag{2.9}$$

where $\varphi^j_{x^i} = \partial\varphi^j/\partial x^i$; $f(t,x,u) = \{f^i_j(t,x,u)\}$, $i = \overline{1,n}$, $j = \overline{1,m}$, is the right-hand side of the process equation (1.5), $f^0(t,x,u)$ is integrand. We further write:

$$G(x(\vartheta)) = F(x(\vartheta)) + \int_S \varphi^j \cos(\bar{n}, t^j) ds, \tag{2.10}$$

where the surface S is the boundary of the domain A, $0 \le t^j \le T^j$, $j = \overline{1,m}$; $t = \vartheta$ are argument values on the surface S; $\bar{n}$ is the external normal vector to the surface S and ds is an element of area on S; $F(x(\vartheta))$ is the boundary functional. In greater detail for the case one has $m = 2$:

$$\begin{aligned} G[x(\vartheta)] &= F[x(\vartheta)] + \\ &+ \int_0^{T^2} \{\varphi^1[T^1, t^2, x(T^1,t^2)] - \varphi^1[0, t^2, x(0,t^2)]\} dt^2 + \\ &+ \int_0^{T^1} \{\varphi^2[t^1, T^2, x(t^1,T^2)] - \varphi^2[t^1, 0, x(t^1,0)]\} dt^1. \end{aligned} \tag{2.11}$$

We also define the functional

$$L(v,\varphi) = G[x(\vartheta)] - \int_A R(t, x(t), u(t)) dt; \tag{2.12}$$

It is easy to prove the properties (2.4) and (2.6) for these constructions.

3 Global Bounds

3.1 Ordinary processes

Let us fill up the row of useful constructions which are related with function $\varphi \in \Pi$. Namely:

$$\mu(t) = \max_{(x,u)\in V(t)} R(t,x,u), \quad \forall t \in (\tau, T); \tag{3.1}$$

$$m = \min G(T,x), \quad x \in V(T); \quad m_0 = \max G_0(x), \quad x \in V(0); \tag{3.2}$$

$$l(\varphi) = m - \int_A \mu(t) dt - m_0; \tag{3.3}$$

where $A = [\tau, T]$; $R(t, x, u)$, $G(T, x)$, $G_0(x)$ have been defined by (2.1), (2.2).

The function φ will be called a *bounding function* when the functional $l(\varphi)$ is defined. We then have: $L(v, \varphi) \geq l(\varphi)$, $\forall v \in E$, $\varphi \in \Pi$. It is easy to obtain this inequality comparing the three terms in the right hand sides (2.3) and (3.3), and due to (2.4) and the inequality :

$$I(v) \geq l(\varphi), \quad \forall v \in D, \quad \varphi \in \Pi. \tag{3.4}$$

The functional $l(\varphi)$ is an invariant of the transformation (2.6). Therefore the bounding function $\varphi(t, x)$ can be defined in such a way that $\mu(t) = 0$, $m = 0$. The function φ is then said to be *normalized*. For the latter we have: $l(\varphi) = -m_0$.

Thus a lower bound for the functional I on D corresponds to any bounding function φ. This inequality can be used to obtain sufficient optimality conditions and directly to obtain global bounds for the respective criterion. These are also useful for the estimation of the proximity of approximate solutions to the optimal ones. From these bounds it is possible to pick up the best:

$$\bar{l} = \sup\ l(\varphi), \quad \varphi \in \Pi. \tag{3.5}$$

In the special case where the initial state is fixed $x(\tau) = x_0$ one has

$$l(\varphi) = m - \int_A \mu(t)dt - \varphi(\tau, x_0) \tag{3.6}$$

Example 2.5.2. Let us consider the family of problems

$$\left.\begin{array}{rcl} I(v,T) & = & \int_\tau^T (u^2 - x^2)dt \to \min \\ \dot{x} & = & u; \quad x(\tau) = x(T) = 0 \end{array}\right\} \tag{3.7}$$

with a parameter T. The Lagrangian $\varphi = \psi(t)x$ is not a bounding one for this problem. Indeed we have

$$R(t, x, u) = \psi u - u^2 + x^2 + \dot{\psi} x.$$

The supremum $\mu(t)$ (3.1) does not exist for any function $\psi(t)$. Let us analyze the bounding function of the following form:

$$\varphi(t, x) = \sigma(t)x^2; \quad \varphi_x \equiv 2\sigma(t)x; \quad \varphi_t = \dot{\sigma}(t)x^2; \tag{3.8}$$

Then

$$R(t, x, u) \equiv \varphi_x u - u^2 + x^2 + \varphi_t = (1 + \dot{\sigma})x^2 + 2\sigma x u - u^2. \tag{3.9}$$

For the existence of a maximum of the function R maximum it is sufficient that the quadratic form $-R$ is positive definite for every $t \in (\tau, T)$. The necessary and sufficient condition for this is

$$-R_{uu} \equiv 2 \geq 0; \quad R_{xx}R_{uu} - R_u^2 \equiv -4(1+\sigma^2+\dot{\sigma}) \geq 0.$$

The first inequality is satisfied identically, whereas the second imposes a constraint on $\sigma(t)$, namely

$$1+\sigma^2+\dot{\sigma} \leq 0; \quad t \in (\tau, T) \tag{3.10}$$

We then have

$$\mu(t) = \max_{x,u} R(t,x,u) = 0; \quad l(\varphi) = \varphi(T,\tau) - \int_\tau^T \mu(t)dt - \varphi(\tau, x_0) = 0,$$

if such a function $\sigma(t)$ does exist. For example, let $T = 1/3$ and $\sigma(t) = -2t$. It is readily seen that (3.10) is satisfied on the segment $[0, 1/3]$, so that $I(v) \geq 0$.

Let us find a differentiable function $\sigma(t)$ which satisfies (3.10) over a maximal interval $[\tau, T]$. This function is obtained by solving the equation

$$1+\sigma^2+\dot{\sigma} = 0. \tag{3.11}$$

We choose a solution which is continuous over the maximal interval (τ, T). The general solution of this equation is

$$\sigma(t) = -\tan(t+c).$$

A particular solution which is continuous over the maximal interval is

$$\sigma = -\tan(t+\pi/2)$$

The continuity interval of this solution is (τ, π). Thus for any $T < \pi$: $I(T,v) \geq 0, \forall v \in D$.

The bounding function of the form (3.8) does not exist when $T > \pi$. But in the latter case a lower bound for the functional $I(v)$ does not exist either, namely: $\inf I(v) = -\propto$, $v \in D$. Indeed, let us consider a sequence of trajectories

$$x_\alpha(t) = \begin{cases} \alpha \sin t, & t < \pi/2 \\ \alpha, & \pi/2 \leq t \leq T - \pi/2 \\ \alpha \sin(T-t), & T - \pi/2 < t \leq T \end{cases}$$

The process $v_\alpha = (x_\alpha(t), u_\alpha(t) = \dot{x}_\alpha(t))$ is admissible for any α. We then have according to (3.7): $I(v_\alpha) = -\alpha^2(T-\pi) \to -\propto, \quad \alpha \to \propto$.

3.2 Multistage processes

The inequality (2.4) is also valid in this case if

$$l(\varphi) = m - \sum_{t=1}^{T-1} \mu(t) - m_0 \qquad (3.12)$$

The terms in the right hand side are defined by (3.1), (3.2), (2.7).

3.3 Multiargument processes

Here the global bound is the following

$$l(\varphi) = m - \int_A \mu(t)dt; \qquad (3.13)$$

where the integrand $\mu(t)$ is defined by (2.9),(3.1), $m = \inf_{x(\vartheta)} G[x(\vartheta)]$, and $G[x(\vartheta)]$ - by (2.10) or (2.11).The inequality (3.4) is valid in also this case. The bound $l(\varphi)$ is then invariant with respect to the transformation (2.6).

Example 2.5.6.(A.W.J. Stoddart [70]). Let us consider the problem:

$$I(v) = \int_A [px^2(t) + u^2(t)]dt \to \min,$$

$$dx(t)/dt = u(t); \quad | u(t) | \leq 1 \text{ on } A, \quad x(t) = 1 \text{ on } S$$

where $t = (t^1, t^2)$; $A = \{(t^1, t^2)\}$: $\{| t^1 | \leq 1, | t^2 | \leq 1\}$; $x \in R^1$ $(n = 1)$; $u = (u_1, u_2)$; $\rho > 0$.

With $\varphi(t, x) = atx$, we obtain the lower bounds for I according to (2.9), (3.1), (2.11). We have

$$R(t, x, u) = atu + 2ax + px^2 + u^2;$$

$$\mu(t) = -a^2/p - a^2t^2/4 \text{ for } | at | \leq 2$$

$$\mu(t) = -a^2/p + 1 - | at | \text{ for } | at | \geq 2$$

$$m = -a$$

and for $| a | \leq \sqrt{2}$:

$$l = -8a - (4/p + 2/3)a^2;$$

while for $\sqrt{2} \leq | a | \leq 2$:

$$l = -8a - 4a^2/p + 4 - 8a^{-2}(\pi/4 - \cos^{-1} | a | /2)/3 - 2\sqrt{(4a^{-2} - 1)} -$$

$$- | a | \sqrt{(4 - a^2)/3} - 4 | a | [\sqrt{2} + \sin h^{-1} 1 - \sinh^{-1} \sqrt{(4a^{-1} - 1)}]/3;$$

for $| a |\geq 2$:

$$l = -8a - 4a^2/p + 4 - 2\pi a^{-2}/3 - 4 \mid a \mid (\sqrt{2} + \sinh^{-1} 1)/3.$$

Appropriate choices of parameter a for various values of p give the following lower bounds l:

p	1	2	3	4	6	8	10	100
l	3.43	6.00	8.02	9.75	12.99	16.11	19.19	156.5
I	3.62	6.67	9.33	11.73	16.00	20.00	24.00	204.0

The last line gives minimal values i of I for those symmetric $x(t)$ with $u(t^1, t^2) = (s, 0)$, where $| t^2 |\leq t^1$, and $u(t^1, t^2) = (0, s)$, where $| t^2 |\geq t^1$. Specifically,

$$I = 4p(18 + p)/(18 + 3p) \text{ with } s = 2p(6 + p) \text{ for } p \leq 6$$

$$I = 4 + 2p \text{ with } s = 1 \text{ for } p \geq 6$$

If I is not sufficiently close to l for the assigned accuracy, there are two possibilities: (a) l is not a sufficiently good lower bound; or (b) I is not close enough to the infimum. One can seek to raise l by considering a more general φ or to lower I by considering a wider class of $v = (x(\cdot), u(\cdot))$ within D.

4 Sufficient Conditions of Optimality

4.1 The case of fixed boundary state

Let us next introduce a useful construction related to a bounding function φ. It is the pair of functions $\tilde{v}(\varphi) = (\tilde{x}(t), \tilde{u}(t))$, that satisfy the condition:

$$R(t, \tilde{x}(t), \tilde{u}(t)) = \max_{(x,u)\in V(t)} R(t, x, u) = \mu(t), \quad t \in A^* \tag{4.1}$$

and more generally, the family of sets

$$\tilde{V}(t) = \text{Arg} \max_{(x,u)\in V(t)} R(t, x, u); \tag{4.2}$$

$$\tilde{E} = \{v \in E : \ (x(t), u(t)) \in \tilde{V}(t), \ \forall t\};$$

where A^* is the interior of the control interval (region) A. We first consider the case of fixed boundary conditions : $x(\vartheta) = x_0(\vartheta), \vartheta \in S = A \backslash A^*$. The optimality conditions have a unified form for all the three types of processes. The function $R(t, x, u)$ is then defined by (2.1), (2.7), (2.9), accordingly.

Theorem 4.1. *Suppose there exists a bounding function $\bar{\varphi}(t, x)$ such that the process $\bar{v} = \tilde{v}(\bar{\varphi})$ is defined and admissible, $\tilde{v}(\bar{\varphi}) \in D$. Then*

$$I(\bar{v}) = \min_{v \in D} I(v) = l(\bar{\varphi}) = \max_{\varphi \in \Pi} l(\varphi) \tag{4.3}$$

I.e. the process $\bar{v}$ minimizes the functional $I(v)$ on the set D, the functional $l(\bar{\varphi})$ is its minimum, and the pair $(\bar{\varphi},\ \bar{v})$ is a solution of the pair of dual problems (4.3).

Proof. By the definition of the process $\tilde{v}$ and by formulae (2.3), (3.3) and these for $L(\varphi, v)$, $l(\varphi)$ we have: $I(\tilde{v}) = L(\bar{\varphi}, \tilde{v}) = l(\bar{\varphi})$. This equation in common with the inequality $I(v) \geq l(\bar{\varphi})$, $\forall v \in D$, gives us (4.3). The theorem is thus proved.

This theorem is a global sufficient condition of optimality. These optimality conditions give us the basis for an approach to the solution of variational problems (Principle of Optimality [2]). It is the following: for different bounding functions one is to find the solutions $\tilde{x}(t), \tilde{u}(t)$ to the family of extremal problems (4.1) with parameter t . Then $\varphi = \bar{\varphi}$ is to be taken such that the process $\tilde{v} = (\tilde{x}(t), \tilde{u}(t))$ would satisfy the process equations and the above mentioned properties of the functions $x(t), u(t)$ in continuous time (the satisfaction of condition (1.1) is a consequence of the construction of $\tilde{v}$). In general the function $\bar{\varphi}$ is essentially nonunique and when specified for different subclasses of Π it leads to different methods of solutions. The function $\bar{\varphi}(t, x) \in \Pi$ is said to be *a solving function* and the set of all these functions is called $\bar{\Pi}$. The specification of the pair $\bar{v} \in D$, $\bar{\varphi} \in \Pi$ indicates that the pair of dual problems (4.3) had been solved.

Example 4.1. Consider the problem from Example 2.4.2. We have: $\tilde{v} = (\tilde{x}(t),\ \tilde{u}(t)) = (0, 0)$, $t \in (0, T)$, for any function $\varphi(t, x) = \sigma(t)x^2$, that satisfies (3.10). Evidently, $\tilde{v} \in D$. Therefore the process $\tilde{v}$ is optimal when $T < \pi$. Let us make a special emphasis on the fact that the optimal process is unique but the solving function $\varphi(t, x)$ is not unique.

Theorem 2.1 gives a nonunique formulation of the global optimality conditions. We have chosen it to achieve maximal laconism and simplicity sacrificing generality. Let us now write down a few more general sufficient optimality conditions that are necessary for our analysis.

Theorem 4.2. *Suppose there exist a bounding function $\bar{\varphi}(t,x)$ and an admissible process $\bar{v} = (\bar{x}(t), \bar{u}(t)) \in D$ such that*

$$R(t, \bar{x}(t), \bar{u}(t)) = \max_{(x,u)\in V(t)} R(t,x,u) = \mu(t), \quad t \in A^*, \tag{4.4}$$

then (4.3) is valid.

Theorem 4.3. *Suppose there exist a bounding function $\bar{\varphi}(t,x)$ and an admissible processes sequence $\{v_s\} = \{x_s(t), u_s(t)\} \subset D$ such that*

$$\int_A \{R[t, x_s(t), u_s(t)]dt - \mu(t)\}dt \to 0 \tag{4.5}$$

Then

$$I(v_s) \to \inf_{v\in D} I(v) = l(\bar{\varphi}) = \max_{\varphi\in\Pi} l(\varphi) \tag{4.6}$$

Theorem 2.3 is applied when an optimal element $\bar{v} \in D$ does not exist but a solving function $\bar{\varphi} \in \Pi$ does exist.

Theorem 4.5. *Assume that sequences $\varphi_s \subset \Pi$ and there exists an admissible processes sequence $\{v_s\} = \{x_s(t), u_s(t)\} \subset D$ such that:*

$$\int_A [R_s(t, x_s(t), u_s(t)) - \mu_s(t)]dt \to 0 \tag{4.7}$$

Then

$$I(v_s) \to \inf_{v\in D} J(v) = \sup_{\varphi\in\Pi} l(\varphi) = \lim l(\varphi_s) \tag{4.8}$$

According to the latter Theorems the pair of functions $\tilde{v}(\varphi) = (\tilde{x}(t), \tilde{u}(t))$ defined by (4.1) is an optimal process for a special function-parameter $\varphi = \bar{\varphi}$. We can add that for any bounding function φ the pair $\tilde{v}(\varphi)$ is a certain approximation of the optimal process. Let $\tilde{v}(\varphi) \in E$ so that this is a piecewise continuous and differentiable pair. The distance ρ of the pair from the set D of admissible processes is defined by (1.10). Let a sequence $\{\varphi_s\} \in \Pi$ be a solution to the dual problem (4.8). Then for some sufficiently weak conditions $\rho[\tilde{v}(\varphi_s)] \to 0$, when $l(\varphi_s) \to d$. Namely, the sequence $\{\tilde{v}_s\}$ is a solution of problems 3 and 4 formulated in section 1.1. New numerical algorithms for computing admissible and optimal processes can be built using this idea. This fact and the computer idea will be proved below (Section 10.4.).

4.2 A nonfixed boundary state

Suppose the boundary state $x(\vartheta)$, $\vartheta \in S$, is variable. Then the conditions of theorems described in the earlier section must be supplemented by the condition for a minimum of the function $G(x(\vartheta))$. This is the function $G(x) = F(T,x) + \varphi(T,x)$, $x = x(T)$, when single argument processes (ordinary or multistage) with a fixed initial state are considered. The latter conditions (4.1), (4.4) must be completed by condition

$$G[\bar{x}(T)] = \min_{x \in V_x(T)} G(x) = m \tag{4.9}$$

The conditions (4.5), (4.7) must be supplemented, respectively, by conditions

$$G[\bar{x}_s(T)] \to \min_{x \in V_x(T)} G(x) = m, \text{ or } G_s[\bar{x}_s(T)] - \min_{x \in V_x(T)} G_s(x) \to 0 \tag{4.10}$$

4.3 Control synthesis

Let us introduce functions $\tilde{u}(t,x)$ and $P(t,x)$ implied by function $\varphi \in \Pi$ through

$$P(t,x) = \max_{u \in V^x(T)} R(t,x,u) = R[t,x,\tilde{u}(t,x)] \tag{4.11}$$

Suppose the process $\tilde{v}(\tau, x_0)$ is associated with the control $\tilde{u}(t,x)$ and the function $\tilde{u}(t,x)$ is a control synthesis for a set B of initial conditions (τ, x_0). This is always true, for example, in the case of a discrete argument when there are no state constraints: $V_x^t = X$, $t = \tau + 1, ...,$. Then for all (τ, x_0) the following optimality estimate is true [4]:

$$\begin{aligned} I(\tilde{v}(\tau,x_0)) - d(\tau,x_0) \leq \Delta(\varphi) = \\ = \int_A [\sup_{x \in V_x(T)} P(t,x) - \inf_{x \in V_x(T)} P(t,x)]dt + \\ + \sup_{x \in V_x(0)} G(x) - \inf_{x \in V_x(0)} G(x), \end{aligned} \tag{4.12}$$

where $d(\tau, x_0) = \inf I(v)$, $v \in D(\tau, x_0)$. I.e. the synthesis $\tilde{u}(t,x)$ is ϵ-optimal, $\epsilon = \Delta(\varphi)$. Thus, a Δ-optimal feed-back policy corresponds to any bounding function φ. This estimation is also valid for multistage processes if we replace the integral sum in (4.12).

Minimizing the functional $\Delta(\varphi)$ it is possible to make it sufficiently small. A group of numerical algorithms for an approximate optimal solution is based on this idea (see below).

Suppose there exists a function $\varphi(t,x) \in \Pi$ that satisfies the conditions:

$$P(t,x) = c(t), \quad \forall x, t < T; \quad \varphi(T,x) = -F(x) + C \tag{4.13}$$

where $c(t)$ is a function and C is a constant. Then, according to (4.12) the synthesis $\tilde{u}(t,x)$ is optimal. If we take $c(t) = 0$ and $C = 0$ (function φ is normalized) the function $\varphi(t,x)$ is the dynamic programming optimal value function with a negative sign: $\varphi(t,x) = -d(t,x)$. In the corresponding cases the equation (4.13) is the Hamilton – Jacobi or the dynamic programming equation .

5 Estimations of Attainability Domains. Strengthened Conditions for Lower Bounds and Sufficient Optimality Conditions

Let us consider an ordinary optimal control problem reduced to the terminal form $f^0 = 0$. Due to condition (3.1) on the admissible trajectory $x(t) \in D_x$ the following inequality is valid

$$d\varphi(t,x(t))/dt = R(t,x(t),u(t)) \le \mu(t), \quad \forall t \in (\tau, T).$$

Integrating the inequality we have:

$$\varphi(t,x(t)) - M(t) \le \varphi(\tau, x(\tau)); \quad \forall t, x(t) \in D_x; \quad M(t) = \int_\tau^t \mu(\xi) d\xi; \tag{5.1}$$

Thus, any bounding function gives us the estimate of the attainability domain, when the initial state is fixed: $x(\tau) = x_0$.

This inequality is also true for a multistage problem if

$$M(t) = \sum_\tau^t \mu(\xi) d\xi.$$

Let $V_\varphi(t)$ be the family of sets restricted by conditions:

$$(x,u) \in V(t); \quad \varphi(t,x) - M(t) \le \varphi(\tau, x_0) \tag{5.2}$$

The problems (I, D_φ), (I, D) are equivalent. Let

$$\mu_\varphi(t) = \sup_{(x,u)\in V_\varphi(t)} R(t,x,u), \quad m_\varphi = \inf_{x \in V_\varphi(T)} G(x),$$

$$l_\varphi = m_\varphi - \int_A \mu_\varphi(t)dt - \varphi(\tau, x_0) \tag{5.3}$$

Obviously,

$$I(v) \geq l_\varphi \geq l(\varphi), \quad \forall v \in D, \quad \varphi \in \Pi. \tag{5.4}$$

We will name the functional l_φ a *strengthened* lower bound. Obviously, a strengthened variant of the theorems 2.1 - 2.5 is true if we substitute $V(t)$, $\mu(t)$, m by $V_\varphi(t)$, e. c.

6 The Principle of Optimality

Let us sum up the mathematical facts related to the bounding functions. These are assigned by defining the functions $R(t, x, u)$ and $G(x)$ through (2.1), (2.2) and by the following values and variables row: $\mu(t)$, $L(\varphi, v)$, $l(\varphi)$, $\tilde{v}(\varphi) = (\tilde{x}(t), \tilde{u}(t))$, $\tilde{v}(t)$, $\tilde{u}(t, x)$, $P(t, x)$, $\Delta(\varphi)$.

All these values and variables have the above discussed properties which are useful for the analysis of optimal control problems.

A totality of these facts forms an analytical environment for the analysis of the optimal control problem and adjacent problems. The sufficient conditions of optimality are the central fact in this analytical environment.

We will further have to master, to prove, to develop and to apply this approach. The first means an elaboration of new analytical and computer methods for the analysis of control problems. The second means a revelation of the relation of our sufficient optimality conditions to the other optimality conditions in variational calculus and optimal control theory and the proof of the Principle of Optimality. The directions of development and application are obvious.

The above mentioned mathematical facts are elementary but they imply nontrivial corollaries. We shall observe that sufficient conditions of optimality include basic equalities and inequalities of variational calculus and optimal control theory like the equations of the maximum principle, the Jacobi conditions, and the Hamilton – Jacobi – Bellman equations. This means that they are quite close to necessary conditions. And in fact, after some natural additional assumptions they do become necessary. This observation had allowed to find new classes of solutions for problems of variational calculus problems and methods of their solution. Some simplest solutions of this type were considered above. Additional cases will be considered below. These facts were the basis for new ideas of constructing numerical algorithms for computing optimal or simply admissible processes. The mathematical methods which use the bounding functions and related constructions were found efficient not only for problems formulated here but in many other problems also in the analysis and synthesis of dynamic control systems.

The presented mathematical facts, and the resulting new possibilities and untraditional directions in solving problems of variational calculus and optimal control were developed in the papers and books surveyed here, starting from [1-5]. But earlier papers containing some elements of this theory should also be mentioned. The Hamilton – Jacobi method in variational calculus and analytical mechanics can be considered as the first application of solving functions. We can also consider that they are used in Bellman's dynamic programming [8] which is a generalization of the Hamilton -Jacobi method to modern problems of control and, in particular, to problems of optimal control of multistage processes. However, these are solving functions of special type, defined by equation (4.13). They do not cover all the possible applications of this theory. Some of Caratheodory's ideas which were described by L.C.Young in his book [9] do approach some of the global optimality conditions described here. But only a particular form of the solving function φ, namely, the Hamilton – Jacobi function (of extremal I-length) is used by this author.

7 Relations to Other Optimality Conditions

The relation of the described optimality conditions to Pontryagin's maximum principle [10] is obvious from the following necessary conditions (4.4) for extremum, see [1,6,7]:

$$R_x(t, \bar{x}(t), \bar{u}(t)) \equiv \dot{\psi} + H_x(t, \psi(t), \bar{x}(t), \bar{u}(t)) = 0, \tag{7.1}$$

$$\bar{u}(t) \in \operatorname{Arg}\max_u \; R(t, \bar{x}(t), u) = \operatorname{Arg}\max_u \; H(t, \psi(t), \bar{x}(t), u), \tag{7.2}$$

$$\begin{aligned} H(t, \psi, x, u) &= \psi f(t, x, u) - \psi_0 f^0(t, x, u) \\ \psi(t) &= \varphi_x(t, \bar{x}(t)), \end{aligned} \tag{7.3}$$

where ψ_0 is any positive constant. Analogous conditions for the discrete - time case have the following form, see [5]:

$$R_x(t, \bar{x}(t), \bar{u}(t)) \equiv H_x[t, \psi(t+1), \bar{x}(t), \bar{u}(t)] -$$

$$- \psi(t) = 0, \quad t \in [\tau, T-1]; \tag{7.4}$$

$$R_u(t, \bar{x}(t), \bar{u}(t)) \equiv H_u[t, \psi(t+1), \bar{x}(t), \bar{u}(t)] = 0 \tag{7.5}$$

Thus, the maximum conditions for the function $R(t, x, u)$ given earlier coincide with the equations of the maximum principle excluding abnormal problems when $\psi_0 = 0$. Together with the process equations (1.3), (1.4) they form a closed system of equations where the solving function $\varphi(t, x)$ is represented only by its gradient $\psi(t)$ along an optimal trajectory. An analogous coincidence of equations is true for appropriate extensions of the maximum principle (Dubovitzky – Milyutin conditions) and for state constraints, see M.Khrustalev [29] and [7, p.120-136]. It turns out the conditions type of (7.1), (7.2) in a strengthened variant coincide with the equations of the maximum principle including also the abnormal case $\psi_0 = 0$[]. Equations (2.11) extend these necessary conditions of optimality to global sufficient conditions which depend on the functions $\varphi(t, x)$ such that $\varphi_x(t, \bar{x}(t)) = \psi(x)$. The simplest conditions of this type can be obtained by taking a linear solving function $\varphi(t, x) = \psi(t)x$. They were considered in [11]. In [3,7] differential equalities for the matrix $\sigma(t) = \varphi_{x^i x^j}(t, \bar{x}(t))$ were given. Their satisfaction guarantees the a strong or weak relative minimum of the functional. The (necessary and sufficient) Jacobi conditions of variational calculus are equivalent to the existence of the matrix $\sigma(t)$ in the appropriate cases. The development of such kinds of conditions for a local optimum is given in the papers by Rozenberg [12] and Zeidan [13].

The Bellman dynamic programming equations [8] and the Hamilton – Jacobi partial differential equations of variational calculus coincide with equation (4.13) that define the solving function of a special type. An extensive analysis of the relations between the value functions and the solving functions has been done by Girsanov [14].

8 Analytical Methods of Solving Singular and Improper Problems

As mentioned earlier the solving function is not completely defined by the optimality conditions (4.4). By considering different classes of problems we may define different methods of finding them. This possibility was first used in those problems where the equations of the maximum principle or more generally the variational methods can not be used or are ineffective, and an application of the dynamic programming equation is too difficult. First, it includes problems where the minimum is beyond the set D. This means that the necessary conditions of variational calculus can not be used. Secondly, it includes the so called singular problems which are unsuitable for traditional methods. These two classes of problems have a nontrivial common part. Many typical minimizing sequences like sliding or discontinues solutions usually originate from singularities. On the other hand, singular problems can have solutions within the set of admissible processes.

Some special methods of dealing with singular problems were developed using the presented approach. This includes an "alternative formalism" for the scalar state problems ($n = 1$), see [2,6], and the method of multiple maxima for multidimensional state problems, see [6,7,15]. The method of multiple maxima and some general approaches it initiated (in variational problems connected with the idea of extension) are the subject of present intensive studies by V.Gurman and his associates [16]. An interested reader can find a detailed description of those methods and their applications in our references. The methods were helpful in finding some new classes of solutions: discontinuous solutions of a special type - the so called (x, u)-objects [17,7], cyclic sliding solutions [18,19], positional control etc.

We shall illustrate this approach on the following problem.

8.1 The degenerate Problem of Euler. Discontinuous solutions

The following problem has been known as the Problem of Euler in Variational Calculus:

$$I(v) = \int_{\tau}^{T} f^0(t, x(t)).u(t))dt \Rightarrow \min \tag{8.1}$$

$$dx(t)/dt = u(t); \quad x(\tau) = x_0; \quad x(T) = x_1; \quad x \in R^1; \quad u \in R^1$$

Let $f^0(t, x, u) = g(t, x) + h(t, x)u$. The state restrictions may be of the type: $a(t) \leq x(t) \leq b(t)$, $\forall t$. The functions $g(t, x,)$, $h(t, x)$, $a(t)$, $b(t)$ are defined and continuous; the function $h(t, x)$ is differentiable; $x_0 \in [a(\tau), b(\tau)]$, $x_1 \in [a(T), b(T)]$. We have:

$$R(t, x, u) = [\varphi_x - h(t, x)]u - g(t, x,) + \varphi_t$$

We define the function $\varphi(t, x)$ by the equality: $\varphi_x = h(t, x)$. Then:

$$\varphi(t, x) = \int_k^x h(t, \xi)d\xi + c(t); \quad \varphi_t = \int_k^x \partial h(t, \xi)/\partial t \; d\xi + \dot{c}(t); \tag{8.2}$$

where $c(t)$ and k are an arbitrary function and a constant, respectively. Substituting it in the expression for $R(t, x, u)$ and taking $c(t) = 0$, $k = 0$ we have

$$R(t, x) = -g(t, x) + \int_k^x \partial h(t, \xi)/\partial t \; d\xi \tag{8.3}$$

The function $R(t, x)$ is continuous and there exists such a function $\tilde{x}(t)$, that

$$R(t, \tilde{x}(t)) = \max_{x \in [a,b]} R(t, x), \quad t \in (\tau, T), \quad \tilde{x}(\tau) = x_0, \quad \tilde{x}(T) = x_1 \tag{8.4}$$

Let the function $\tilde{x}(t)$ be piecewise continuous. We have

$$L(v) = \varphi(T, x_1) - \int_{\tau}^{T} R(t, x(t))dt \; - \; \varphi(\tau, x_0), \tag{8.5}$$

where φ and R are defined by (8.2), (8.3). The admissible trajectories set D_x is a set of all continuous, piecewise differentiable functions $x(t)$ restricted by conditions

$$x(\tau) = x_0; \;\; x(T) = x_1; \;\; a(t) \leq x(t) \leq b(t), \;\; \forall t, \tag{8.6}$$

Let us introduce the set E of piecewise continuous pairs of functions $v = (x(t), u(t))$ restricted only by (8.6) but not by the equation $u(t) = dx(t)/dt$. For any $v = (x(t), u(t)) \in E$ a sequence $\{x_s(t)\} \subset D_x$ and a corresponding sequence $\{v_s\} = \{x_s(t),\, u_s(t) = dx_s(t)/dt\} \subset D$ exist such that $I(v_s(t)) = L(x_s(t)) \to L(x(t))$. These sequences may be defined as $v_\epsilon = (x_\epsilon(t), u_\epsilon(t) = dx_\epsilon(t)/dt)$, $\epsilon \to 0$, where $x_\epsilon(t) = x(t)$ everywhere except at the discontinuity points in the ϵ-neighborhoods A_ϵ. In these neighborhoods the sign of the control $u_\epsilon(t)$ does not change. We have by (8.2), (2.5):

$$L(x(t)) = I(v) + \sum_{t \in \beta} \varphi(t, x(t))|_{t-0}^{t+0} =$$

$$= I(v) + \sum_{t \in \beta} \int_{x(t-0)}^{x(t+0)} h(t, \xi)d\xi, \tag{8.7}$$

where β is the set of discontinuity points of the function $x(t)$. Further:

$$I(v_\epsilon) \;=\; L(v_\epsilon) \;=\; \int_{A \setminus \cup_{t \in \beta} A_\epsilon(t)} f^o(t, x(t), u(t))dt \; +$$

$$+ \sum_{t \in \beta} \int_{A_\epsilon(t)} f^o(t, x_\epsilon(t), u_\epsilon(t))dt \tag{8.8}$$

It is easy to show: $I(v_\epsilon) \to L(x(t))$, *epsilon* $\to 0$. Indeed, the first term tends to $I(v)$ and the second one tends to the second term in (8.7).

An approximate sequence v_ϵ is completely defined by a trajectory $x(t)$. In other words an approximate sequence v_ϵ corresponds to the family of processes $v = (x(t), u(t)) \in E$ with a fixed trajectory $x(t)$ and an arbitrary control $u(t)$. We can single out an element from this family by imposing the condition $u(t) = dx(t)/dt$ at the differentiability points of the trajectory $x(t)$. We shall name this subclass of the class E as D'.

The trajectory $\tilde{x}(t)$ defined by (8.4) is generally a discontinuous function. As a rule, the points $t = \tau$ and $t = T$ are the points of discontinuity. The existence of other discontinuity points is also possible. Let $\tilde{x}(t) \in E_x$. The sequence v_ϵ of the above, associated with the trajectory $\tilde{x}(t)$, satisfies Theorem 2.3 and is the solution of the problem (8.1). It is obvious that $u_\epsilon(t) \to \infty$ for $t \in A_\epsilon$. The problem has been solved. A solution of this type will be referred to as the *discontinuous solution.*

For the functional (8.1) to have a minimum at $x = x(t)$, it is necessary and sufficient that the function $R(t, x)$ would have a maximum for every fixed $t \in (\tau, T)$. Then the equation of the extremal is

$$R_x = g_x(t, x) - h_t(t, x) = 0, \quad t \in (0, T), \tag{8.9}$$

if the function R is differentiable. We have thus obtained necessary and sufficient conditions of minimum of a functional (8.1). Equation (8.9) can also be obtained also by the method of variations, in the form of a degenerate Euler equation. Using Green's theorem [], we can derive sufficient conditions for a strong local minimum on its solution $x^0(t)$. It coincides with the condition for a local maximum of $R(t, x)$ in the neighborhood of $x^0(t)$ for every fixed t. If equation (8.9) has several solutions, the absolute minimal consists of pieces of these solutions and of pieces of the boundary, connected by jumps.

The construction of the absolute minimal from these pieces by the method of variations combined with Green's theorem is highly complicated. Our method provides an attractively simple solution to this problem: by using (8.3) we construct the function $R(t, x)$ and for every fixed $t \in (\tau, T)$ we find the value of x on which $R(t, x)$ attains its maximum on the segment $a(t) \leq x \leq b(t)$.

8.2 The Problem of Euler with a bounded nonlinearity

Consider the problem

$$I(v) = \int_A [g(t, x, u) + h(t, x)u]dt \to \min \tag{8.10}$$

$$\dot{x} = u, \quad x(\tau) = x_0, \quad x(\tau) = x_1, \quad a(t) \leq x(t) \leq b(t),$$

where $A = (\tau, T)$, $n = r = 1$, the functions $g(t, x, u)$, $h(t, x)$, $a(t)$, $b(t)$ are defined and continuous for all values of their arguments; the function

$g(t, x, u)$ is bounded with respect to u; the function $h(t, x)$ is differentiable. We have:

$$R(t, x, u) = [\varphi_x - h(t, x)]u - g(t, x, u) + \varphi_t$$

Defining the function $\varphi(t, x)$ by (8.2) we get:

$$R(t, x, u) = -g(t, x, u) + \int_k^x \partial h(t, \xi)/\partial t \, d\xi \qquad (8.11)$$

Suppose there exists a pair of functions $\tilde{x}(t)$, $\tilde{u}(t)$ such that

$$R(t, \tilde{x}(t), \tilde{u}(t)) = \max_{x \in [a,b],\ u} R(t, x, u), \quad t \in (0, T), \qquad (8.12)$$

$$\tilde{x}(\tau) = x_0, \quad \tilde{x}(T) = x_1$$

Let $(\tilde{x}(t), \tilde{u}(t)) \in E$. We have

$$L(x(t), u(t)) = \varphi(T, x(T)) - \int_A R(t, x(t), u(t))dt - \varphi(x(\tau)), \qquad (8.13)$$

where φ and R are defined by (8.2), (8.11).

Similar to the problem (8.1) it is not difficult to demonstrate by using representations (8.7), (8.8) that $I(v_\epsilon) \to L(v)$, $\epsilon \to 0$ for any $v \in D'$, where $\{v_\epsilon\}$ is the approximate sequence described above.

Suppose the pair $v = (x(t), u(t)) \in E$ is given. Let us take an integer $s > 0$ and construct the pair of functions $v_s = (x_s(t), u_s(t)) \in D'$ in the following way. The interval (τ, T) is divided by the sequence: $\tau < \tau_1 < \tau_2 < \ldots < \tau_s < T$, which contains all the points of discontinuity of the trajectory $\tilde{x}(t)$. On each subinterval (τ_p, τ_{p+1}), $p = 1, 2, \ldots, S-1$, the function $x_s(t)$ is defined by the following formula:

$$x_s(t) = \tilde{x}(\tau_p + 0) + \tilde{u}(\tau_p + 0)(t - \tau_p)$$

and on the subintervals (τ, τ_1) and (τ_s, T) by:

$$x_s(t) = x_0 + \frac{\tilde{x}(\tau_1) - x_0}{\tau_1 - \tau}(t - \tau) \text{ for } \tau \le t \le \tau_1$$

$$x_s(t) = x_1 + \frac{\tilde{x}(\tau_s) - x_1}{\tau_s - T}(t - T) \text{ for } \tau_s \le t \le T$$

The function $u_s(t)$ is defined by: $u_s(t) = \dot{x}_s(t)$. Consider the sequence $\{v_s\}$ above defined with an additional condition: $\Delta_s = \max(\tau_1 - \tau, \tau_2 - \tau_1, \ldots, \tau_s - \tau_{s-1}, T - \tau_s) \to 0$. The sequence of the state trajectories

$\{x_s(t)\}$ converges to the function $\tilde{x}(t)$ while the sequence of its derivatives $\{\dot{x}_s(t) = u_s(t)\}$ converges (in measure) not to $\dot{\tilde{x}}(t)$ but to the function $\tilde{u}(t) \neq \dot{\tilde{x}}(t)$. The sequence $\{v_s\} = \{x_s(t), u_s(t)\}$ is bounded. The integrand (8.13) $R_s(t) = R(t, x_s(t), u_s(t))$ is piecewise continuous and bounded. Due to Lebesgue's theorem we have $L(v_s) \to L(x(t), u(t))$, $\forall (x(t), u(t)) \in E$.

There also exists a sequence $\{v_\eta\} \in D$ such that $L(v_\eta) \to L(x(t), u(t))$, $\forall (x(t), u(t)) \in E$. It is a combination of the sequences $\{v_s\}$ and $\{v_\epsilon\}$. Suppose, for example, that we want to construct an admissible process $v = (x(t), u(t)) \in D$ such that $| I(v) - L(v_0) | < \delta$, where the pair $v_0 = (x_0(t), u_0(t)) \in E$ is preassigned. Let us take an element $v_s \in \{v_s\}$, $| I(v_s) - L(v_0) | < \delta/2$ and approximate the latter by the process $v_\eta \in \{v_\epsilon\}$, $| I(v_\eta) - L(v_S) | < \delta/2$. The process v_η is a solution to the problem.

The intervals where $\dot{\tilde{x}}_s(t)$ is close to $\tilde{u}(t)$ alternate infinitely often with impulses of infinite height. These kinds of sequences we shall name as (x, u)-*policy* or *impulse sliding policy*. It is completely defined by the pair of functions $\tilde{x}(t), \tilde{u}(t)$. The former, $\tilde{x}(t)$, will be called the *function of closure zero*, and the latter, $\tilde{u}(t)$, the *basic control*. An algorithm for solving the problem consists in finding the functions $\tilde{x}(t)$ and $\tilde{u}(t)$ from a maximization of the function $R(t, x, u)$ for a fixed t.

Example. Consider the degenerate problem of Euler (2.43), when

$$g = 1 - e^{-(t-u)^2}, \qquad h = -tx$$

We have:

$$\varphi = -x^2 t/2; \quad \varphi_t = -x^2/2; \quad \varphi_x = -xt;$$

$$R(t, x, u) = -1 + e^{-(t-u)^2} - x^2/2$$

Maximizing this function R we get: $\tilde{x}(t) = 0$, $\tilde{u}(t) = t$. The optimal solution is the (x, u)-policy.

The discontinuous solutions of the above and the (x, u) -policies were given in [17]. They have interesting applications in mechanical and geometrical problems [6,7].

The solutions of these kinds have even more interesting and diverse constructions and realizations for $n > 1$, like, for example, the cyclic sliding policies or the positional controls. They have intuitive realizations in jet propulsion control problems. If the amount of fuel is taken as the process argument, then to the motion with the engine turned off corresponds to vertical segments in the state diagram. In these points the trajectory is discontinuous. The above optimal solutions can be interpreted as a sequence of infinitely short turning on of the engine at the given points of the state space with the given thrusts and then the cyclic bendings of the trajectory corresponding to the motion of the system with the engine turned off. As indicated in V.Gurman [19] and I.Ioslovich [20,21], the optimal change of space orbits and the control for stopping rotation of a solid body with a fixed point are of this kind.

9 The Extension of the Class of Solving Functions and Optimality Conditions

Function-theoretical properties of admissible processes were established to satisfy two goals which are generally contradictory. These are the simplicity and the sufficient completeness of the class D. To satisfy the latter it would be the best to define D in such a way that the minimum of the functional $J(v)$ would exist. However, at first, this is not always possible, and at second, even if it would be possible, then the problem formulation might be too cumbersome. To achieve simplicity it would be convenient to assume that the functions $x(t)$ and $u(t)$ are continuous or sufficiently smooth. But the solution would then very often be in the form of a minimizing sequence. The properties of the admissible processes that are taken here form a compromise between these concepts.

A similar situation exists for the properties of the solving functions $\varphi \in \bar{\Pi}$. A continuous differentiability is assumed entirely for the sake of simplicity and brevity. However, even in linear problems this is burdensome. The solving function in these problems has the form $\varphi = \psi(t)x$, where $\psi(t)$ is the adjoined vector function which is defined by the maximum principle equations. The latter is, as a rule, piecewise differentiable. To include such kind of functions into the class Π we have to weaken the related assumptions, changing the assumption of continuous differentiability to that of continuity on $T \times X$ and to continuous differentiability everywhere on $T \times X$ except for a finite number of cross-sections for fixed t's. This was already done in [1].

The problem of extending the class Π and the extension of the sufficient conditions for optimality to necessary conditions is not reduced, however, only to weakening of the function-theoretical restrictions on Π. There the following aspects can be chosen:

1. The existence of an extending sequence $\varphi_s \in \Pi$ which satisfies (2.4).
2. A weakening of the requirements on Π to such an extent that the equality (2.4) could be strengthened to:

$$\inf_{D} J(v) = \max_{\Pi} l(\varphi) \tag{9.1}$$

3. The use of a strengthened variant of the sufficient optimality conditions so that:

$$\inf_D J(v) = \max_\Pi l_\varphi(\varphi) \tag{9.2}$$

4. The extension of the class of solving and bounding functions either by weakening the function-theoretical restrictions in such a way that 2. holds or by going over to other representations in order to extend the possibilities of choosing the optimization tools.

It was supposed in [2] that the sufficient conditions for optimality can also be the necessary conditions under some additional assumptions. This was formulated as an optimality principle.

There are two approaches to prove this principle. They use quite different mathematical tools. In one of these a proof of the existence of the Hamilton – Jacobi – Bellman equation under given conditions and in a given class of functions is used. In the other the direct existence of the maximum of the functional $l(\varphi)$ and its equivalence with the minimum of the functional $J(v)$, $v \in D$, is demonstrated.

The first approach was applied for proving the necessity of the sufficient conditions (4.4) for the discrete time processes [5]. It was also applied for continuous time processes without state constraints and terminal conditions [22]. In this case it was found necessary to take the class Π as the class of locally Lipschitz functions. Then (5.1) is satisfied.

The second approach, namely the proof of the existence of $\max l(\varphi)$, $\varphi \in \Pi$, was used by Ioffe [23].

Let us suppose that φ is normalized and that the inequality

$$R(t, x, u) \leq 0, \quad (x, u) \in V(t)$$

is changed to the integral form by using (1.3), (1.4), $\mu(t) = 0$, which is actually more general:

$$\varphi(t_1^*, x_1^*) - \varphi(t_0^*, x_0^*) \leq d'(t_0^*, t_1^*, x_0^*, x_1^*) =$$

$$= \min \int_{t_0^*}^{t_1^*} f^0(t_1, x(t), u(t))dt, \quad \forall t_0^*, t_1^*, x_0^*, x_1^*$$

This integral extension of the class Π often turns to be convenient for extending the sufficient conditions to be also the necessary ones. However, this is not a constructive way because it contains the family $d'(\cdot)$ of $\min J(v)$ for different boundary constraints. For the Euler problem of variational calculus it was shown that there exists a continuous function φ which satisfies the equality $\max l(\varphi) = d$ in the integral extension of the class Π for the lower semicontinuous functionals.

This approach was developed further by Vinter [24,25] who extended the sufficient conditions of variational calculus in the optimal terminal control problems without state constraints to be also necessary. He proved the existence of a function φ that satisfies (5.1) where Π is the class of locally Lipschitz functions with an additional assumption called strong calmness. This assumption is obviously quite weak but not constructive. In this respect it very much resembles the unconstructivity of Ioffe's condition.

These unconstructive elements are absent in an approach to the extension of sufficient conditions where solving sequence $\{\varphi_s\} \subset \Pi$ is instrumental. Vinter & Lewis [26] proved existence of solving sequences of smooth bounding functions which satisfy (2.4) for the end-point problem without state constraints. Klötzler [27] and Pickenhain [28] proved that such sequences also exist for the state constrained problems.

M.Khrustalev [29,30] has proved the equality (9.2), where Π is the class of locally Lipschitz functions, for ordinary processes without state constraints including the terminal conditions.

An extension of the classes of bounding functions Π and solving functions $\bar{\Pi}$ from the set needed in the extension of sufficient optimality conditions to the necessary conditions went in the following directions.

Khrustalev [7,31] introduced solving functions φ that are discontinuous on the cross-sections of $T \times X$ for a fixed t. At the same time the functions $\varphi(t, \bar{x}(t))$ are continuous along the optimal trajectory $\bar{x}(t)$. For these functions φ the equality $J(\bar{v}) = L(\varphi, \bar{v})$ holds on an optimal process $\bar{v}$. Besides that, the directions of discontinuities are oriented in such a way that $L(\varphi, v) \geq l(v)$, $v \in E$ holds and also therefore that the sufficient conditions (4.4) are satisfied. For this extension of the class $\bar{\Pi}$, our first order necessary maximum conditions for the function $R(t, x, u)$ with respect to $x, u \in V(t)$ are the same as the necessary conditions of Dubovitzky – Milutin [32] including jump conditions for the problems with state constraints. For the convex problems of this kind of the solving functions have the form $\varphi = \psi(t)x$.

Klötzler [27,33,34] formed a similar reciprocal problem with a multidimensional argument using discontinuous solving functions but requiring neither continuity of the function $\varphi(t, \bar{x}(t))$ nor the special orientation of discontinuities. Instead, definitions of the functionals $L(\varphi, v)$ and $l(\varphi)$ were extended for discontinuous functions using some special formula. These sufficient conditions were successfully used by Klötzler, Pickenhain, and Andreeva for investigating of an interesting class of isoperimetric problems [35,36,37,28].

Gurman [16] and Konstantinov [38] introduced and applied vector solving functions $\varphi(t, x)$, $T \times X \to R^k$, $k > 1$. Finally, Moskalenko [39] and Gurman [16] used a more general formulation of the auxiliary problem (L, E) which led to generalized sufficient conditions.

The fact that the sufficient conditions for global optimality can be extended to necessary and sufficient conditions opens the possibility to realize Hilbert's idea of eliminating the method of variations from the calculus of variations, but from another perspective. If Hilbert saw this possibility in the development of direct methods, then the question is in the reduction of a spectrum of necessary and sufficient conditions of variational calculus to necessary and sufficient conditions of global and relative maximum of the function $R(t, x, u)$ with respect to $(x, u) \in V(t)$ and to conditions for the existence of the solving functions $\varphi(t, x)$. For this the difference between the analytical properties of the solving functions which are taken in the existence theorems and the analytical properties needed in the corresponding conditions of variational calculus should be eliminated. For example, the equations (7.1) of the maximum principle may be obtained from the condition (4.4) under the assumption that the function $\varphi(t, x)$ is twice differentiable along the optimal trajectory $\bar{x}(t)$. But, as shown earlier, only weaker properties are proved. This gap can be filled with such results as that of Subbotin & Subbotina [40] who showed that the maximum principle equations are necessary consequences of the sufficient conditions with a solving function from the class of Lipschitz functions. The above is also true for multistage processes. In [5] it was proved without any assumptions on analytical properties that the unique map φ: $T \times X \to R^1$ exists. From Ioffe's results [23] it follows that this map is continuous when $X = R^n$. But for (7.4) to hold the function $\varphi(t, x)$ should be differentiable at $x = \bar{x}(t)$.

10 Computer Algorithms Based on the Techniques of Global Bounds

Computer algorithms based on the technique of global bounds may be divided into 3 groups:

(i) Methods of successive improvements of the sweeping type similar to the traditional techniques presented for example by Kelley [41], Eneev [42], Krylov & Chernousko [43], Bryson & Ho [44]. However, the choice of an improving function allows to optimize not in a local (gradient) direction but in a global one.
(ii) Dual methods which are connected with a construction of sequences of solving functions $\{\varphi_s\} \subset \Pi$ maximizing the functional $l(\varphi)$ given by

(3.3). This way we obtain an increasing sequence of lower bounds for the functional J on the set D which converges under appropriate conditions to $\inf_D J(v)$. Yet a solution to the problem is not this sequence but the sequence $\{v_s\} \subset E$ which satisfies (1.1) and $\rho(v_s) \to 0$, (1.10). The role of this sequence is played by $\{\tilde{v}_s\} = \{\tilde{x}_s(t), \tilde{u}_s(t)\} \subset \tilde{E}_s$ which is related to $\{\varphi_s(t,x)\}$ through (4.1). This way we get an approximation to $\inf J(v)$, $v \in D$, by an "outside" approximation of an admissible process. We thus solve not only the problem (P) but also the problem (I).
(iii) Methods where the ϵ-optimal feedback control $u(t,x)$ is constructed using the bound (2.5). This leads to a minimization of the functional $\Delta(\varphi)$ until it is not greater then a given ϵ.

10.1 The methods of successive improvement of the control for ordinary and multistage processes

We start by the description of the mentioned methods of local improvement of control in terms of the improving function. Let us assume that we know an admissible process $v_0 = (x_0(t), u_0(t)) \in D$. We want to improve it, i.e. to find $av = (x(t), u(t)) \in D$ such that $J(v) < J(v_0)$. We replace the optimization of the functional $J(v)$ by an optimization of $L(v, \varphi)$ given by (2.3), (2.8) with a suitably chosen function φ. We shall look for v which is sufficiently close to v_0 in such a way that the sign of $\Delta J = J(v) - J(v_0)$ is the same as of its main linear part:

$$\delta J = \delta L = G_x(x_0(T))\delta x(T) - \int_A (R_x \delta x(t) + R_u \delta u(t))dt; \qquad (10.1)$$

$$\delta x = x - x_0, \qquad \delta u = u - u_0$$

It is tacitly assumed above that the functions $R(t,x,u)$ and $G(x)$ are differentiable. The formula for δL is given for the function $\varphi(t,x)$. We require that it complies with the equalities:

$$R_x(t, x_0(t), u_0(t)) = 0 \qquad (10.2)$$

$$G_x(x(T)) = \psi(T) + F_x(x(T)) = 0 \qquad (10.3)$$

These equations contain only the gradient of the function $\varphi(t,x)$ along the trajectory $x_0(t)$. The value of $\psi(t) = \varphi_x(t, x_0(t))$ and the values of (7.1) and (7.4) are determined after replacing $\bar{x}(t), \bar{u}(t)$ by $x_0(t), u_0(t)$. This means that the equations (10.2) and (10.3) are satisfied by functions of the form $\varphi(t,x) = \psi_i(t)x^i$, where the vector $\psi(t) = \{\psi_i(t)\}$ is determined by (10.2) and (10.3). We call this function a local improving function for control. Then:

$$\delta J(v_0) = \delta L(v_0, \varphi) = \int_A R_u(t, x_0(t), u_0(t))\delta u(t)dt; \tag{10.4}$$

where $R_u(t, x_0(t), u_0(t))$ is equal to $H_u(t, \psi(t), x_0(t), u_0(t))$ or $H_u(t, \psi(t+1), x_0(t), u_0(t))$ for the continuous and the discrete variants of the problem, respectively.

Let there be given a function $\delta u(t)$ and an arbitrarily small parameter $\epsilon > 0$ such that:

1. The right-hand side of (10.4) is positive,
2. $u(t, \epsilon) = u_0 + \epsilon \delta u \in V_u$, $t \in T$,
3. $x(t, \epsilon) \in V_x$, where $x(t, \epsilon)$ is the trajectory determined by the open loop control $u(t, \epsilon)$, the equation of motion, and the initial conditions (let the latter be assigned).
4. $v(\epsilon) = (x(t, \epsilon), u(t, \epsilon)) \in D$.

Then there exists an $\epsilon > 0$ such that:

$$J(v) < J(v_0), \quad v = v(\epsilon) \tag{10.5}$$

Without the state constraints, i.e. for $V_x^t = X$, $t \in (0, T]$, the improvement of the given open loop control $u_0(t)$ reduces to the following steps:

(i) Find the trajectory $x_0(t)$ by solving the Cauchy problem (1.3) or (1.4) with $u = u_0(t)$, $x(0) = x_0$. The open loop control $u_0(t)$ should satisfy $v_0 = (x_0(t), u_0(0)) \in D$.

(ii) Find $\psi(t)$ and $R_u(t, x_0(t), u_0(t))$ by solving the linear Cauchy problem (10.2) with the initial condition (10.3) which determines a local improving function $\varphi = \psi(t)x$.

(iii) Set a variation $\delta u(t)$ of the control program which makes the right-hand side of (10.4) positive.

(iv) For different $\epsilon > 0$ solve the problem (I) with $u = u_0 + \epsilon \delta u$. The value of ϵ should be taken in such a way that (10.5) would hold.

The basic part of this algorithm is the "sweeping" solution of the pair of Cauchy problems: the equation of motion from τ to T and the adjoint equation from τ to T. A consecutive repetition of these operations allows to find the improving sequence $\{v_s\} \subset D$.

The expression (10.4) gives the gradient of the functional $J(u)$ in the space of control functions $u(t)$. The presented method can then be considered as an application of the gradient techniques to the above class of problems. A week point of this is the local character of the improvement which is guaranteed only for small variations of the control $u(t)$. This is not only troublesome because the convergence is slow but also because the small variations may not be realizable (for example when the set V_u^t is finite). This deficiency may be avoided when globally improving functions are used.

It was shown in [45] that the function $\varphi(t,x)$ is globally improving for a given process $v_0 = (x_0(t), u_0(t)) \in D$ if it satisfies the following conditions:

$$R(t, x_0(t), u_0(t)) = \min_x R(t, x, u_0(t)), \quad t \in T \tag{10.6}$$

$$G(x_0(t)) = \max_x G(x), \quad t = T$$

A process $v = (x(t), u(t))$ which is determined by the control $\tilde{u}(t,x) = \arg\max_u R(t,\ x, u)$ satisfies the inequality $J(v) < J(v_0)$ if the process v_0 is not optimal. For continuous processes it also holds that $\tilde{u}(t,x) = \arg\max_u H[t, \varphi_x(t,x), x, u]$. That is when previously the local improvement was realized by a small variation of control in order to increase the function $R(t, x_0(t), u)$, then the new control is chosen as a global maximum of R with respect to u. The condition (10.6) which is satisfied by an improving function φ can be slightly weakened:

$$R(t, x_0(t), u_0(t)) \le R(t, x(t), u_0(t)), \quad t \in \bar{T} \tag{10.7}$$

$$G(x_0(T)) \ge G(x(T))$$

where $x(t)$ is the trajectory determined by $\tilde{u}(t,x)$.

To satisfy the inequalities (10.6) it is enough to consider improving functions of the form:

$$\varphi(t,x) = \psi_i(t)x^i + \sigma_{ij}(t)(x^i - x_0^i(t))(x^j - x_0^j(t))$$

where the coefficients $\psi(t) = \{\psi_i\}$, $\sigma_{ij}(t)$, $i,j = \overline{1,n}$ have to be found. It is not difficult to observe that the equations for $\psi(t)$ implied by (10.6) are the same as (10.2) and (10.3). The matrix $\sigma(t)$ is determined not uniquely. Here one possibility is to consider the additional equations:

$$R_{x^i x^j}(t, x_0(t), u_0(t)) = \delta_{ij}\eta, G_{x^i x^j}(x_0(t)) = -\delta_{ij}\alpha, \quad i,j = \overline{1,n} \tag{10.8}$$

Here δ_{ij} is the Kronecker delta: $\delta_{ij} = 0$ for $i \ne j, \delta_{ii} = 1$; η and α are positive constants. The equations (10.8) form the system of $(n+1)n/2$ linear differential (or difference) equations with unknowns $\sigma_{ij}(t) = \sigma_{ji}(t)$ and the given boundary condition at $t = t_1$ These equations together with (10.2), (10.3), and arbitrary positive η, α determine the coefficients of the function $\varphi(t,x)$ such that $x = x_0(t)$ is a relative minimum of $R(t,x,u_0(t))$ and maximum of $G(x)$. By an appropriate choice of η we are able to satisfy the inequalities (10.7) and therefore the relation (10.5). In this way we obtain the following algorithm for improving a solution:

(i) Set $\eta \geq 0$, $\alpha \geq 0$ and find $\psi(t), \sigma(t), \varphi(t,x), \tilde{u}(t,x)$ by solving the linear Cauchy problem (10.2), (10.3), (10.8) from T to τ,

(ii) Find the process $v = (x(t), u(t) = u[t, x(t)]) \in D$ by solving the Cauchy problem for the equation of motion with $u = \tilde{u}(t,x)$, $x(\tau) = x_0$, from τ to T and verify the inequality $J(v) < J(v_0)$. If this is not satisfied, then select another η, α and repeat the calculations. This procedure improves any process that does not satisfy the equations of the maximum principle or its discrete equivalent. By consecutively repeating the above algorithm we finally find an improving sequence $\{v_s\} \subset D$. However, in general it may not converge to $\inf J(v)$, $v \in D$.

Example 10.11 [46]. The problem is:

$$J = -x^2(2) \to \min$$

$$\text{s.t.} \qquad \begin{aligned} &x^1(t+1) = x^1(t) + 2u(t) \\ &x^2(t+1) = -(x^1(t))^2 + x^2(t) + u^2(t), \quad t = 0, 1 \\ &x^1(0) = 3, \; x^2(0) = 0, \; | \, u(t) \, | \leq 5 \end{aligned}$$

The optimal solution is $\bar{u}(0) = -2$, $\bar{u}(1) = \mp 5$, $J = -19$. For this problem Pontryagin's maximum principle does not hold. For $t = 0$ at $\bar{u}(0) = -2$ the Hamiltonian $H(t,u) = \psi_i(t) f^i(t, \bar{x}(t), u)$ attains the minimum rather than the maximum. Taking $\sigma_{12}(t) = \sigma_{22}(t) = 0$, t, $\sigma_{11} = \sigma$ we thus obtain

$$\varphi(t,x) = \psi_1(t)x^1 + \psi_2(t)x^2 + \sigma(t)(x^1 - x_0^1(t))^2/2$$

The functions R and G then take the form:

$$R(t,x,u) = \psi_i(t+1)[x + 2u] + \psi(t+1)[-x^2 + x + u^2] +$$

$$+0.5\sigma(t+1)[x + 2u - x_0^1(t+1)]^2 - \psi_i(t)x^i - 0.5\sigma(t)[x^1 - x_0^1(t)]^2$$

$$G(x) = -x^2 + \psi_i(2)x^i + 0.5\sigma(2)[x^1 - x_0^1(2)]^2$$

The adjoint equation and the equation for $\sigma(t)$ are as follows:

$$\psi(t) = \psi_1(t+1) - 2x^1(t)\psi_2(t+1), \quad \psi_1(2) = 0$$

$$\psi_2(t) = \psi_2(t+1), \quad \psi_2(t) = 1$$

$$\sigma(t) = -2\psi_2(t+1) + \sigma(t+1) - \eta, \quad \sigma(2) = \alpha$$

See Table 1 for the results.

10.2 A special class of nonlinear optimal control problems. Control of quantum systems by laser radiation

We choose a class of problems for which the global improving function that satisfies (10.6) has the form $\varphi(t,x) = \psi_i(t)x^i$. In this case the algorithm presented above is substantially simplified because there is no need to adjust the coefficients η, α or to solve the system of equations (10.8). The functions that define the problem now have the form:

$$f(t,x,u) = A(t,u)x + b(t,u); \quad F(x) = \lambda_i x^i + \gamma_{ij}x^i x^j; \tag{10.9}$$

$$u \in [u_1, u_2]; \quad f^0(t,x,u) = a^0(t,u)x + b^0(t,u);$$

where $A(t,u)$, $b(t,u)$, $a^0(t,u)$, $b^0(t,u)$ are continuous matrix and vector-functions, λ, Γ are a preassigned vector and a nonpositive matrix; u_1, u_2 are some assigned values.

An interesting subclass of these problems is connected with the control of quantum systems by means of a laser radiation. It was investigated and algorithmized using the method described above in [47]. It was developed in [48] for some interesting problems of synthesizing new materials with the aid of lasers at the quantum level. The distinctive feature of the problems is the very large dimension of the state vector which reaches the order of tens of thousands. In a simulated experiment a good convergence and effectivity of the method was obtained.

10.3 The multivariate Knapsack problem

The above method seems to be also effective for the following problem of integer programming:

$$J(v) = \sum_{t=0}^{N} c_t u_t \to \min \quad u_t \in [0, \beta_t] \tag{10.10}$$

$$\sum_{t=0}^{N} a_t^i u_t \leq b^i \quad i = \overline{1,n} \tag{10.11}$$

u_t is integer. Introducing the following sequence $\{x_t\} \subset R^n$, $t = 0, 1, \ldots,$ N:

$$x_{t+1}^i = x_t^i + a_t^i u_t, \quad x(1) = 0, \quad x^i(N+1) \leq b^i, \quad i = \overline{1,n}, \tag{10.12}$$

we can transform the problem (10.10)-(10.11) to the multistage optimization problem where:

$$t_0 = 0, \quad T = N+1, \quad X = R^n, \quad U - \text{ is a set of integers},$$

$$f(t,x,u) = x + a(t)u, \quad f^0 = c(t)u, \quad V_x = R^n \text{ for } t < T,$$
$$V_x = \{x : x^i < b^i \text{ for } t = T\}, \quad V^x = [0, b_t].$$

We have:

$$\varphi(t,x) = \psi_i(t)x^i + \sigma_{ii}(t)(x^i - x_0^i(t))^2/2, \quad \sigma_{ij} = 0, \quad i \neq j$$

$$R(t,x,u) = \psi_i(t+1)(x^i + a^i(t)u) + \sigma_{ii}(t+1)(x^i + a^i(t)u - x_0^i(t+1))^2 -$$
$$-\psi_i(t)x^i - \sigma_{ii}(t)(x^i - x_0^i(t))^2 - c(t)u$$
$$G(x) = \psi_i(T)x^i + \sigma_{ii}(T)(x^i - x_0^i(T))^2$$

Taking $\eta = 0$ and solving the equations (10.2), (10.3), and (10.8) we obtain

$$\psi_i(t) = \text{ constant } = \psi_i = 0 \text{ if } x_0^i(T) < b^i \text{ and}$$
$$\psi_i(t) \geq 0 \text{ if } x_0^i(T) = b^i, \quad i = \overline{1,n},$$
$$\sigma_{ii}(t) = \text{ constant } = -\alpha_i, \quad \alpha_i > 0$$
$$R(t,x,u) = -A(t)u^2/2 + B(t,x)u + C(t,x)$$
$$A(t) = \alpha_i(a^i(t))^2 > 0$$
$$B(t,x) = -\alpha_i a^i(t)(x^i - x_0^i(t+1)) + \psi_i a^i(t) - c(t)$$
$$C(t,x) = \alpha_i[x_0^i(t+1) - x_0^i(t)] + \alpha_i[x_0^{i2}(t+1) - x_0^{i2}(t)]$$

The expression for $R(t,x,u)$ satisfies (10.6). The control $\tilde{u}(t,x)$ is taken as an integer from the interval $[0, \beta_t]$ which is closest to the value $u^*(t,x)$ $= B(t,x)/A(t)$. The values α_i are chosen in such a way that the improved trajectory satisfies the inequalities (10.11).

Example 10.2.1 [49]. The problem is:

$$J = -[6u_1 + 4u_2 + u_3] \rightarrow \min$$

$$\text{s.t.} \qquad \begin{aligned} u_1 + 2u_2 + 3u_3 &\leq 5 \\ 2u_1 + u_2 + u_3 &\leq 4 \\ u_t &= \{0,1\} \end{aligned}$$

and the optimal solution, see Table 2:

$$\bar{u}_1 = \bar{u}_2 = 1, \quad \bar{u}_3 = 0, \quad J = -10$$

Example 10.2.2 [50]. The problem is :

$$J = -[3u_1 + 3u_2 + 13u_3] \rightarrow \min$$

$$\text{s.t.} \qquad -3u_1 + 6u_2 + 7u_3 \le 8$$
$$6u_1 - 3u_2 + 7u_3 \le 8$$
$$0 \le u_t \le 5, \quad u_t \quad - \text{ integer}$$

and the optimal solution, see table 3:

$$\bar{u}_1 = \bar{u}_2 = 0, \quad \bar{u}_3 = 1, \quad J = -13$$

In the paper [45] the version with a global improving function quadratic in x is discussed. There exist other versions of this method which are presented in [51].

10.4 The methods of successive improvement of the bounding function

The method is presented according to [52,53]. This is elaborated for ordinary, multistage and multiargument processes. Here we consider only ordinary processes.

Suppose there exists a function $\varphi_0(t,x) \in \Pi$. We now set the improvement problem which is to find a function $\varphi \in \Pi$ such that $l(\varphi) > l(\varphi_0)$, where $l(\varphi)$ is the lower bound from Section 4. We assume that it has the form: $\varphi = \varphi_0 + \lambda\gamma$, where λ, $\gamma(t,x)$ are a coefficient and a function which should be determined. We introduce a functional:

$$\delta(v) = \int_A r(t, x(t), u(t))dt + \gamma(t, x(t)) \mid_\tau^T; \quad r(t,x,u) = \gamma_x f; \tag{10.13}$$

We denote by $R(t,x,u,\lambda)$, $\tilde{E}(\lambda)$ etc. the appropriate constructions (4.1), (4.2), associated with $\varphi = \varphi_0 + \lambda\gamma$, and also $R_0(t,x,u) = R(t,x,u,0)$ etc. Taking into account (10.13) and (3.6) the increment $\Delta l = l(\lambda) - l_0$ can be written in the form:

$$\Delta l = \lambda\delta(v) + [L(v, \varphi_0) - l_0], \quad v \in \tilde{E}(\lambda) \tag{10.14}$$

where $L(v, \varphi_0)$ is defined by (). From here it follows that $l(\varphi) > l(\varphi_0)$ if at least for one $v \in \tilde{E}(\lambda)$:

$$\lambda\delta(v) > 0 \tag{10.15}$$

The specification of λ and γ that satisfy the above inequality will be further said to be *an elementary bound improving operation.* In the sequel for simplicity we shall consider only the case when the set $\tilde{E}$ contains a single element $\tilde{v}(\lambda) = (\tilde{x}(t,\lambda), \tilde{u}(t,\lambda))$, i.e. the function $R(t,x,u,\lambda)$ that has a single maximum. We also denote $\delta(\lambda) = \delta(\tilde{v}(\lambda))$. Under sufficiently general conditions the function $\delta(\lambda)$ is lower semicontinuous at $\lambda = 0$. Thus if we define the function γ to satisfy:

$$\delta(0) = \int_A r(t, \tilde{x}_0(t), \tilde{u}_0(t))dt + \gamma(t, \tilde{x}_0(t)) \mid_\tau^T > 0 \tag{10.16}$$

then for a sufficiently small $\lambda > 0$ the inequality (10.15) holds and therefore $l(\varphi) \geq l(\varphi_0)$. We then see that the elementary operation can be done in two steps. In the first step $\gamma(t, x)$ is chosen according to (10.16) and in the second $a\lambda > 0$ is chosen.

It is easier to interpret the idea of elementary operation when the improving component is taken in the form $\gamma = \nu_i(t)x^i$ and the functional $\delta(v)$ in the form:

$$\delta(\lambda) = \int_\tau^T \nu(t)\tilde{z}(t)dt + \sum_{t\in\beta} \nu(t)\tilde{x}(t) \mid_{t-1}^{t+0}; \tag{10.17}$$

$$z(t) = dx(t)/dt - f(t, x(t), u(t));$$

where $\tilde{z}(t)$ is related to the process $\tilde{v}$, β is the set of points of discontinuity of the function $\tilde{x}(t)$.

It follows from (10.17) that if there exists a value $t = \vartheta$ such that $\tilde{z}(\vartheta) \neq 0$ or in the continuous time $\tilde{x}(\vartheta + 0) - \tilde{x}(\vartheta - 0) \neq 0$, then the improvement of the function $\varphi_0(t, x)$ can be achieved by adding a linear term $\gamma(t, x) = \nu_i(t)x^i$ where the function $\nu(t)$ is taken to keep the right hand side of (10.17) positive for $\tilde{v} = \tilde{v}_0(\tilde{x}_0(t), \tilde{u}_0(t))$.

In more complicated cases the incorporation of $\gamma(t, x)$ is necessary only if the maximum of $R_0(t, x, u)$ is not unique.

A weak point in this method is the necessity to maximize $R(t, x, u)$ for every t in order to form the process $\tilde{v} = (\tilde{x}(t), \tilde{u}(t))$ or more generally, the set $\tilde{E}$. Therefore the method can be applied only to problems where this maximization can be performed analytically or when there exist efficient numerical procedures.

Repeating the elementary improving operations we come to the sequence $\{\varphi_s\}$ for which the value $l(\varphi)$ does increase.

There exist theorems where it is shown that under some stronger conditions for γ and λ the above sequences ensure a solution to the optimization problem for a broad class of systems. Namely, the sequence $\{\tilde{v}_s\} = \{\tilde{x}_s(t), \tilde{u}_s(t)\}$ corresponding to $\{\varphi_s\}$ by (4.1) is a generalized solution to the Problem 3 in the sense of (1.10) and to the Problem 4 .

Example 10.4.1. Find a solution to the system:

$$dx(t)/dt = u(t); \quad x(0) = x_0 > 0; \quad x(1) = 0;$$

that minimizes the functional:

$$J = \int_0^1 (u^2(t) + x^2(t))dt$$

Here x and u are scalar functions. We look for a solution in the form of a sequence:

$$\varphi_s = \psi_s(t)x, \qquad \gamma_s = \nu_s(t)x$$

We have:

$$R_s(t,x,u) = \psi_s u - u^2 - x^2 + \dot{\psi}_s x$$

$$\tilde{x}_s = \dot{\psi}_s/2$$

$$\tilde{u}_s = \psi_s/2$$

$$\mu_s(t) = R_s(t,\tilde{x}_s,\tilde{u}_s) = (\psi_s^2 + \dot{\psi}_s^2)/4$$

$$l_s = -\psi_s(0)x_0 - \int_0^1 (\psi^2 + \dot{\psi}^2)dt$$

$$\tilde{z} = \dot{\tilde{x}}_s - \tilde{u}_s$$

$$\Delta S = \Delta_2^1 + \Delta_2^2$$

$$\Delta_s^1 = \int_0^1 \mid \tilde{z}_s \mid dt$$

$$\Delta_s^2 = \mid \tilde{x}(1) \mid + \mid \tilde{x}_s(0) - x_0 \mid$$

$$\delta_s(x,u) = -\nu_s(1)x_s(1) - \nu_s(0)(x_0 - x(0)) - \int_0^1 \nu_s(t)z_s dt$$

$$\tilde{\delta}_s = \delta_s(\tilde{x}_s,\tilde{u}_s)$$

$$\Delta\tilde{x}_s = \dot{\nu}_s/2$$

$$\Delta\tilde{u}_s = \nu_s/2$$

$$R_s(t,x,u,\lambda) = R_s(t,x,u) + \lambda r_s(t,x,u)$$

$$r_s(t,x,u) = \nu_s u + \dot{\nu}_s x$$

$$\tilde{x}_s(\lambda) = \tilde{x}_s + \lambda\Delta\tilde{x}_s$$

$$\tilde{u}_s(\lambda) = \tilde{u}_s + \lambda\tilde{u}_s$$

The value λ_s is taken to satisfy the condition:

$$\tilde{\delta}_s(\lambda) \equiv \delta_s(\tilde{x}_s + \lambda\Delta\tilde{x}_s, \tilde{u}_s + \lambda\Delta\tilde{u}_s) = 0$$

which in this case is an elementary improving operation [51]. We have:

$$\tilde{\delta}_s(\lambda) = \delta_s(\tilde{x}_s,\tilde{u}_s) + \lambda\delta_s(\Delta\tilde{x}_s,\Delta\tilde{u}_s)$$

$$\lambda_s = -\delta_s(\tilde{x}_s, \tilde{u}_s)/\delta_s(\Delta\tilde{x}_s, \Delta\tilde{u}_s)$$

$$\psi_{s+1} = \psi_s + \lambda_s \nu_s$$

The function $R(t, x, u)$ has a unique maximum at $\tilde{x}(t)$, $\tilde{u}(t)$ for any ψ. The elementary operation is thus solvable in the class of linear functions $\gamma(t, x) = \nu(t)x$ for any ψ that does not ensure a strict optimum. We provide the specific iterations starting from $\psi_0 = 0$.

Iteration 1

We have:

$$\tilde{x}_0(t) = \tilde{u}_0(t) = 0$$

$$\Delta_0^1 = 0, \quad \Delta_0^2 = x_0 - \tilde{x}_0(0) = x_0$$

$$\tilde{z}(t) \equiv 0, \qquad l_0 = 0$$

$$\tilde{\delta}_0 = \delta_0(\tilde{x}_0, \tilde{u}_0) = -\nu_0(0)x_0 + \int_0^1 \nu_0(t)\tilde{z}_0(t)dt = -\nu_0(0)x_0$$

The condition (10.16) is satisfied for $\nu_0(0) = -1$. For other values of t the function $\nu_0(t)$ can be defined arbitrary. We define it in a simple way: $\nu_0(t) = -1$. We have $\tilde{\delta}_0 = x_0$, $\Delta\tilde{x}_0 = 0$, $\Delta\tilde{u}_0 = -1/2$, $\delta_0(\Delta\tilde{x}_0, \Delta\tilde{u}_0) = -1/2$, $\lambda_0 = 2x_0$. Hence $\psi_1(t) = 0 + \lambda_0\nu_0 = -2x_0$. Moreover $\tilde{x}_1(t) \equiv 0$, $\tilde{u}_1(t) = -y_0$, $\Delta_1^1 = x_0$, $l_1 = x_0^2 > l_0 = 0$.

Thus in the first iteration the value of l did increase but the pair $\tilde{x}, \tilde{u}$ did not move closer to D neither in the boundary conditions, i.e. in the norm Δ^2, nor in the integral norm Δ^1.

Iteration 2

We have:

$$\tilde{\delta}_1 = \delta_1(\tilde{x}_1, \tilde{u}_1) = -\nu_1(0) + x_0 \int_0^1 \nu_1(t)dt$$

According to (10.16) and the requirements of normalization [53] (1st way):

$$\nu_1(0) = -1, \quad \nu_1(t) \equiv 1 \text{ for } t \in (0, 1)$$

This function is discontinuous and does not comply with the conditions of the elementary operation. Therefore we take as $\nu_1(t)$ a continuous function from the approximating sequence $\{1 - 2(t-1)^k\}$, $k = 2, 4, 6, \ldots$. We select the function which is simplest for computing, i.e. $\nu_1(t) = 1 - 2(t-1)^2$.

We have:

$$\tilde{\delta}_1 = 4/3x_0$$

$$\Delta x_1 = 2(1-t) \qquad \Delta u_1 = 1/2 - (t-1)^2$$

$$\delta_1(\Delta x_1, \Delta u_1) = -29$$

$$\lambda_1 = \tilde{\delta}_1/\delta_1(\Delta x_1, \Delta u_1) = 40/87x_0$$

$$\tilde{x}_2(t) = \lambda_1 \Delta u_1 = 80/87x_0(1-t)$$

$$\tilde{u}_2(t) = \tilde{u}_1 + \lambda_1 \Delta u = -1/87x_0[67 + 40(t-1)^2]$$

The estimate of the distance from D is:

$$\Delta_1^1 = 10/87x_0 \approx 1/9x_0, \quad \Delta_2^2 = x_0 - \tilde{x}_2(0) = 7/87x_0 \approx 7/90x_0$$

The lower bound is $l_2 \approx x_0^2$. Therefore in the second iteration the pair $\tilde{x}, \tilde{u}$ was moved substantially closer to D, approximately 10 times in each norm.

The above method was applied in order to develop algorithms for solving integral assignment problems, as well as those of scheduling, traveling salesman type [54], and also various optimization problems of space maneuvres [55] and problems for distributed parameters systems [56].

10.5 Methods of ϵ-optimal control synthesis

Our problem is to find an ϵ-optimal control synthesizing function $\bar{u}(t,x)$. We consider the case when there are no state constraints, including boundary constraints, i.e. $V_x(t) = R^n \ \forall t, x \in T \times R^n$. Other problems can be solved by this method using penalty functions. We have shown in the above that this problem can be solved by using the bounding expression (2.5) and by minimizing the functional $\Delta(\varphi)$ until it reaches the value $\Delta(\bar{\varphi}) = \epsilon$. Then the synthesizing function $\tilde{u}(t,x) = \arg\max_{u \in V^x(t)} R(t,x,u)$ is ϵ-optimal. The problem of finding an optimal control synthesis is therefore reduced to the minimization of the functional $\Delta(\varphi)$. The lower bound for the latter is zero. This bound is attained when in the class Π or its above mentioned refinements there exists a solution of the dynamic programming equation (2.6) or a sequence which approximates this solution in the sense of $\Delta(\varphi)$.

There exist numerical algorithms that do use this approach. One of such algorithms [4,6,57,58] is as follows. The desired function $\varphi(t,x)$ is taken as an interpolating polynomial in the space $X = R^n$. Its parameters depend on t and are determined from the equations:

$$P(t, x_l(t)) = 0, \quad G(x_l(T)) = 0 \tag{10.18}$$

where $\{x_l\}$ is a given set of interpolation knots, $P(t,x)$ and $G(t,x)$ are given by (4.11) and (2.2). The equations (10.18) form a system of normal differential (difference) equations in the function $\varphi(t,x)$ parameters with the given boundary conditions for $t = T$. Solving this system we come to the function $\varphi(t,x)$, the corresponding control synthesis $\tilde{u}(t,x)$, and the bound $\Delta(\varphi)$. If the latter is too large, the computations are then repeated with a "better" set $\{x_l\}$. This is reiterated until we get $\Delta(\varphi) < \epsilon$.

A second algorithm that was used in some interesting applied problems [7,pp.349-357] consists in solving the problem $\Delta(\varphi) \to \min$ by the Ritz method. Then a class of functions $\varphi(t,x) = \xi(t,x,a)$ that depend on a parameter a is taken. The functional $\Delta(a) = \Delta(\xi(t,x,a))$ is computed and the minimal value of $\Delta(\varphi)$ is found using the method of mathematical programming for this class situations.

The possibilities of using methods of the above are limited because of the operations $\sup_x P(t,x)$ and $\inf_x P(t,x)$ that are in (2.5). For many specific problems [59,60,61] these operations can be performed analytically. In the mentioned cases it is quite easier to realize and justify algorithms for solving the problem $\Delta(\varphi) \to \min$ in the class of bounding functions that are quadratic in x. Moreover, in order to obtain exact solutions in the form of minimizing sequences of synthesizing controls.

11 Application of the Global Bounds Techniques to Other Control Problems for Dynamic Systems

The following problems of the control for dynamic systems within the framework of optimal control theory have been the objects of this technique. These are control synthesis under incomplete information on the process equations and the state with a game-type criterion [62], [63], [64], [7] and with invariance criteria [65]; the control of systems with an aftereffect [66], [67]; the optimal control of noninertial systems with delays [68]; the estimation of the attainability domains [16],[31]; the problems of integer programming [54], [69].

References

[1] Krotov V.F. *Methods of solution of variational problems on the basis of sufficient conditions for absolute minimum. I.* Avtomat. i Telemeh.†, 23 (1962) 12, 1571-1583.

[2] Krotov V.F. *Methods of solving variational problems. II. Sliding regimes.* Avtomat. i Telemeh.†, 24 (1963) 5, 581-598.

[3] Krotov V.F. *Methods for variational problem solution based on absolute minimum sufficient conditions. III.* Avtomat. i Telemeh.†, 25 (1964) 7, 1037-1046.

[4] Krotov V.F. *Approximate synthesis of optimal control.* Avtomat. i Telemeh.†, 25 (1964) 11, 1521-1527.

[5] Krotov V.F. *Sufficient optimality conditions for discrete control systems.* Dokl. AN SSSR, 172 (1967) 1, 18-21.

[6] Krotov V.F., Bukreev V.Z., Gurman V.I. *New Variational Methods in Flight Dynamics.* Mashinostroyenie, Moskva, 1969. (English transl.: NASA, Transl. TTF-657, 1971).

[7] Krotov V.F., Gurman V.I. *Methods and Problems of Optimal Control.* Nauka, Moskva, 1973 (in Russian).

[8] Bellman R. *Dynamic Programming.* Princeton University Press, New Jersey, 1957.

[9] Ioffe A.D., Tihomirov V.M. Theory of Optimal Problems. Nauka, Moskva, 1974 (English transl.: North Holland, Amsterdam, 1973).

[10] Pontryagin L.S., Boltianski V.G., Gamkrelidze P.V., Mishchenko E.F. *Mathematical Theory of Optimal Processes.* Fizmagtiz, Moskva, 1961 (English transl.: Interscience Publishers Inc., New York, 1962).

[11] Mereau P.M., Powers W.F. *A direct sufficient condition for free final time optimal control problems.* SIAM J. Control Optim., 14 (1976) 4, 613-622.

[12] Rozenberg G.S. *On the necessity and the sufficient conditions for minimum in variational problems.* Differencialnye Uravnenija, (1968) 2 (transl. to English as Differential Equations).

[13] Zeidan V. *1st and 2nd order sufficient conditions for optimal control and the calculus of variations.* Appl. Math. Optim., 11 (1984) 3, 209-226.

[14] Girsanov I.V. *On a relation between Krotov and Bellman functions in the dynamic programming method.* Vestnik Moskov. Univ., Ser. I, Mat. Meh., (1968), 56-59 (in Russian).

[15] Gurman V.I. *Singular Problems in Optimal Control.* Nauka, Moskva, 1977 (in Russian).

[16] Gurman V.I. *Principle of Extensions in Control Problems.* Nauka, Moskva, 1985 (in Russian).

[17] Krotov V.F. *A fundamental problem of variational calculus for a simple functional on the set of discontinuous functions.* Dokl. AN SSSR, 137 (1961) 1, 31-34 (in Russian).

[18] Gurman V.I. *A method for investigating one class of optimal sliding regimes.* Avtomat i Telemeh.†, 26 (1965) 7, 1169-1176.

[19] Gurman V.I. *On optimal trajectories between nonoriented orbits in the central field.* Kosmicheskiye Issledovania, (1966) 3 (in Russian).

[20] Ioslovich I.V. *The optimal stabilization of a satellite in an inertial system of coordinates.* Astronaut. Acta, 13 (1962), 37-47 (in Russian).

[21] Borshchevski M.Z., Ioslovich I.V. *Some problems in optimal stabilization of cosymmetric satellite.* Kosmicheskiye Issledovania, (1966) 3 (in Russian).

[22] Khrustalev M.M. *On sufficient conditions of global maximum.* Dokl. AN SSSR, Ser. Mat., 174 (1967) 5 (in Russian).

[23] Ioffe A.D. *Convex functions occuring in variational problems and the absolute minimum problem.* Mat. Sb., 88 (1972), 194-210.

[24] Vinter R.B. *Weakest conditions for existence of Lipshitz continuous Krotov functions in optimal control theory.* SIAM J. Control Optim., 21 (1983) 2, 215-234.

[25] Vinter R.B. *New global optimality conditions in optimal control theory.* SIAM J. Control Optim., 21 (1983) 2, 235-245.

[26] Vinter R.B., Lewis R.M. *A verification theorem which provides a necessary and sufficient condition for optimality.* IEEE Trans. Autom. Control, AC-25 (1980) 1, 84-89.

[27] Klötzler R. Starke *Dualität in der Steurungstheorie.* Math. Nachr., 95 (1980), 253-263.

[28] Pickenhain S. Starke *Dualität bei verallgemeinerten Steuerungsproblemen.* Zeit. f. Analysis u. ihre Anwendungen, 1 (1982) 4, 15-24.

[29] Khrustalev M.M. *Exact Description of Reachable Sets and Global Optimality Conditions for Dynamic Systems.* Avtomat. i Telemeh.†, N5, pp.62-70, May, 1988.

[30] Khrustalev M.M. *Exact Description of Reachable Sets and Conditions of Global Optimality for Dynamic Systems.* Avtomat. i Telemeh.†, N7, pp.874-881, December, 1988.

[31] Khrustalev M.M. *On sufficient optimality conditions in the problem with phase coordinates constraints.* Avtomat. i Telemeh.†, (1967) 4, 18-29 (in Russian).

[32] Dubovitzky A.Ya., Milyutin A.A. *Extremum problems under constraints.* Zh. Vychisl. Mat. i Mat. Fiz., 5 (1965) 3, 395-453 (in Russian).

[33] Klötzler R. *Dualität in der Steuerungstheorie.* Sonderdruck Heft, (1981) 2/4.

[34] Klötzler R. *Globale Optimierung in der Steurungstheorie.* Hauptvortrag zur GAMM-Tagung, 1982.

[35] Klötzler R. *Models and applications of duality in optimal control.* IFIP Conf., Hanoi, Viet-Nam, 10-14 January 1983.

[36] Klötzler R. *Geometrical applications of duality in optimal control.* Int. Conf. Math. Methods in OR, Sofia, 1983, 52-69.

[37] Andreeva I.A., Klötzler R. *Zur analytische Lösung geometrischer Optimierugsaufgaben mittels Dualität bei Steuerungsproblem.* Teil I. Z. Angew. Math. Mech., 64 (1984), 35-44.

[38] Konstantinov G.N. *Normalization of Controls in Dynamic Systems.* Izd. Irkutskogo Univ., Irkutsk, 1983 (in Russian).

[39] Moskalenko A.I. *Methods of Linear Transformations in Optimal Control.* Nauka, Moskva, 1983 (in Russian).

[40] Subbotin A.I., Subbotina N.N. *On proving the dynamical programming method in an optimal control problem.* Izv. AN SSSR, Tehnich. Kibern., (1983) 2, 24 (in Russian).

[41] Kelley H.J. *Gradient theory of optimal flight path.* ARS J., 30 (1960) 10.

[42] Eneev T.M. *On application of gradient method in optimal control theory problems.* Kosmicheskiye Issledovania, 4 (1966) 5 (in Russian).

[43] Krylov I.A., Chernousko F.L. *On a method of consecutive approximations for solving optimal control problems.* Zh. Vychisl. Mat. i Mat. Fiz., 2 (1962) 6, 1132-1139 (in Russian).

[44] Bryson A.E., Ho Y.Ch. *Applied Optimal Control.* Hemisphere Publ. Corp., Washington, 1975.

[45] Krotov V.F., Feldman I.N. *An iterative method for solving problems of optimal control.* Izv. AN SSSR, Tehnich. Kibern., (1983) 2, 162-167 (in Russian).

[46] Propoi A.I. *Elements of the Theory of Optimal Discrete Systems.* Nauka, Moskva, 1981 (in Russian).

[47] Kazakov V.A., Krotov B.F. *Optimal control of resonance interaction of the light and a substance.* Avtomat i Telemeh.†, (1987) 4, 9-15.

[48] Kazakov V.A., Somoloi J., Tannor D.J. *Optimal Control of Quantum Systems. I. A Shortsighted Optimization Procedure.* J. Chem. Phys., 1993 (in appear).

[49] Emelichev V.A., Komlik V.I. *A Method for Constructing a Sequence of Plans for Solving Discrete Optimal Problems.* Nauka, Moskva, 1981 (in Russian).

[50] Wagner G. *Principles of Operations Research with Application to Managerial Decisions.* Prentice Hall, Englewood Cliffs, 1969.

[51] Gurman V.I., Baturin V.A., Danilina E.V. *New Methods of Improving Controlled Process.* Nauka, Novosibirsk, 1987 (in Russian).

[52] Krotov V.F. *Computing algorithms for solution and optimization of a controlled system of equations.* I. Izv. AN SSSR, Tehnich. Kibern., (1975) 5, 3-15 (in Russian).

[53] Krotov V.F. *Computing algorithms for solution and optimization of a controlled system of equations. II.* Izv. AN SSSR, Tehnich. Kibern., (1975) 6, 3-13 (in Russian).

[54] Krotov V.F., Sergeev S.I. *Computing algorithms for solution of some problems in linear and linear-integer programming.* Avtomat i Telemeh.†, Part I (1980) 12, 86-96, Part II (1981) 1, 86-96, Part III (1981) 3, 83-94, Part IV (1981) 4, 103-112.

[55] Egorov V.A., Gusev L.I. *The Dynamics of the Flight between the Earth and the Moon.* Nauka, Moskva, 1980 (in Russian).

[56] Gukasian M.H. *On computing algorithms for optimization of linear controlled systems with distributed parameters.* In: Investigations in Mechanics of a Solid Deforming Body. Erewan, 1981, 93-97.

[57] Bukreev V.Z. *Concerning certain method of approximate synthesis of optimal control.* Avtomat i Telemeh.†, (1968) 11, 5-13.

[58] Bukreev V.Z., Rozenblat G.M. *Approximate design of optimal control for discrete-time systems.* Avtomat i Telemeh.†, (1976) 1, 90-93.

[59] Bakhito R.U., Krapchetkov N.P., Krotov V.F. *The synthesis of an approximately optimal control for one class of controlled systems.* Avtomat i Telemeh.†, (1972) 10, 33-43.

[60] Trigub M.B. *Approximately optimal stabilization of one class of non-linear systems.* Avtomat i Telemeh.†, (1987) 1, 34-47.

[61] Lekkerke P.I., Derleman T.W. *Calculation method for dynamic optimization using quadratic approximations to minimum surface.* Automatica, 7 (1971) 6, 713.

[62] Krotov V.F. *Optimization methods of control processes with minimax criteria.I,* Automat. i Telemeh.†,No.12 (1973).

[63] Krotov V.F. *Optimization methods of control processes with minimax criteria.II,* Avtomat. i Telemeh.†,No.1 (1974).

[64] Krotov V.F. *Optimization methods of control processes with minimax criteria.III.* Avtomat. i Telemeh.†,No.2 (1974).

[65] Khrustalev M.M. *Necessary and sufficient conditions of invariance.* Avtomat. i Telemeh.†,No.4 (1968).

[66] Andreeva A.A., Kolmanovskii V.B., Shaychet L.E. *Control of the systems with aftereffect.* Moscow, Nauka, 1992, pp.64-98 (in Russian).

[67] Kolmanovskii V.B., Myshkis A.D. *Functional – differential equations.* Kluwer Ac. Publ. N. Y. 1992, pp.162-173.

[68] Khrustalev M.M. *Global Optimality Conditons for Controlled Non-inertial Systems.* Control and Cybernetics, 17, N02-3, 1988.

[69] Krotov V.F. *A Technique of Global Bounds in Optimal Control Theory.* Control and Cybernetics, 17, N02-3, 1988.

[70] Stoddard A.W.J. *Estimation of Optimality for Multidimensional Control Systems.* J. of Opt. Theory and Appl., 1969, v.3, N03, pp.385-391.

†Avtomatika i Telemehanika is translated to English as Automation and Remote Control.

Institute for Control Problems
Russian Academy of Sciences
Profsoyuznaya st. 65
Moscow 117342, Russia

It.no	u(0)	u(1)	x^1 (1)	x^2 (1)	x^2 (1)	ψ_2 (1)	ψ_2 (1)	J
1	0	0	3	-9	-18	-6	1	18
2	-2	-5	-1	-5	-19	2	1	19

Table 1: The numerical results of example 1

Iter.	Control			Vector α		
number	u_1	u_2	u_3	α_1	α_2	Functional
0	0	0	0	1	1	0
1	1	0	0	1	1	-6
2	1	1	0			-10

Table 2: The numerical results of example 2

Iter.	Control			Vector α		
number	u_1	u_2	u_3	α_1	α_2	Functional
0	0	0	0	0.3	0.3	0
1	0	0	1			-13

Table 3: The numerical results of example 3

On the Theory of Trajectory Tubes – A Mathematical Formalism for Uncertain Dynamics, Viability and Control

A.B. Kurzhanski and T.F. Filippova

Abstract. This paper is a survey on the theory of trajectory tubes for differential inclusions, which appears to be a relevant tool for modeling uncertain dynamics. It is motivated by results in nonlinear analysis, particularly, in viability theory for differential inclusions, as well as by recent achievements in the theory of estimation and control for systems with unknown but bounded uncertainties. The motivations for these studies come from applied areas and the rapidly increasing number of applications range from computer sciences and engineering to economics as well as ecological and biomedical modeling. The theory is constructive and has led to the development of effective tools of computational and graphic animation.

Key words: differential inclusions, semigroup, dynamic system, state constraints, viability theory, trajectory tube, attainability set, funnel equation, bilinear systems, singular perturbations, stochastic filtering, state estimation, uncertain system, nonlinear control synthesis.

1 Introduction

This paper is a survey of recent results in the theory of tubes (assemblies) of solutions to differential inclusions. Of particular interest is the description of the behaviour of these tubes when the system is subjected to state constraints. The objects under investigation are then known as the "viability tubes" and their crossections turn to be the attainability domains for the original differential inclusion with the state constraint (the "viability" restriction [1]). Therefore a crucial item is the investigation of the evolution of these domains ,especially if we take into account the following. Starting at a specified set, the overall system generates a multivalued map that satisfies the semigroup property and therefore, defines a generalized dynamic system. One of the basic problems discussed here is the ability to describe either the infinitesimal generator of the respective semigroup or some equivalent notion. This issue is treated here in dual form, namely, either with the aid of an evolution equation of the "funnel" type with set-valued trajectories, or through a generalized partial differential equation for the support functions of the crossections of the viability tubes. (The latter is possible when the crossections are convex).

The other principal facts given here are: the calculation of viability tubes through a parametrized variety of tubes for a system without state constraints; the description of trajectory tubes for bilinear viable systems with star-shaped set-valued trajectories; the generalized set-valued Lagrangian scheme.

A separate issue is related to the theory of singular perturbations for differential inclusions which allows to calculate viability tubes for state constraints that are nonsmooth and even discontinuous in time. The emphasis of this paper is therefore such that it hardly overlaps with the results of [1, 3] .

As already mentioned above, the mathematical motivations for the topics of this survey come from problems in nonlinear analysis as well as in system modeling, in dynamics, control and game-type issues for evolutionary systems and related topics [34, 36, 38].

The applications discussed in this paper are in guaranteed state estimation for systems with unknown but bounded errors and in nonlinear control synthesis [38, 70].

It should be understood that the investigation of set-valued uncertain systems on a finite time interval requires to study the trajectory tubes in all the details. This is in contrast with other theories like those of robust stability and stabilization, which deal with an infinite time horizon with the goal being to achieve only a favourable asymptotic behaviour of the trajectories and where therefore the objects of description are the more general asymptotic properties of the system, like those of being stable or asymptotically stable uniformly in some appropriate sense.

Moreover, the formulation of the final results is given through relations that involve only the simplest operations like the sums and intersections of sets, rather than the calculation of tangent cones and related constructions (although these most important notions are of course indispensable in the proofs). The results may therefore allow a rather effective algorithmization and parallelization.

The problems of this paper therefore require a serious and motivate development of new techniques of set-valued analysis, particularly its constructive methods — ellipsoid–valued calculus, for example [13, 57, 58]. The last issue is however beyond the scope of this paper.

In order to commence the survey it is necessary to introduce some essential notations.

2 Notations

Here we give a list of basic notations used throughout the paper.

R^n –
the n–dimensional Euclidean space

(x, y), $x'y$ –
the inner product of $x, y \in R^n$ with the prime as a transpose

$\| x \|$ –
$\| x \| = (x, x)^{1/2}$

int A –
the interior of $A \subseteq R^n$

co A –
the convex hull of $A \subseteq R^n$

comp R^n –
the variety of all compact subsets $A \subseteq R^n$

conv R^n –
the variety of all compact convex subsets $A \subseteq R^n$

cl conv R^n –
the variety of all closed convex subsets $A \subseteq R^n$

$d(x, A)$ –
the distance $d(x, A) = \inf \{\| x - y \| \mid y \in A\}$

$h(A, B)$ –
the Hausdorff distance for $A, B \subseteq R^n$, i.e. $h(A, B) = \max \{h^+(A, B), h^+(B, A)\}$, $h^+(A, B) = \inf\{d(x, B) \mid x \in A\}$

$\mathfrak{R}^{n,m}, \mathfrak{R}^n$ –
the space of all $n \times m$–matrices, $\mathfrak{R}^n = \mathfrak{R}^{n,n}$

$\mathfrak{R}^n_+$ –
the variety of all symmetric positively definite matrices $L \in \mathfrak{R}^n$

I, I_n –
the identity matrix in $\mathfrak{R}^n$

det L –
the determinant of $L \in \mathfrak{R}^n$

$C^n[t, \tau]$ –
the space of continuous functions $z : [t, \tau] \to \mathfrak{R}^n$

$C^n_k[t,\tau]$ –
the space of k - times continuously differentiable functions z : $[t,\tau] \to \Re^n$, $0 < k \leq \infty$

$L^n_p[t,\tau]$ –
the space of functions $z : [t,\tau] \to \Re^n$ with $\| z(s) \|^p$ integrable on $[t,\tau]$

$\Re^{n,m}[t,\tau]$ –
the space of continuous functions $L : [t,\tau] \to \Re^{n,m}[t,\tau]$

$\Re^n[t,\tau]$ –
$\Re^n[t,\tau] = \Re^{n,n}[t,\tau]$

$\Re^{n,m}_k[t,\tau]$ –
the space of k - times continuously differentiable functions L : $[t,\tau] \to \Re^n$ $(0 < k \leq \infty)$

$\Re^n_k[t,\tau]$ –
$\Re^n_k[t,\tau] = \Re^{n,n}_k[t,\tau]$

$\rho(l \mid A)$ –
the support function of $A \subseteq R^n$ at point $l \in R^n$: $\rho(l \mid A) = \sup\{l'a \mid a \in A\}$

co $\varphi(l)$ –
the closed convex hull of function $\varphi : R^n \to R^1$

$\partial\varphi(l)$ –
the subdifferential of a convex function $\varphi : R^n \to R^1$ at point $l \in R^n$

3 Uncertain Systems. Differential Inclusions and the Viability Problem

One of the principal motivations for studying the topics of this survey comes from the theory of systems with unknown, but bounded uncertainties. These may be described as follows. Consider the ordinary differential equation

$$\dot{x} = f(t, x, v) \tag{3.1}$$

with function $f : T \times R^n \times R^n \to R^m$ measurable in t and continuous in the other variables. Here x stands for the state space ("phase") vector, t stands for time and v is the input, which ,for example, may be a control or disturbance. The variables v are assumed to be *unknown but bounded*

$$v \in Q(t, x) \tag{3.2}$$

where $Q(t,x)$ is a multivalued map $(Q : T \times R^n \to \text{comp}R^m)$ measurable in t and continuous in x. The given data allows to consider a multivalued function

$$\mathcal{F}(t,x) = \bigcup \{ \, f(t,x,v) \mid v \in Q(t,x) \, \}$$

and further on , a differential inclusion

$$\dot{x} \in \mathcal{F}(t,x) \tag{3.3}$$

that reflects the variety of all models (3.1) possible under uncertainty (3.2).

Otherwise,

$$\mathcal{F}(t,x) = f(t,x,Q(t,x))$$

so that the multivalued function $\mathcal{F} : T \times R^n \to \text{comp}R^n$ turns to be measurable in t and continuous in x in the Hausdorff metric [12].

In particular, we may have

$$f(t,x,v) = A(t,v)x + h(t),$$

$$v \in Q(t) \in \text{comp}R^m,$$

so that

$$\mathcal{F}(t,x) = A(t,Q(t))x + h(t), \tag{3.4}$$

and (3.1) describes a *linear system with uncertain coefficients*, or

$$\begin{aligned} f(t,x,v) = &\quad A(t)x + B(t)v, \\ v \in &\quad Q(t) \in \text{comp}R^m, \end{aligned} \tag{3.5}$$

when (3.1) is a *"proper" linear system.*

The multifunction $\mathcal{F}(t,x)$ may also reflect the uncertainty in the model f (3.1) itself, when

$$f = f(t,x) \in \mathcal{F}$$

and $\mathcal{F}$ is a prespecified variety of functions f.

One of the first points of interest of this survey is to study the set of all solutions $x[t] = x(t,t_0,x^0)$ to a differential inclusion (the "contingent differential equation")

$$\dot{x} \in \mathcal{F}(t,x), \quad t \in T, \tag{3.6}$$

of type (3.1), that are emitted by the initial set X^0 so that

$$x(t_0) = x^0, \quad x^0 \in X^0 \in \text{comp}R^n \tag{3.7}$$

A further problem that concerns the set of these solutions is to single out a subset of those trajectories $x[t] = x(t, t_0, x^0)$ that satisfy both (3.6) and a restriction on the state vector (the "viability" constraint)

$$Gx[t] \in Y(t), \tag{3.8}$$

where G is $p \times n$–matrix ($p \leq n$), $Y(\cdot)$ ($Y(t) \in \text{conv}R^p$) is a convex compact valued multifunction absolutely continuous in t in the sense that its support function

$$f(l,t) = \rho(l \mid Y(t)) = \max\{(l,y) \mid \; y \in Y(t)\;\}$$

is absolutely continuous in t, uniformly in l (further on, with regard to specific versions of the results , we will either strengthen this condition to an $f(l,t)$ continuously differentiable in t or relax it ,allowing $f(l,t)$ to be only measurable in t).

The viability constraint (3.8) may particularly arise in the "guaranteed" state estimation problem [43] when it is induced by a measurement equation

$$y(t) = G(t)x + w, \tag{3.9}$$

where y is the measurement, $G(t)$ — a continuous matrix function, w — the unknown but bounded "measurement noise" and

$$w \in Q(t), \quad Q(t) \in \text{comp}R^p$$

with Q continuous in t. In a more general form this equation may be written as

$$y \in G(t,x), \tag{3.10}$$

where $G(t,x)$ is a multivalued map ($G : T \times R^n \to \text{conv}R^p$) or taking

$$G^*(t,x) = G(t,x) - y(t)$$

and omitting the asterisk, as

$$0 \in G(t,x) \tag{3.11}$$

(the other requirements on $G(t,x)$ will be indicated below, in conjunction with the specific theorems to be discussed).

The "guaranteed" estimation problem will thus consist in describing the set $X[\cdot]$ of solutions to the system (3.6), (3.7), (3.11). The crossection $X[t]$ of this set will be the set-valued estimate itself.

The problem of jointly solving the system of inclusions (3.6), (3.7), (3.11) could be also treated as a generalized version of the viability problem [2]. The solution set to this system will be further referred to as the *viability bundle* or *viability tube* and defined more precisely below.

As we shall observe, the differential inclusion (3.6) and the viability constraints (3.7) may generate a class of generalized dynamic systems that are relevant for describing solutions to problems of dynamics and control.

Particularly, the crossections of the viability tubes are actually the *attainability domains* of system (3.1), (3.2), (3.7) and are therefore important for the investigation of problems in estimation and control. When taken in inverse time, they are crucial for the solutions of problems in nonlinear control synthesis.

We shall now proceed with a precise formulation and a constructive theory for the problems under discussion.

4 Trajectory Bundles and Trajectory Tubes. The Generalized Dynamic System

Following Section 3 we further assume that the notions of continuity and measurability of single-valued and multivalued maps are taken in the sense of [12, 26].

Consider the nonlinear differential inclusion (3.6), where $x \in R^n$, $\mathcal{F}$ is a continuous multivalued map ($\mathcal{F} : [t_0, t_1] \times R^n \rightarrow \text{conv} R^n$) that also satisfies the Lipschitz condition with constant $L > 0$, namely

$$h(\mathcal{F}(t,x), \mathcal{F}(t,y)) \leq L \parallel x - y \parallel, \quad \forall x, y \in R^n$$

Assuming a set $X_0 \in \text{comp} R^n$ to be given, denote $x[t] = x(t, t_0, x_0)$, $t_0 \in T = [t_0, t_1]$, to be a solution to (3.6) (an isolated trajectory) that starts at point $x[t_0] = x_0 \in X_0$.

Here we define the solution as a Caratheodory–type trajectory $x[\cdot]$, i.e. as an absolutely continuous function $x[t]$ $(t \in T)$ that satisfies the inclusion

$$\frac{d}{dt} x[t] = \dot{x}[t] \in \mathcal{F}(t, x[t]) \tag{4.1}$$

for almost every $t \in T$.

Further on, we require all the solutions $\{ x[t] = x(t, t_0, x_0) \mid x_0 \in X_0 \}$ to be extendable up to the instant t_1 [3, 21].

Let $Y(t)$ be a continuous multivalued map ($Y : T \to \text{conv}R^n$), $X_0 \subseteq Y(t_0)$.

Definition 4.1 [1, 44]. *A trajectory $x[t] = x(t, t_0, x_0)$ ($x_0 \in X_0$, $t \in T$) of the differential inclusion (4.1) will be defined to be viable on $[t_0, \tau]$ if*

$$x[t] \in Y(t) \quad \textit{for all} \;\; t \in [t_0, \tau]. \tag{4.2}$$

The further investigations actually follow under the assumption that there exists at least one solution $x^*[t] = x^*(t, t_0, x_0^*)$ of (4.1) (together with a starting point $x^*[t_0] = x_0^* \in X_0$) that satisfies condition (4.2) with $\tau = t_1$. The specified isolated trajectory $x^*[t]$ is therefore assumed to be viable on the whole segment $[t_0, t_1]$. The conditions for the existence of such trajectories may be given in terms of generalized duality concepts [38]. The known theorems on viability also provide the existence of such trajectories $x^*[t]$ even for any starting point $x_0^* \in X_0$ [1, 2]. Here however our requirements are weaker. We therefore consider a somewhat more general relationship between the mappings $\mathcal{F}$ and Y than in the above investigations.

Let $\mathcal{X}(\cdot, t_0, X_0)$ be the set of all solutions to the inclusion (4.1) that emerge from X_0(the "solution bundle"). Denote $\mathcal{X}[t] = \mathcal{X}(t, t_0, X_0)$ to be its crossection at instant t and the set

$$\mathcal{Q} = \bigcup \{ \; \mathcal{X}(t, t_0, X_0) \quad | \quad t_0 \leq t \leq t_1 \; \}$$

to be the integral funnel to (4.1) [21, 62]. One may observe that under our assumptions $\mathcal{Q} \in \text{comp}R^n$ [3].

The subset of $\mathcal{X}(\cdot, t_0, X_0)$ that consists of all solutions to (4.1) viable on $[t_0, \tau]$ will be further denoted as $X(\cdot, \tau, t_0, X_0)$ with its s – crossections as $X(s, \tau, t_0, X_0)$, $s \in [t_0, \tau]$. We introduce symbol $X[\tau]$ for these crossections at instant τ, namely

$$X[\tau] = X(\tau, t_0, X_0) = X(\tau, \tau, t_0, X_0)$$

As indicated above, it is not difficult to observe that the sets $X[\tau]$ are actually the attainability domains (or "reachable sets") at instant τ for the differential inclusion (4.1) with state constraint (4.2).

The map $X(t, t_0, X_0)$ ($X : T \times T \times \text{comp}R^n \to \text{comp}R^n$) satisfies the *semigroup property*

$$X(t, \tau, X(\tau, t_0, X_0)) = X(t, t_0, X_0),$$

$$t_0 \leq \tau \leq t \leq t_1,$$

and therefore defines a generalized dynamic system with set-valued trajectories $X[t] = X(t, t_0, X_0)$, [8, 68]. The multivalued functions $\mathcal{X}[t]$, $X[t]$, $t \in T$, will be referred to as the *trajectory tube* and *viable trajectory tube* respectively.

The main objective of this survey is to study the properties and means of calculating the viable trajectory tubes and their crossections (the attainability domains) together with the impact of these properties on the theory of control.

5 Some Basic Assumptions and Basic Geometrical Invariants

Further on we mainly assume one of the two following groups of hypotheses to be fulfilled with exceptions from this rule being always pointed out.

Let us denote the graph of the map $\mathcal{F}(t,\cdot)$ as $gr_t\mathcal{F}$ (t is fixed):

$$gr_t\mathcal{F} = \{ \{x, y\} \in R^n \times R^n : \ y \in \mathcal{F}(t,x) \}$$

Assumption A1

(i) For some $D \in \text{conv}R^n$ such that $\mathcal{Q} \subseteq \text{int}D$ the set $(D \times R^n) \cap gr_t\mathcal{F}$ is convex for every $t \in T$.

(ii) There exists a solution $x_*[\cdot]$ to (4.1) such that $x_*[t_0] \in X_0$ and

$$x_*[t] \in \text{int}\, Y(t), \quad \forall\, t \in T.$$

(iii) The set $X_0 \in \text{conv}R^n$.

In order to formulate the next group of assumptions we need to recall the following notion.

Definition 4.2. *A set $Z \subseteq R^n$ will be defined to be star–shaped (with a center at c) if $c \in Z$ and $c + \lambda Z \subseteq Z$ for all $\lambda \in (0,1]$.*

The variety of all star–shaped compact sets $Z \subseteq R^n$ with center at point c will be denoted as $\text{St}(c, R^n)$ and with center at $0 \in R^n$ as $\text{St}(R^n)$.

Assumption A2.

(i) For every $t \in T$, the graph $gr_t\mathcal{F} \in \text{St}(R^{2n})$.

(ii) There exists an $\epsilon > 0$ such that

$$\epsilon S \subseteq Y(t)$$

for all $t \in T$ where $S = \{x \in R^n : \| x \| \leq 1\}$.

(iii) The set $X_0 \in \mathrm{St}(R^n)$.

One may observe that the assumptions A1, A2 are overlapping. However in general neither of them covers the other.

One of the important properties of trajectory tubes and viable trajectory tubes is to preserve some geometrical characteristics of the sets $\mathcal{X}[t]$, $X[t]$ along the system trajectories. These are given by the following assertions.

Lemma 5.1. *Under assumption A1 the crossections $\mathcal{X}[t]$, $X[t]$ are convex and compact for all $t \in T$ ($\mathcal{X}[t]$, $X[t]$ $\in convR^n$).*

Lemma 5.2. *Under assumption A2 the crossections $\mathcal{X}[t]$, $X[t]$ are star–shaped and compact for all $t \in T$ ($\mathcal{X}[t]$, $X[t]$ $\in St(R^n)$).*

Loosely speaking, a differential inclusion with a convex graph $gr_t\mathcal{F}$ generates convex–valued tubes $\mathcal{X}[t]$, $X[t]$, while a star–shaped graph $gr_t\mathcal{F}$ generates tubes with star–shaped crossections $\mathcal{X}[t]$, $X[t]$. Convexity and star–shapedness are therefore the two simplest basic geometrical invariants for the crossections of the trajectory tubes and viable trajectory tubes.

6 The Evolution Equation

Having defined the notion of viable trajectory tube and having observed that the mapping $X[t] = X(t, t_0, X_0)$ defines a generalized dynamic system, we arrive at the following natural question: does there exist some sort of evolution equation that would describe the tube $X[t]$ and therefore also the related generalized dynamic system?

It should be emphasized here that the space compR^n of all compact subsets of R^n to which the "states" $X[t]$ do belong is only a metric space with a rather complicated nonlinear structure. In particular, there does not exist even an appropriate universal definition of the difference of sets $A, B \in$ compR^n. It is mainly due to this reason that the problem of constructing the infinitesimal generator for the generalized system of the above still remained open.

In this paper we present an approach to the solution of this problem through the technique of evolution "funnel equations" rather than by constructing an evolution "differential" system. This approach is motivated by the ideas of investigations [62, 73] and developed further for systems with state constraints (3.8).

We shall further demand that one of the following assumptions concerning the mapping $Y(\cdot)$ would be fulfilled.

Assumption B1. The graph $grY \in \text{conv}R^{n+1}$.

Assumption B2. For every $l \in R^n$ the support function $f(l,t) = \rho(l \mid Y(t))$ is differentiable in t and its derivative $\partial f(l,t)/\partial t$ is continuous in $\{l,t\}$.

Theorem 6.1. *Suppose that the assumptions A1 or A2 hold and that the map Y satisfies either assumption B1 or B2. Then the multifunction* $X[t] = X(t, t_0, X_0)$ *is the set–valued solution to the following evolution equation*

$$\lim_{\sigma \to +0} \sigma^{-1} h(\, X[t+\sigma], \bigcup_{x \in X[t]} (x + \sigma \mathcal{F}(t,x)) \bigcap Y(t+\sigma) \,) = 0, \qquad (6.1)$$

$$X[t_0] = X_0, \quad t_0 \le t \le t_1.$$

The proof of this main theorem will follow from a sequence of technical lemmas given further.

In what follows, a function denoted $o(\sigma)$ without or with any type of indices (i.e., $o^*(\sigma)$, $o_i(\sigma)$, etc.) will always be presumed to satisfy $\sigma^{-1} o(\sigma) \to 0$ if $\sigma \to 0$.

Let $\tau \in T = [t_0, t_1]$ be fixed, $S = \{x \in R^n : \| x \| \le 1\}$, $X[\tau] = X(\tau, t_0, X_0)$. First we have the

Lemma 6.1. *Under either the assumption A1 or A2 there exists a* $\sigma_* > 0$ *such that*

$$X[\tau + \sigma] \subseteq \bigcup_{x \in X[\tau]} (x + \sigma \mathcal{F}(\tau, x)) \bigcap Y(\tau + \sigma) + o(\sigma) S \qquad (6.2)$$

for every $\sigma \in (0, \sigma_*]$.

Proof. Since $X[\tau+\sigma] = X(\tau+\sigma, \tau, X[\tau])$, the definition of viable trajectories yields

$$X[\tau + \sigma] \subseteq \mathcal{X}(\tau + \sigma, \tau, X[\tau]) \bigcap Y(\tau + \sigma) \qquad (6.3)$$

Being the crossection at instant $\tau + \sigma$ of the solution tube $\mathcal{X}[t] = \mathcal{X}(t, \tau, X[\tau])$ generated by the differential inclusion (3.6) and "starting position" $\{\tau, X[\tau]\}$, the set $\mathcal{X}[t]$ satisfies the funnel equation [62] :

$$\lim_{\sigma \to +0} \sigma^{-1} h(\, \mathcal{X}[t+\sigma], \bigcup_{x \in \mathcal{X}[t]} (x + \sigma \mathcal{F}(t,x)) \,) = 0,$$

$$\mathcal{X}[\tau] = X[\tau], \quad \tau \le t \le t_1.$$

Therefore

$$\mathcal{X}(\tau+\sigma,\tau,X[\tau]) \subseteq \Big(\bigcup_{x\in X[\tau]} (x+\sigma\mathcal{F}(\tau,x)) \Big) + o(\sigma)S \tag{6.4}$$

Provided P, Q, W are given subsets of R^n with $Q = -Q$, it is possible to verify the inclusion

$$(P+Q)\cap W \subseteq P\cap(Q+W)+Q$$

From here and from (6.3), (6.4) it follows

$$X[\tau+\sigma] \subseteq R(\sigma,\tau)\bigcap(Y(\tau+\sigma)+o(\sigma)S)+o(\sigma)S \tag{6.5}$$

where

$$R(\sigma,\tau) = \bigcup_{x\in X[\tau]} (x+\sigma\mathcal{F}(\tau,x)).$$

One should observe that the set $R(\sigma,\tau)$ is either compact and convex (due to assumption A1) or compact and star–shaped (due to assumption A2) for every $\sigma > 0$.

The next step is to verify the inclusion

$$R(\sigma,\tau)\bigcap(Y(\tau+\sigma)+o(\sigma)S) \subseteq R(\sigma,\tau)\bigcap Y(\tau)+o_1(\sigma)S \tag{6.6}$$

for some function $o_1(\sigma)$.

From the accepted assumptions it follows that there exist vectors $x_* \in X[\tau]$ and $v_* \in \mathcal{F}(\tau,x_*)$ (in the case of Assumption A2 we take $x_* = 0$, $v_* = 0$) and numbers $r > 0$, $\sigma_* > 0$, $K > 0$ such that for every $\sigma \in [0,\sigma_*]$ one has

$$x_*+\sigma v_*+rS \subseteq Y(\tau+\sigma), \quad x_*+\sigma v_* \in R(\sigma,\tau),$$
$$R(\tau,\sigma) \subseteq KS, \quad 0 \le r^{-1}o(\sigma) \le 1.$$

We proceed further by proving the inclusion

$$R(\sigma,\tau)\bigcap(Y(\tau+\sigma)+o(\sigma)S) \subseteq R(\sigma,\tau)\bigcap Y(\tau)+2Kr^{-1}o(\sigma)S \tag{6.7}$$

Indeed, suppose that given are a number $\sigma \in [0,\sigma_*]$ and a vector $z \in R(\sigma,\tau)\cap(Y(\tau+\sigma)+o(\sigma)S)$. Selecting vector

$$y = (1-r^{-1}o(\sigma))x + r^{-1}o(\sigma)(x_*+\sigma v_*)$$

one may observe that $y \in R(\sigma,\tau)$ and

$$\| y - z \| \le 2Kr^{-1}o(\sigma)$$

From the above we arrive at two inclusions

$$(1 - r^{-1}o(\sigma))x \in (1 - r^{-1}o(\sigma))(Y(\tau+\sigma) + o(\sigma)S)$$

$$\subseteq (1 - r^{-1}o(\sigma))Y(\tau+\sigma) + o(\sigma)S,$$

$$r^{-1}(x_* + \sigma v_* + rS) \subseteq r^{-1}o(\sigma)Y(\tau+\sigma)$$

Adding these inclusions we come to

$$y + o(\sigma)S \subseteq Y(\tau+\sigma) + o(\sigma)S$$

and, after reducing the term $o(\sigma)S$, to the inclusion

$$y \in Y(\tau+\sigma)$$

(having also made use of the convexity of the set $Y(\tau+\sigma)$.)

The last relation immediately yields (6.7) and the inclusion (6.6) is therefore established. The result given in Lemma 6.1 now follows from (6.5), (6.6).

Proceeding further, introduce an auxiliary system

$$\begin{array}{ll} \dot{z}(t) \in \mathcal{F}(\tau, x_0), & z(\tau) = x_0, \\ z(t) \in Y(t), & \tau \le t \le t_1 \end{array} \tag{6.8}$$

where $x_0 \in X_0$ is a fixed vector.

Denote $Z(\tau+\sigma, \tau, x_0)$ to be the crossection at instant $\tau+\sigma$ of the tube of viable solutions to (6.8) and

$$Z(\tau+\sigma) = \bigcup_{x_0 \in X[\tau]} Z(\tau+\sigma, \tau, x_0).$$

Lemma 6.2. *Under either Assumption A1 or A2 for every $\epsilon > 0$ there exists a $\sigma_* > 0$, such that for all $\sigma \in [0, \sigma_*]$ the following inclusions are true*

$$X[\tau+\sigma] \subseteq Z(\sigma, \tau) + \epsilon\sigma S, \tag{6.9}$$

$$Z(\sigma,\tau) \subseteq \bigcup_{x \in X[\tau]} (x + \sigma\mathcal{F}(\tau, x)) \bigcap Y(\tau+\sigma) + \sigma\epsilon S. \tag{6.10}$$

Proof. This result is a detailed version of Lemma 6.1 and may be proved through a similar scheme.

The next lemma gives us the estimate opposite to (6.9).

Lemma 6.3. *With Assumption A1 or A2 fulfilled it is possible for any $\epsilon > 0$ to indicate a $\sigma_* > 0$, such that for every $\sigma \in (0, \sigma_*]$ one has*

$$Z(\sigma, \tau) \subseteq X[\tau + \sigma] + \sigma\epsilon S. \tag{6.11}$$

Proof. Suppose $z^* \in Z(\sigma, \tau)$. Then there exists a pair

$$x_0 \in X[\tau], \quad v(t) \in \mathcal{F}(\tau, x_0), \quad \tau \leq t \leq \tau + \sigma,$$

such that the respective solution

$$z[t] = x_0 + \int_\tau^t v(s)ds, \quad \tau \leq t \leq \tau + \sigma,$$

to the inclusion (6.8) satisfies the conditions

$$z[\tau + \sigma] = z^*, \quad z[t] \in Y(t) \ (\tau \leq t \leq \tau + \sigma)$$

Recall that the map $\mathcal{F}(t, x)$ is assumed to be Lipschitz in x and continuous in t uniformly in $x \in Q$. It is then not difficult to verify that

$$\mathcal{F}(\tau, x_0) = \mathcal{F}(\tau, z[t] - \int_\tau^t v(s)ds) \subseteq$$

$$\subseteq \mathcal{F}(\tau, z[\tau]) + LK\sigma S \subseteq \mathcal{F}(t, z[t]) +$$
$$+ (O(\sigma) + LK\sigma)S,$$

where

$$K = \max\{ \| x \| \colon \ x \in \mathcal{F}(t, x), \ x \in Q, \ t \in T \}$$

and the function $O(\sigma) \to 0$ with $\sigma \to 0$.

Therefore

$$\dot{z}[t] \in \mathcal{F}(t, z[t]) + (O(\sigma) + LK\sigma)S,$$
$$z[\tau] = x_0, \quad \tau \leq t \leq \tau + \sigma$$

From the Gronwall Lemma for differential inclusions [11, 73] it follows that there exists a solution $y[t]$ to

$$\dot{y} \in \mathcal{F}(t, y), \quad y[\tau] = x_0, \quad \tau \leq t \leq \tau + \sigma$$

that satisfies

$$\| y[t] - z[t] \| \leq (\exp L\sigma)(O(\sigma) + LK\sigma)\sigma = o^*(\sigma), \tag{6.12}$$

$$\tau \leq t \leq \tau + \sigma$$

Hence $y[t] \in Y(t) + o^*(\sigma)S$ for every $t \in [\tau, \tau+\sigma]$. Following the scheme for Lemma 6.1 it is possible to construct a function $w(t)$ that satisfies

$$\dot{w} \in \mathcal{F}(t, w), \quad w(\tau) \in X[\tau], \quad \tau \leq t \leq \tau + \sigma$$

$$\| y(t) - w(t) \| \leq o_1^*(\sigma)$$

for a certain function $o_1^*(\sigma)$. Then

$$y(\tau+\sigma) \in X(\tau+\sigma, \tau, X[\tau]) + o_1^*(\sigma)S =$$

$$= X(\tau+\sigma, t_0, X_0) + o_1^*(\sigma)S$$

and finally in view of (6.12) we have

$$z^* \in X[\tau+\sigma] + (o^*(\sigma) + o_1^*(\sigma))S = X[\tau+\sigma] + o_2^*(\sigma)S$$

where the function $o_2^*(\sigma)$ does not depend upon the vector $z^* \in Z(\sigma, \tau)$.

The last lemma leads to

Corollary 6.1. *Under Assumptions A1 or A2 one has*

$$\lim_{\sigma \to +0} \sigma^{-1} h(\ X[\tau+\sigma], Z(\sigma, \tau)\ ) = 0. \tag{6.13}$$

We proceed by establishing an inclusion opposite to either (6.2) or (6.10). This however will require some additional properties of the map $Y(\cdot)$.

Lemma 6.4. *Under Assumption B1 for any $\sigma > 0$ one has*

$$R(\sigma, \tau) \bigcap (Y(\tau+\sigma) \subseteq Z(\sigma, \tau). \tag{6.14}$$

Proof. Consider the set $Z^*(\sigma+\tau, \tau, x_0)$ of solutions $\{z[\cdot]\}$ to (6.8) with constant "velocities" $\dot{z}[t] \equiv v \in \mathcal{F}(\tau, x_0)$ ($\dot{z}[t] = \text{const}$).

Denote

$$Z^*(\sigma, \tau) = \bigcup \{\ Z^*(\sigma+\tau, \tau, x_0) \ \mid\ x_0 \in X[\tau]\ \}$$

Clearly $Z^*(\sigma, \tau) \subseteq Z(\sigma, \tau)$. If we now assume $z \in R(\sigma, \tau) \cap (Y(\tau+\sigma)$, then there exists a pair of vectors $x \in X[\tau]$, $v \in \mathcal{F}(\tau, x)$ such that

$$x + \sigma v \in Y(\tau+\sigma), \quad x \in Y(\tau)$$

Since $grY \in \text{conv}R^{n+1}$ we have

$$x + sv = (1 - s\sigma^{-1})x + s\sigma^{-1}(x + \sigma v) \in Y(s)$$

for any $s \in [0, \sigma]$.

Therefore $z \in Z^*(\sigma, \tau)$ and (6.14) is proved.

Relations (6.14) and (6.13) yield

Corollary 6.2. *Assume either of the Assumptions A1 or A2 and also B1 of this Section. Then*

$$\lim_{\sigma \to +0} \sigma^{-1} h(\ Z(\sigma, \tau), R(\sigma, \tau) \bigcap Y(\tau + \sigma)\ ) = 0. \tag{6.15}$$

We are now able to prove the basic theorem. However, this will require some auxiliary constructions. Thus, at this point it is necessary to recall the notion of *contingent cone* $T_P(x)$ for an arbitrary closed set $P \subseteq R^n$ at a point $x \in P$, namely

$$T_P(x) = \{\ v \in R^n : \ \liminf_{\sigma \to +0} \sigma^{-1} d(x + \sigma v, P) = 0\ \}$$

and of a *contingent derivative* $DY(t, y)(\alpha)$ to a multivalued mapping Y [1, 5] :

$$DY(t, y)(\alpha) = \{\ v \in R^n : \ (\alpha, v) \in T_{grY}(t, y)\ \}, \quad \forall \{t, y\} \in \mathrm{gr} Y, \ \alpha \in R^1$$

Denote $V(t, y) = DY(t, y)(1)$ for $\{t, y\} \in \mathrm{gr} Y$. Under Assumption B2 the set $V(t, y)$ is closed and convex in R^n for all $\{t, y\} \in \mathrm{gr} Y$ [3].

Following [16, 60] consider a local approximation $Y_\tau(\sigma)$ for the set–valued map Y in the neighbourhood of a fixed point τ:

$$Y_\tau(\sigma) = \bigcap_{y \in Y(\tau)} (y + \sigma V(t, y)), \quad Y_\tau(0) = Y(\tau)$$

Lemma 6.5 [16]. (i) *Under Assumption B2 the following equality is true for all $\sigma > 0$:*

$$Y_\tau(\sigma) = \{\ x \in R^n : \ l'x \le \rho(l \mid Y(\tau)) + \sigma \partial f(l, \tau)/\partial \tau, \quad \forall l \in R^n\ \}.$$

(ii) *Under Assumptions A1(ii) or A2(ii) for every $\epsilon > 0$ there exists a $\sigma_* > 0$ such that for all $\sigma \in (0, \sigma_*]$*

$$Y(\tau + \sigma) \subseteq Y_\tau(\sigma) + \epsilon \sigma S, \tag{6.16}$$

$$Y_\tau(\sigma) \subseteq Y(\tau + \sigma) + \epsilon \sigma S. \tag{6.17}$$

Note that the graph of the map $Y_\tau(\sigma)$, regarded as a function of σ, is convex.

Proof of Theorem 6.1. Supposing at first that Assumptions A1 or A2 and B1 are true and combining the results of Corollaries 6.1, 6.2, we arrive immediately at the equality (6.1).

We now have to prove the theorem under Assumption A1 or A2 and B2.

In view of Lemma 6.1 we need to verify only that

$$R(\sigma,\tau)\bigcap Y(\tau+\sigma)\subseteq X(\tau+\sigma)+o(\sigma)S \tag{6.18}$$

for some function $o(\sigma)$.

From our assumptions and from Lemma 6.5 one may observe that

$$R(\sigma,\tau)\bigcap Y(\tau+\sigma)\subseteq R(\sigma,\tau)\bigcap(Y_\tau(\sigma)+o_1(\sigma)S) \tag{6.19}$$

and that Assumption A1(ii) or A2(ii) respectively remains true for $Y_\tau(\sigma)$. Then, following the reasoning of Lemma 6.1, we will have

$$R(\sigma,\tau)\bigcap(Y_\tau(\sigma)+o_1(\sigma)S)\subseteq R(\sigma,\tau)\bigcap Y_\tau(\sigma)+o_2(\sigma)S \tag{6.20}$$

(It is essential here that $grY\in \mathrm{conv}R^{n+1}$).

From (6.19), (6.20) we come to the inclusion

$$R(\sigma,\tau)\bigcap Y(\tau+\sigma)\subseteq R(\sigma,\tau)\bigcap Y_\tau(\sigma)+o_2(\sigma)S \tag{6.21}$$

It remains to prove that

$$R(\sigma,\tau)\bigcap Y_\tau(\sigma)\subseteq X(\sigma+\tau)+o_3(\sigma)S \tag{6.22}$$

Assume $z\in R(\sigma,\tau)\bigcap Y_\tau(\sigma)$. Then $z=x+\sigma v\in Y_\tau(\sigma)$ for some $x\in X[\tau]$, $v\in\mathcal{F}(\tau,x)$. Since $grY\in \mathrm{conv}R^{n+1}$ we have

$$x+sv\in Y_\tau(s),\quad \forall s\in[0,\sigma]$$

As in Lemma 6.3 it is possible to establish the existence of a solution $y(t)$ to the inclusion

$$\dot y\in\mathcal{F}(t,y), \tag{6.23}$$

$$y(\tau)=x,\quad t\in[\tau,\tau+\sigma]$$

that satisfies the inequality

$$\| x+sv-y(\tau+s)\|\leq\tilde o(\sigma),\quad s\in[0,\sigma]$$

for a certain $\tilde{o}(\sigma)$. Therefore

$$y(\tau+s) \in Y_\tau(s) + \tilde{o}(\sigma)S, \quad s \in [0,\sigma] \tag{6.24}$$

$$\| y(\tau+\sigma) - z \| \leq \tilde{o}(\sigma) \tag{6.25}$$

Due to Lemma 6.5(ii) we derive from (6.24) that $y(\tau) \in Y(\tau+\sigma) + \tilde{o}_1(\sigma)S$.

Then, following the schemes of Lemma 6.1, we may find a solution $y_*(t)$ to (6.23) that satisfies the relations

$$y_*(\tau+\sigma) \in X[\tau+\sigma],$$

$$\| y_*(\tau+\sigma) - y(\tau+\sigma) \| \leq \tilde{o}_2(\sigma)$$

The latter inequality together with (6.25) leads to the inclusion

$$z \in X[\tau+\sigma] + o(\sigma)S$$

for $o(\sigma) = \tilde{o}(\sigma) + \tilde{o}_2(\sigma)$. The inclusion (6.18) is therefore true. This finalized the proof of Theorem 6.1.

We will now pass to the discussion of the uniqueness of solution to the evolution equation (6.1).

7 The Unicity Theorem

Let us denote $\mathcal{Z}[t_0,t_1]$ to be the set of all multivalued functions $Z[t]$ $(Z : [t_0,t_1] \to \mathrm{comp}R^n)$ such that

$$\lim_{\sigma \to +0} \sigma^{-1} h\Big(Z[t+\sigma], \bigcup_{x \in Z[t]} (x + \sigma \mathcal{F}(t,x)) \bigcap Y(t+\sigma) \Big) = 0, \tag{7.1}$$

$$Z[t_0] = X_0, \quad t_0 \leq t \leq t_1,$$

where the limit in (7.1) is uniform with respect to $t \in [t_0,t_1]$.

Following the schemes of the proof of Theorem 6.1 we may observe that under the Assumptions of the Theorem we have

$$X[\cdot] = X(\cdot, t_0, X_0) \in \mathcal{Z}[t_0,t_1]$$

Let us begin however with the more general case when these assumptions are not required. Consider some properties of the maps $Z[\cdot] \in \mathcal{Z}[t_0,t_1]$.

Lemma 7.1. *Assume that the set-valued map Y satisfies the Lipschitz condition (with a constant $k > 0$):*

$$h(Y(\tau_1), Y(\tau_2)) \leq k \mid \tau_1 - \tau_2 \mid, \quad t_0 \leq \tau_1, \ \tau_2 \leq t_1$$

Then for every $Z[\cdot] \in \mathcal{Z}[t_0, t_1]$ *the following inclusion holds*

$$Z[\tau] \subseteq X[\tau] = X(\tau, t_0, X_0)$$

for every $\tau \in [t_0, t_1]$.

Proof. Let the instant $\tau \in [t_0, t_1]$ be arbitrary and $z \in Z[\tau]$. Consider the subdivision $\{ t_i \; ; \; i = 1, \ldots, N \}$ of the interval $[t_0, \tau]$ with a uniform step $\sigma_N = (\tau - t_0)/N$:

$$t_i = t_0 + i\sigma, \quad i = 1, \ldots, N, \quad t_N = \tau.$$

Let

$$o(\sigma, Z) = \sup_{t_0 \leq t \leq t_1} h(\; Z[t+\sigma], \; \bigcup_{x \in Z[t]} (x + \sigma \mathcal{F}(t,x)) \bigcap Y(t+\sigma) \;)$$

From the definition of $Z[\cdot]$ we obtain

$$\sigma^{-1} o(\sigma, Z) \to 0 \quad (\sigma \to +0) \tag{7.2}$$

and for a certain finite sequence of vectors $\{z_i, f_i\}_{i=0,1,\ldots,N}$ we have

$$z_i \in Z[t_i], \quad f_i \in \mathcal{F}(t_i, z_i), \quad z_N = z,$$

$$z_0 \in X_0, \quad z_i + \sigma_N f_i \in Y(t_{i+1}),$$

$$z_i = z_{i-1} + \sigma_N f_{i-1} + l_i, \quad \| \, l_i \, \| \leq o(\sigma_N, Z),$$

$$i = 1, \ldots, N-1$$

Consider the piecewise–linear interpolation $z_{(N)}(t_i)$ $(t_0 \leq t \leq \tau)$:

$$z_{(N)}(t_i) = z_i, z_{(N)}(t) = z_i + (z_{i+1} - z_i)(t - t_i)\sigma_N^{-1}$$

$$t_i \leq t \leq t_{i+1}, \quad i = 0, 1, \ldots, N-1$$

Then for every $t \in [t_i, t_{i+1}]$ $(i = 0, 1, \ldots, N-1)$ we have

$$z_{(N)}(t_i) = z_i \in Y(t_i) + l_i \subseteq Y(t) + (k\sigma_N + o(\sigma N, Z))S$$

$$z_{(N)}(t_{i+1}) = z_{i+1} \in Y(t_{i+1}) + l_{i+1} \subseteq Y(t) + (k\sigma_N + o(\sigma N, Z))S$$

Since the set $Y(t)$ is convex, the last relations yield the following inclusion for the interpolating function $z_{(N)}(t)$:

$$z_{(N)}(t) \in Y(t) + (k\sigma_N + o(\sigma N, Z))S \tag{7.3}$$

for all $t \in [t_0, \tau]$.

It is not difficult to prove that the sequence $\{z_{(N)}(\cdot)\}_{N\to\infty}$ has a limit point $x_*(\cdot)$ in the space $C_n[t,\tau]$ and that the function $x_*(\cdot)$ is a solution to the following differential inclusion

$$\dot{x}_*(t) \in \mathcal{F}(t,x), \quad t_0 \le t \le \tau,$$

$$x_*(t_0) \in X_0, \quad x_*(\tau) = z$$

Relations (7.3), (7.2) yield

$$x_*(t) \in Y(t), \quad t_0 \le t \le \tau,$$

since $Y(t)$ is compact in R^n for all $t \in [t_0, t_1]$.

Therefore $x_*(\cdot) \in X(\cdot,\tau,t_0,X_0)$ and $x_*(\tau) = z \in X[\tau]$. Lemma 7.1 is thus proved.

Corollary 7.1. *Under the Assumptions of Lemma 7.1 the following relations are true whatever is the tube* $Z(\cdot) \in \mathcal{Z}[t_0.t_1]$*:*

(i) *The inclusion*

$$Z(t) \subseteq Y(t)$$

is true for every $t \in [t_0, t_1]$

(ii) *There is a constant* $M > 0$ *such that*

$$Z(t+\sigma) \subseteq Z(t) + M\sigma S$$

where $t_0 \le t \le t+\sigma \le t_1$.

Example 7.1. Consider the following system in R^n ($0 = t_0 \le t \le t_1$):

$$\begin{cases} \dot{x}_1 = x_1 x_2^2 - x_1 \\ \dot{x}_2 = x_1^2 x_2 - x_2 \end{cases}$$

with the initial condition

$$x_0 \in X_0 = \{\ x = (x_1, x_2)\ :\ |\ x_1\ | \le 1,\ x_2 = 1\ \}$$

and the state constraints

$$x(t) \in Y = \{\ x = (x_1, x_2)\ :\ |\ x_1\ | \le 2,\ 1 \le x_2 \le 2\ \}$$

It is not difficult to verify that for every $\tau \in (0, t_1]$ the set

$$X[\tau] = X(\tau, t_0, X_0) = \{x^{(1)}\} \cup \{x^{(2)}\}$$

where $x^{(1)} = (1,1)$ and $x^{(2)} = (-1,1)$.

One may observe however that we can find at least three multivalued solutions $Z[\cdot] \in \mathcal{Z}[t_0, t_1]$ to the evolution equation (7.1), namely

$$Z_1[\cdot] = X[\cdot], \quad Z_i(t) = \begin{cases} X_0, & t = 0 \\ \{x^{(i)}\}, & 0 < t \le t_1 \end{cases} \qquad (i = 2, 3)$$

Here we emphasize that both viable trajectories $x^{(1)}(t) \equiv x^{(1)}$ and $x^{(2)}(t) \equiv x^{(2)}$ lie on the boundary of Y.

The next result indicates that for the "interior" trajectory $x_*(t)$ the situation mentioned above, namely, that $x_*(t) \notin Z[t]$ will be impossible.

For every $\tau \in [t_0, t_1]$ denote

$$X_{int}[\tau] = X_{int}(\tau, t_0, X_0) = \{ \; z \in R^n : \; \exists x(\cdot) \in X(\cdot, \tau, t_0, X_0)$$

$$x(\tau) = z, \quad x(t) \in \mathrm{int} Y(t) \quad \forall t \in [t_0, \tau] \; \}$$

Lemma 7.2. *Let Assumption A1(ii) or A2(ii) be fulfilled. Then for every $\tau \in [t_0, t_1]$ the following inclusion holds*

$$X_{int}[\tau] \subseteq Z(\tau)$$

where $Z(\cdot)$ is any multifunction from the class $\mathcal{Z}[t_0, t_1]$.

The proof of this Lemma is similar to that of Lemma 6.6.

Corollary 7.2. *Under the Assumptions of Lemmas 7.1 and 7.2 one has*

$$clX_{int}[\tau] \subseteq Z(\tau) \subseteq X[\tau], \quad t_0, \le t \le t_1,$$

for any $Z(\cdot) \in \mathcal{Z}[t_0, t_1]$.

We are now able to formulate the Unicity Theorem.

Theorem 7.2. *Let the Assumptions of Theorem 3.1 be true. Then the multivalued function $X[\tau] = X(\tau, t_0, X_0)$ is the unique solution to the funnel equation (6.1) in the class $\mathcal{Z}[t_0, t_1]$ of all the set–valued mappings $Z(\cdot)$ that satisfy this equation uniformly in $t \in [t_0, t_1]$.*

Proof. Following the lines of the proof of Theorem 6.1 it is possible to establish the equality

$$\mathrm{cl}X_{int}[\tau] = X[\tau],$$

whatever is the instant $t \in [t_0, t_1]$. From Corollary 7.2 we may then conclude that for each $Z(\cdot) \in \mathcal{Z}[t_0, t_1]$. The set–valued function $X[t]$ is therefore the unique solution to the evolution equation (6.1) and Theorem 6.2 is thus proved.

8 Funnel Equations. Specific Cases

In this Section we demonstrate some examples of differential inclusions for which the Assumptions A1, A2 given above are fulfilled, presenting also the specific versions of the funnel equations.

(i) Linear systems. Consider the linear differential inclusion (3.5)

$$\dot{x} \in A(t)x + B(t)Q(t), \tag{8.1}$$

$$x(t_0) = x_0 \in X_0, \quad t_0 \leq t \leq t_1,$$

where $x \in R^n$, $A(t)$ and $B(t)$ are continuous $n \times n-$ and $n \times m-$ matrices, $Q(t)$ is a continuous map ($Q : [t_0, t_1] \to \text{conv}R^m$), $X_0 \in \text{conv}R^n$.

Here the Assumptions A1(i), A1(iii) will be fulfilled automatically. Hence to retain the Assumption A1(ii) we introduce

Assumption A*. There exists a solution $x_*[\cdot]$ of (8.1) such that $x_*[t_0] \in X_0$ and

$$x_*[t] \in \text{int}Y(t), \quad \forall t \in [t_0, t_1].$$

The following result is a consequence of Theorem 6.1.

Theorem 8.1. *Suppose Assumptions A* and B1 or B2 are fulfilled. Then the set-valued function $X[t] = X(t, t_0, X_0)$ is the solution to the evolution equation*

$$\lim_{\sigma \to +0} \sigma^{-1} h(\ X[t+\sigma],\ ((I+\sigma A(t))X[t]+\sigma P(t)) \bigcap Y(t+\sigma)\ ) = 0, \tag{8.2}$$

$$X[t_0] = X_0, \quad t_0 \leq t \leq t_1.$$

A separate question is how to solve the equation (8.2). We will demonstrate this in the Sections 10-11 by introducing a certain multivalued "convolution integral".

(ii) A Bilinear Differential Inclusion. Consider the differential inclusion of type (3.4) with $A(t, Q(t)) = \mathcal{A}(t)$ and $h(t) \in P(t)$ unknown but bounded ($P(t) \in \text{conv}R^n$ being continuous in t). This yields

$$\begin{array}{ll} \dot{x} \in \mathcal{A}(t)x + P(t), & x(t_0) = x_0 \in X_0, \\ x(t) \in Y(t), & t_0 \leq t \leq t_1, \end{array} \tag{8.3}$$

and the right–hand side $\mathcal{F}(t,x) = \mathcal{A}(t)x + P(t)$ depends bilinearly upon the state vector x and the set–valued map $\mathcal{A}$.

We assume that $\mathcal{A}(\cdot)$ is a continuous mapping from $[t_0, t_1]$ into the set conv$\Re^n$ of convex and compact subsets of the space $\Re^n$ of $n \times n$–matrices.

The equation (8.3) is a well–known important model of an uncertain bilinear dynamic system with set–membership description of the unknown matrices $A(t) \in \mathcal{A}(t)$, $h(t) \in P(t)$ and $x_0 \in X_0$ [39].

It is not difficult to demonstrate that the Assumption A1 does not hold here, so that the sets $\mathcal{X}[t] = \mathcal{X}(t, t_0, X_0)$ (and also $X[t] = X(t, t_0, X_0)$) need not be convex.

Example 8.1. Indeed, consider a differential inclusion in R^2:

$$\begin{cases} \dot{x}_1 \in [-1, 1]x_2, \\ \dot{x}_2 = 0, \quad 0 \leq t \leq 1 \end{cases}$$

with the initial condition

$$x(0) = x_0 \in X_0 = \{ \, x = (x_1, x_2) \; : \; x_1 = 0, \; | \, x_2 \, | \leq 1/2 \, \}$$

Then

$$\mathcal{X}[1] = \mathcal{X}(1, 0, X_0) = X^1 \bigcup X^2$$

where

$$X^1 = \{ \, x = (x_1, x_2) \; : | \, x_1 \, | \leq x_2 \leq 1/2 \, \},$$

$$X^2 = \{ \, x = (x_1, x_2) \; : | \, x_1 \, | \leq -x_2 \leq 1/2 \, \}.$$

It is obvious that the set $\mathcal{X}[1]$ is not convex.

However one can verify that though A1 is not fulfilled for the system (8.3), the Assumption A2(i) is indeed satisfied for this system, provided $0 \in X_0$ and $0 \in P(t)$ for all $t \in [t_0, t_1]$.

The Assumption A2 may hence be rewritten in the following reduced form

Assumption A** .

(i) The condition $0 \in P(t)$ is true for all $t \in [t_0, t_1]$.

(ii) There exists an $\epsilon > 0$ such that

$$\epsilon S \subseteq Y(t), \quad \forall t \in [t_0, t_1].$$

(iii) The set $X_0 \in \mathrm{St}(R^n)$ and $0 \in X_0$.

To formulate the analogy of Theorem 6.1 we need to introduce an additional notation:

$$\mathcal{M} * X = \{ \, z \in R^n : \; z = Mx, \; M \in \mathcal{M}, \; x \in X \, \}$$

where $\mathcal{M} \in$ conv$\Re^n$, $X \in$ convR^n.

Theorem 8.2. *Under Assumptions A^{**} and B1 or B2 the set–valued map map $X[t] = X(t, t_0, X_0)$ is the solution to the following evolution equation*

$$\lim_{\sigma \to +0} \sigma^{-1} h(X[t+\sigma], ((I+\sigma A(t)) * X[t] + \sigma P(t)) \bigcap Y(t+\sigma)) = 0, \quad (8.4)$$

$$X[t_0] = X_0, \quad t_0 \le t \le t_1.$$

(iii) A Nonlinear Example. We finalize this Section by still one more example, which is given by a multifunction $\mathcal{F}(t,x)$ with a star–shaped graph $gr_t\mathcal{F}$.

Namely, let $\mathcal{F}(t,x)$ be of the form

$$\mathcal{F}(t,x) = G(t,x)U + P(t) \quad (8.5)$$

where the $n \times n$–matrix function $G(t,x)$ is continuous in t, Lipschitz–continuous in x and positively homogeneous in x, $U \in \mathrm{conv}R^n$. The map $P(t)$ is the same as in the previous subsection (ii). One can immediately verify that the condition A2(i) holds in this case so that the crossection $X[t]$ of the solution tube to the equation (8.4) is star–shaped.

Remark 8.1. Concluding this Section we wish to mention an approach similar to this one but with the evolution equation (6.1) written down with the Hausdorff semidistance

$$h^+(A,B) = \max_{a\in A} \min_{b\in B} \| a-b \|, \quad A, B \in \mathrm{comp}R^n,$$

rather than with the Hausdorff metric

$$h(A,B) = \max\{h^+(A,B), h^+(B,A)\}$$

The solution to (6.1) then ceases to be unique and the preference out of the solutions may then be given to the "maximal" one (with respect to inclusion) [30, 51, 52].

The h^+–technique may seem to be more adequate for dealing with discontinuous set–valued functions $X[t] = X(t, t_0, X_0)$.

The evolution equation (6.1) delivers a formalized model for the generalized dynamic system generated by the mapping $X[t] = X(t, t_0, X_0)$. However, once the sets $X[t]$ are convex, there exists an alternative presentation of this evolutionary system. Namely, each of the sets $X[t]$ could be described by its support function

$$\varphi(l,t) = \rho(l \mid X[t]) = \max_{x\in X[t]} (l,x)$$

The dynamics of $X[t]$ would then be reflected by a generalized partial differential equation for $\varphi(l,t)$. The description of this equation is the next issue to be discussed.

9 The Evolution Equation as a Generalized Partial Differential Equation

In this Section we restrict our attention only to the system for which the Assumption A1 is satisfied. In this case the assembly $X(\cdot, \tau, t_0, X_0)$ is a convex compact subset of $C^n[t_0, t_1]$ and the set $X[\tau] = X(\tau, t_0, X_0) \in \text{conv}R^n$ for every $\tau \in [t_0, t_1]$.

Since there is an equivalence between the inclusion $x \in X$ and the system of inequalities

$$(l, x) \leq \rho(l \mid X), \quad \forall l \in R^n,$$

we may now try to describe the evolution of sets $X[\tau]$ through respective differential relations for the support function $\varphi(l,t) = \rho(l \mid X[t])$. The primary problem of discovering the law of evolution of the multivalued "states" $X[t]$ is thus replaced by the problem of evolution in time of a distribution $\phi_t(l) = \varphi(l,t)$ defined over all $l \in R^n$, being positively homogeneous and convex in l.

Let us calculate the directional derivative

$$\partial^+\varphi(l,t)/\partial t = \lim_{\sigma \to +0} (\varphi(l, t+\sigma) - \varphi(l,t))$$

for every instant $t \in [t_0, t_1]$ and for any fixed $l \in R^n$.

The following result will be proved as a direct consequence of Theorem 6.1.

Theorem 9.1. *Suppose Assumptions A1 and either B1 or B2 to be true. Then the support function $\varphi(l,t) = \rho(l \mid X[t])$ is differentiable in t from the right and the directional derivative in time is*

$$\partial^+\varphi(l,t)/\partial t = \min_{q \in Q(l,t)} \max_{x \in \partial_l \varphi(l,t)} \psi(q, x, l, t) =$$

$$= \max_{x \in \partial_l \varphi(l,t)} \min_{q \in Q(l,t)} \psi(q, x, l, t), \tag{9.1}$$

where

$$\psi(q, x, l, t) = \rho(q \mid \mathcal{F}(t,x)) + \partial \rho(l - q \mid Y(t))/\partial t,$$

$$Q(l,t) = \{q \in R^n : \ \varphi(l,t) - \varphi(q,t) = \rho(l - q \mid Y(t)) \}, \tag{9.2}$$

$$\partial_l \varphi(l,t) = \{x \in X[t] : \ (l,x) = \varphi(l,t) \}.$$

Proof. From Theorem 6.1 we deduce

$$\varphi(l, t+\sigma) = \rho(l \mid R(\sigma, t, X[t]) \bigcap Y(t+\sigma)) + o(\sigma) \parallel l \parallel,$$

where

$$R(\sigma, t, X[t]) = \bigcup_{x \in X[t]} (x + \sigma \mathcal{F}(t, x)),$$

$$t_0 \leq t \leq t + \sigma \leq t_1, \ \sigma^{-1} o(\sigma) \to 0 \ (\sigma \to +0)$$

and $R(\sigma, t, X[t])$ is convex due to the Assumption A1.

Taking the inf–convolution of support functions of respective sets in the above intersection (under assumptions of Theorem 9.1 the inf–convolution is exact [59]) we come to

$$\varphi(l, t + \sigma) = \min_{q \in rS} \{\rho(q \mid R(\sigma, t, X[t])) + \rho(l - q \mid Y(t + \sigma))\} + o(\sigma) \parallel l \parallel$$

for a certain $r > 0$.

Hence we have to verify only the differentiability in σ (from the right) of the following function

$$H(\sigma, l, t) = \min_{q \in rS} \max_{x \in X[t]} h(\sigma, t, l, q, x), \tag{9.3}$$

where

$$h(\sigma, t, l, q, x) = (q, x) + \sigma \rho(q \mid \mathcal{F}(t, x)) + \rho(l - q \mid Y(t + \sigma)) \tag{9.4}$$

with $\{l, t\}$ fixed.

The differentiation of this minmax relation may be realized due to the rules given in [15, Theorem 5.3]. Applying this Theorem all the conditions of which are satisfied we arrive at (9.1).

Let us elaborate on this result. Due to the properties of $X[t]$ we have $X[t] \subseteq Y(t)$, so that (9.2) yields

$$\varphi(l, t) - \varphi(q, t) = \rho(l - q \mid Y(t)) \geq \varphi(l - q, t) \tag{9.5}$$

On the other hand, since the function $\varphi(l, t)$ is convex and positively homogeneous in $l \in R^n$ with t fixed, we also observe

$$\varphi(l, t) \leq \varphi(q, t) + \varphi(l - q, t) \tag{9.6}$$

Comparing this with (9.5), we come to the equality

$$\varphi(l, t) - \varphi(q, t) = \varphi(l - q, t) \tag{9.7}$$

Without loss of generality we could have assumed $0 \in X[t]$. Further on, we will require

$$0 \in \text{int} X[t], \quad \forall t \in [t_0 . t_1] \tag{9.8}$$

which is an additional assumption, however. It is not difficult to prove the following sufficient condition for the sets $X[t]$ to be symmetric (with respect to 0) (we recall that the set $A \subseteq R^n$ is defined to be symmetric with respect to 0 if $A = -A$).

Lemma 9.1. *Suppose the assumption (9.8) holds and the sets X_0, $gr_t\mathcal{F}$, $Y(t)$ are symmetric for every $t \in [t_0, t_1]$. Then the set $X[t]$ is symmetric whatever is the instant $t \in [t_0, t_1]$.*

Under the assumptions of Lemma 9.1 the inequality (9.6) turns into an equality if and only if the vectors l, q are collinear ($l = \alpha q$ and $\alpha \geq 0$). Then (9.2) turns into

$$(1-\alpha)(\rho(l \mid X[t]) - \rho(l \mid Y(t))) = 0$$

so that with $\rho(l \mid X[t]) < \rho(l \mid Y(t))$ we have $\alpha = 1$ and in (9.1) the set $Q(l,t) = \{l\}$. Otherwise, $q = \alpha l$, $\alpha \in [0,1]$ and

$$\psi(q,x,l,t) = \alpha\rho(l \mid \mathcal{F}(t,x)) + (1-\alpha)\partial\rho(l \mid Y(t))/\partial t,$$

so that the minimum in (9.1) is to be taken over $\alpha \in [0,1]$.

Finally, we have

$$\partial^+\varphi(l,t)/\partial t = \begin{cases} \max\{\ \rho(l \mid \mathcal{F}(t,x)) \mid x \in \partial_l\varphi(l,t)\ \}, \\ \qquad\qquad\qquad \text{if } \varphi(l,t) < \rho(l \mid Y(t)) \\ \min\{\max\{\rho(l \mid \mathcal{F}(t,x)) \mid x \in \partial_l\varphi(l,t)\}, \\ \quad \partial\rho(l \mid Y(t))/\partial t\},\ \text{if } \varphi(l,t) = \rho(l \mid Y(t)) \end{cases} \tag{9.9}$$

For a linear system this turns into

$$\partial^+\varphi(l,t)/\partial t = \begin{cases} \rho(A'l \mid \partial_l\varphi(l,t)) + \rho(l \mid B(t)Q(t)), \\ \qquad\qquad\qquad \text{if } \varphi(l,t) < \rho(l \mid Y(t)) \\ \min\{\rho(l \mid \partial_l\varphi(l,t)) + \rho(l \mid B(t)Q(t)), \\ \quad \partial\rho(l \mid Y(t))/\partial t\},\ \text{if } \varphi(l,t) = \rho(l \mid Y(t)) \end{cases} \tag{9.10}$$

Each of the relations (9.9), (9.10) is actually a *generalized partial differential equation* which has to be solved with the boundary condition

$$\varphi(l,t_0) = \rho(l \mid X[t_0]) \tag{9.11}$$

The result given in Theorem 9.1 will later be used in Section 17 for describing a feedback solution strategy for the problem of control synthesis under state constraints.

10 Viability Through Parametrization. The General Case

In this Section the description of viability to trajectory tubes is reduced to the treatment of trajectory tubes for a variety of specially designed new differential inclusions without state constraints. These new inclusions are designed depending upon certain parameters and have a relatively simple structure. The overall solution is then presented as an intersection over the parameters of the parallel solution tubes to the new inclusions.

Let us start with the nonlinear case, namely, with a differential inclusion of the general type

$$\dot{x} \in \mathcal{F}(t, x) \tag{10.1}$$

$$x(t_0) \in X_0, \quad t_0 \leq t \leq t_1$$

with constraints

$$x(t) \in Y(t), \quad t_0 \leq t \leq \tau, \tag{10.2}$$

supposing also that the basic conditions of Section 3 on the maps $\mathcal{F}$, Y and the set X_0 are satisfied.

In this Section we do not require however that either of the Assumptions A1, A2 or B1, B2 are true.

We further need to introduce the notion of the restriction $\mathcal{F}_Y(t, x)$ of the map $\mathcal{F}(t, x)$ to a set $Y(t)$ (at time t). This is given by

$$\mathcal{F}_Y(t, x) = \begin{cases} \mathcal{F}(t, x), & x \in Y(t) \\ \emptyset, & x \notin Y(t) \end{cases}$$

The next property follows directly from the definition of viable trajectories.

Lemma 10.1. *An absolutely continuous function $x(t)$ defined on the interval $[t_0, \tau]$ with $x_0 \in X_0$ is a viable trajectory to (10.1) for $t \in [t_0, \tau]$ if and only if the inclusion*

$$\dot{x}(t) \in \mathcal{F}_Y(t, x)$$

is true for almost all $t \in [t_0, \tau]$.

We will now represent $\mathcal{F}_Y(t, x)$ as an intersection of certain multifunctions. The first step to achieve that objective will be to prove the following auxiliary assertion.

Lemma 10.2. *Suppose A is a bounded set, B a convex closed set, both in R^n. Then*

$$\bigcap \{A + LB \mid L \in \Re^n\} = \begin{cases} A, & 0 \in B \\ \emptyset, & 0 \notin B \end{cases}$$

Proof. First assume $0 \in B$. Then

$$A \subseteq \bigcap\{A + L\{0\} \mid L \in \Re^n\} \subseteq \bigcap\{A + LB \mid L \in \Re^n\} \subseteq A + 0 \cdot B = A$$

Hence, $A = \bigcap\{A + LB \mid L \in \Re^n\}$. If $0 \notin B$, then there exists a hyperplane in R^n that strictly separates the origin from B. By the boundedness of A, for a sufficiently large number $\lambda > 0$, we then have

$$(A + \lambda B) \bigcap (A - \lambda B) = \emptyset$$

Setting $L^+ = \lambda I$ and $L^- = -\lambda I$ we conclude that

$$\bigcap\{A + LB \mid L \in \Re^n\} \subseteq (A + L^+ B) \bigcap (A + L^- B) = \emptyset$$

The lemma is thus proved.

From Lemmas we obtain the following characterization of viable trajectories.

Theorem 10.1. *An absolutely continuous function $x(\cdot)$ defined on an interval $[t_0, t_1]$ with $x(t_0)$ is a viable trajectory to (10.1) for $[t_0, \tau]$ iff the inclusion*

$$\dot{x}(t) \in \bigcap \{ \ (\mathcal{F}(t,x) - Lx(t) + LY(t)) \ \mid \ L \in \Re^n \ \}$$

is true for almost all $t \in [t_0, \tau]$.

We introduce a variety of differential inclusions that depend on a matrix parameter $L \in \Re^n$. These are given by

$$\dot{z} \in \mathcal{F}(t, z) - Lz + LY(t), \tag{10.3}$$

$$z(t_0) \in X_0, \quad t_0 \leq t \leq t_1$$

By $z[\cdot] = z(\cdot, \tau, t_0, z_0, L)$ denote the trajectory to (10.3) defined on the interval $[t_0, \tau]$ with $z[t_0] = z_0 \in X_0$. Also denote

$$Z(\cdot, \tau, t_0, X_0, L) = \bigcup \{Z(\cdot, \tau, t_0, z_0, L) \mid z_0 \in X_0\}$$

where $Z(\cdot, \tau, t_0, z_0, L)$ is the bundle of all the trajectories $z[\cdot]{=}z(\cdot, \tau, t_0, z_0, L)$ issued at time t_0 from point z_0 and defined on $[t_0, \tau]$. The crossections of the set $Z(\cdot, \tau, t_0, X_0, L)$ at time t are then denoted as $Z(\tau, t_0, X_0, L)$.

Theorem 10.2. *For each $\tau \in [t_0, t_1]$ one has*

$$X(\cdot, \tau, t_0, X_0) = \bigcap \{Z(\cdot, \tau, t_0, X_0, L) \mid L \in \Re^n\}.$$

Moreover, the following inclusion is true

$$X[\tau] = X(\tau, t_0, X_0) \subseteq \bigcap \{Z(\tau, t_0, X_0, L) \mid L \in \Re^n\}.$$

Proof. Theorem 10.2 is a direct consequence of Theorem 10.1.

Let us now replace the constant matrix L in (10.3) by a continuous function $L(\cdot) \in \Re^n[t_0, t_1]$, coming thus to the differential inclusion

$$\dot{z} \in \mathcal{F}(t, z) - L(t)z + L(t)Y(t), \tag{10.4}$$

$$z(t_0) \in X_0, \quad t_0 \leq t \leq t_1$$

and keeping the earlier notation for its trajectory bundle $Z(\cdot, \tau, t_0, X_0, L(\cdot))$ and $Z(\tau, t_0, X_0, L(\cdot))$ for the τ–crossection of the bundle.

What follows is a more precise version of Theorem 10.2.

Theorem 10.3. *For each $\tau \in [t_0, t_1]$ one has*

$$X(\cdot, \tau, t_0, X_0) = \bigcap \{Z(\cdot, \tau, t_0, X_0, L) \mid L \in \Re^n[t_0, \tau]\}. \tag{10.5}$$

Moreover,

$$X[\tau] = X(\tau, t_0, X_0) \subseteq \bigcap \{Z(\tau, t_0, X_0, L) \mid L \in \Re^n[t_0, \tau]\}. \tag{10.6}$$

We will further prove that for a linear differential inclusion the relation (10.6) is actually an equality. This partly explains the reasons for a separate announcement of Theorem 10.3.

We now proceed with a generalization of Theorems 10.2, 10.3 for a constraint map Y of the form

$$Y(t) = \{\ x \in R^n : \ 0 \in G(t, x)\ \} \tag{10.7}$$

where the multivalued map $G(t, x)$ ($G : [t_0, t_1] \times R^n \to \text{conv}R^m$) is continuous.

As before, we shall assume that the sets $Y(t)$ of (10.7) are convex for all $t \in [t_0, t_1]$. This property is automatically fulfilled one assumes for example that the graph $gr_t G$ is convex.

Consider the differential inclusion

$$\dot{z} \in \mathcal{F}(t, z) - L(t)G(t, z), \tag{10.8}$$

$$z(t_0) \in X_0, \quad t_0 \leq t \leq t_1$$

with $L \in \Re^{n,m}[t_0, t_1]$ and standing $Z(\cdot, \tau, t_0, X_0, L)$ for the solution bundle to (10.8).

Following the schemes for proving Theorem 10.3, we come to

Theorem 10.4. *For each $\tau \in [t_0, t_1]$ one has*

$$X(\cdot, \tau, t_0, X_0) = \bigcap \{Z(\cdot, \tau, t_0, X_0, L) \mid L \in \Re^{n,m}\} =$$

$$= \bigcap \{Z(\tau, t_0, X_0, L) \mid L \in \Re^{n,m}[t_0, \tau]\}.$$

Moreover,

$$X[\tau] = X(\tau, t_0, X_0) \subseteq \bigcap \{Z(\tau, t_0, X_0, L) \mid L \in \Re^{n}[t_0, \tau]\} \subseteq$$

$$\subseteq \bigcap \{Z(\tau, t_0, X_0, L) \mid L \in \Re^{n,m}\}.$$

11 Viability Through Parametrization. The Linear–Convex Case

This Section is devoted to the linear differential inclusion

$$\dot{x} \in A(t)x + P(t), \tag{11.1}$$

$$x(t_0) \in X_0, \quad t_0 \le t \le t_1$$

under the state constraints

$$x(t) \in Y(t), \quad t_0 \le t \le \tau, \tag{11.2}$$

where the assumptions on $A(t)$, $P(t)$, $Y(t)$, X_0 are the same as in Section 10.

In the linear case it is possible to pursue a direct calculation of the support function $\varphi(l, t) = \rho(l \mid X[\tau])$ of the crossection $X[\tau] = X(\tau, t_0, X_0)$ of viable trajectory bundle $X(\cdot, \tau, t_0, X_0)$. The calculation is based on methods of convex analysis and the set–valued analogies of Lagrangian techniques.

Let us further introduce still another additional notation, namely let $\mathcal{M}^n[t_0, \tau]$ stand for the set of all n–vector–valued polynomials of finite degree, defined on $[t_0, \tau]$. Obviously, $g(\cdot) \in \mathcal{M}^n[t_0, \tau]$ if

$$g(s) = \sum_{i=1}^{k} l^{(i)} s^i, \quad s \in [t_0, \tau], \quad l^{(i)} \in R^n$$

and $\mathcal{M}^n[t_0, \tau] \subseteq C^n_\infty[t_0, \tau]$.

Applying the duality concepts of convex analysis [64, 66, 67] as given in the form presented in [38] we come to the following relations.

For any $l \in R^n$, $\lambda(\cdot) \in C^n[t_0, \tau]$ denote

$$\Phi_\tau(l, \lambda(\cdot)) = \rho(l'S(t_0, \tau) - \int_{t_0}^{\tau} \lambda'(\xi)S(t_0, \xi)d\xi \mid X_0)+ \tag{11.3}$$

$$+\int_{t_0}^{\tau} \rho(l'S(\xi, \tau) - \int_{\xi}^{\tau} \lambda'(s)S(\xi, s)ds \mid P(\xi))d\xi + \int_{t_0}^{\tau} \rho(\lambda'(s) \mid Y(s))ds$$

Here, in the first variable the function $S(t, \tau)$ is the matrix solution for the equation

$$\dot{s} = -sA(t), \quad S(\tau, \tau) = I$$

In [38, §6] , it was proved that

$$\rho(l \mid X[\tau]) = \inf\{ \Phi_\tau(l, \lambda(\cdot)) \mid \lambda(\cdot) \in L^n_p[t_0, \tau] \} \tag{11.4}$$

A slight modification of the mentioned proof indicates that the class of functions L^n_p in the last formula may be substituted by C^n or C^n_∞ or even by $\mathcal{M}^n[t_0, \tau]$. Hence one can writes

$$\begin{aligned} &\inf\{ \Phi_\tau(l, \lambda(\cdot)) \mid \lambda(\cdot) \in L^n_p[t_0, \tau] \} = \\ &= \inf\{ \Phi_\tau(l, \lambda(\cdot)) \mid \lambda(\cdot) \in C^n[t_0, \tau] \} = \\ &= \inf\{ \Phi_\tau(l, \lambda(\cdot)) \mid \lambda(\cdot) \in C^n_\infty[t_0, \tau] \} = \\ &= \inf\{ \Phi_\tau(l, \lambda(\cdot)) \mid \lambda(\cdot) \in \mathcal{M}^n[t_0, \tau] \} \end{aligned} \tag{11.5}$$

From relations (11.4), (11.5) it is possible to derive the following assertion

Lemma 11.1. *The following equality is true*

$$\begin{aligned} X[\tau] &= \bigcap \{ R(\tau, M(\cdot)) \mid M(\cdot) \in \mathfrak{R}^n[t_0, \tau] \} = \\ &= \bigcap \{ R(\tau, M(\cdot)) \mid M(\cdot) \in \mathfrak{R}^n_\infty[t_0, \tau] \} = \\ &= \bigcap \{ R(\tau, M(\cdot)) \mid M(\cdot) \in \mathcal{M}^{n\times n}[t_0, \tau] \} \end{aligned} \tag{11.6}$$

Here

$$R(\tau, M(\cdot)) = (S(t_0, \tau) - \int_{t_0}^{\tau} M(\xi)S(t_0, \xi)d\xi)X_0 + \int_{t_0}^{\tau} (S(\xi, \tau) -$$

$$- \int_{\xi}^{\tau} M(s)S(\xi, s)ds)P(\xi)d\xi + \int_{t_0}^{\tau} M(s)Y(s)ds \tag{11.7}$$

and the last two terms of (11.7) are the multivalued integrals of the Lebesgue type [12].

Proof. Here we shall verify only the first equality in (11.6), as the proofs of the others are similar.

Indeed, first observe that

$$\begin{aligned} &\inf\{\ \Phi_\tau(l,\lambda(\cdot))\ \mid\ \lambda(\cdot)\in C^n[t_0,\tau]\ \} = \\ &= \inf\{\ \Phi_\tau(l,\lambda(\cdot))\ \mid\ \lambda(\cdot)\in \mathcal{M}^n[t_0,\tau]\ \} \end{aligned} \tag{11.8}$$

for every $l \in R^n$.

The equality (11.8) is obviously true for $l = 0$ since $\Phi(0,0) = 0$ and

$$\inf\{\ \Phi_\tau(l,\lambda(\cdot))\ \mid\ \lambda(\cdot)\in C^n[t_0,\tau]\ \} = \rho(0 \mid X[\tau]) = 0$$

Let $l \neq 0$. Then for any $\lambda(\cdot) \in C^n[t_0,\tau]$ it is possible to find $M(\cdot) \in \Re^n[t_0,\tau]$ such that

$$\lambda(t) = l'M(t), \quad t_0 \le t \le \tau$$

On the opposite, the inclusion $M(\cdot) \in \Re^n[t_0,\tau]$ yields $\lambda(\cdot) = l'M(\cdot) \in C^n[t_0,\tau]$. Therefore we may substitute $\lambda(\cdot) = l'M(\cdot)$ in (11.8) without any change of infimum value. The equality (11.8) is thus proved.

The next step is to observe that $\Phi_\tau(l, l'M(\cdot)) = \rho(l \mid R(\tau, M(\cdot))$. Hence from (11.2), (11.4), (11.8) we have

$$\rho(l \mid X[\tau]) = \inf\{\ \rho(l \mid R(\tau, M(\cdot))\ \mid\ \ M(\cdot) \in \Re^n[t_0,\tau]\ \}$$

for every $l \in R^n$. Due to well–known theorems of convex analysis [64] the last equality yields (11.6).

Lemma 11.1 acquires a specific form when $X_0 = R^n$. In this case there is no initial restriction on $x_0 = x[t_0]$.

Corollary 11.1. *Assume $X_0 = R^n$. Then for all $\tau \in [t_0, t_1]$ one has $X[\tau] = J[\tau]$ where*

$$J(\tau) = \bigcap\{J(\tau, M(\cdot)) \mid M(\cdot) \in \Re^n[t_0,\tau], \int_{t_0}^{\tau} M(s)S(\tau,s)ds = I\} \tag{11.9}$$

and

$$\begin{aligned} J(\tau, M(\cdot)) = \int_{t_0}^{\tau} (S(\xi,\tau)- \\ - \int_{\xi}^{\tau} M(s)S(\xi,s)ds)P(\xi)d\xi + \int_{t_0}^{\tau} M(s)Y(s)ds \end{aligned} \tag{11.10}$$

Relations (11.9), (11.10) define the set $J(\tau)$ which will be referred to as the *set–valued convolution integral* by analogy with the convolution integral for single–valued functions (see, for example, [26]). We shall also extend this term to the right–hand part of (11.6).

The main result that we have obtained here is therefore the assertion that in the linear problem the sets $X[\tau]$ happen to be equal to the multivalued convolution integrals defined by relations (11.10), (11.11).

12 The Set–Valued Lagrangian Technique

The assertions of the above Section yield a "standard" duality formulation for calculating $\gamma_0(l) = \rho(l \mid X[\tau]$ (see [38]). This formulation is as follows.

The Primal Problem:

$$\textit{maximize } (l, x[\tau])$$

over all

$$u(\cdot) \in P(\cdot), \quad x_0 \in X_0$$

where $x[t]$ is the solution to the equation

$$\dot{x}[t] = A(t)x[t] + u(t), \quad x[t_0] = x_0 \tag{12.1}$$

In other words,

$$\gamma_0(l) = \max\{\Psi_\tau(l, x_0, u(\cdot)) \mid x_0 \in R^n, \ u(\cdot) \in L_2^n[t_0, \tau]\}$$

under restriction (12.1), where

$$\Psi_\tau(l, x_0, u(\cdot)) = (l, x[\tau]) - \delta(x_0 \mid X_0) -$$

$$- \int_{t_0}^{\tau} (\delta(x[t] \mid Y(t)) + \delta(u(t) \mid P(t)))dt$$

Here

$$\delta(x \mid X) = \begin{cases} 0, & x \in X \\ +\infty, & x \notin X \end{cases}$$

The primal problem generates

The Dual Problem:

$$\textit{determine } \gamma^0(l) = \inf\{\Phi_\tau(l, \lambda(\cdot)) \mid \lambda(\cdot) \in C^n[t_0, \tau]\}$$

along the solutions $s[t]$ to the equation

$$\dot{s}[t] = -s[t]A(t) + \lambda(t), \ s[\tau] = l$$

The function $\Phi_\tau(l, \lambda(\cdot))$ is defined in (11.7) and may be rewritten as

$$\Phi_\tau(l, \lambda(\cdot)) = \rho(s[t_0] \mid X_0) + \int_{t_0}^{\tau} (\rho(s[t] \mid P(t)) + \rho(\lambda(t) \mid Y(t)))dt$$

Relations (11.8), (11.9) indicate that $\gamma_0(l) = \gamma^0(l)$. The optimal values of the primal and dual problems thus coincide.

Here the "standard" Lagrangian formulation yields

Lemma 12.1. *The value $\gamma_0(l) = \gamma^0(l) = \gamma(l)$ may be achieved as a solution to the problem*

$$\gamma(l) = \inf_{\lambda(\cdot)} \max_{u(\cdot), x_0} L(\lambda(\cdot), u(\cdot), x_0)$$

where

$$L(\lambda(\cdot), u(\cdot), x_0) = (s[\tau], x_0) + \int_{t_0}^{\tau} ((s[t], u(t)) + \rho(\lambda(t) \mid Y(t)))dt$$

and

$$\lambda(\cdot) \in C^m[t_0, \tau], \quad u(\cdot) \in P(\cdot), \quad x_0 \in X_0.$$

Lemma 11.1 provides another form of presenting $X[\tau]$. Namely, denote $S[\tau]$ to be the solution to the matrix differential equation

$$\dot{S}[t] = -S[t]A(t) + M(t), \quad S[\tau] = I$$

Also denote

$$\Lambda(x_0, u(\cdot), M(\cdot)) = S[\tau]x_0 + \int_{t_0}^{\tau} (S[t]u(t) + M(t)Y(t))dt$$

Obviously

$$R(\tau, M(\cdot)) = \bigcup \{\Lambda(x_0, u(\cdot), M(\cdot)) \mid x_0 \in X_0, \ u(\cdot) \in P(\cdot)\ \} =$$

$$= S[\tau]X_0 + \int_{t_0}^{\tau} (S[t]P(t) + M(t)Y(t))dt \qquad (12.2)$$

We now reformulate Lemma 10.1 as

Lemma 12.2. *The set $X[\tau]$ may be determined as*

$$X[\tau] = \bigcap_{M(\cdot)} \bigcup_{x_0, u(\cdot)} \Lambda(x_0, u(\cdot), M(\cdot))$$

over all $M(\cdot) \in \Re^n[t_0, \tau]$, $x_0 \in X_0$ and $u(\cdot) \in P(\cdot)$.

This result may be treated as a *generalization of the standard Lagrangian formulation.*

To conclude the discussion on the direct calculation of $X[\tau]$ we would like to indicate that the sets $R(\tau, M(\cdot))$ defined by (12.2) may not be the appropriate elements for describing $X[\tau]$, especially when one deals with the evolution of $X[\tau]$.

The reasons for this are in the following. Assuming function $M(\cdot)$ to be fixed, redenote $R(\tau, M(\cdot))$ as $R_M(\tau, t_0, X_0)$. Then we observe that in general

$$R_M(\tau, t_0, X_0) \neq R_M(\tau, s, R_M(\tau, t_0, X_0))$$

This means, in other words, that the map $R_M(\tau, t_0, X_0)$ does not generate a semigroup of transformations that could define a generalized dynamic system. In this sense the "dynamic" properties of the multifunction $R_M(\tau, t_0, X_0)$ worse than those of $X[\tau] = X(\tau, t_0, X_0)$.

This leads us to an alternative variety of parametrized maps that would be "better" in the sense that they would now possess the semigroup property.

13 An Alternate Presentation of the Tubes $X[\tau]$

Denote $\Re_0^n[t_0, \tau]$ to be the subclass of $\Re^n[t_0, \tau]$ that consists of all continuous matrix functions $M(\cdot)$ that satisfy the condition

$$\det\ (S(t, \tau) - \int_t^\tau M(s)S(t, s)ds) \neq 0 \tag{13.1}$$

for any $t \in [t_0, \tau]$.

In equivalent form this means that the set $M(\cdot) \in \Re_0^n[t_0, \tau]$ must be such that the solution $K[\cdot]$ to the matrix equation

$$\dot{K}[t] = -K[t]A(t) + M(t), \tag{13.2}$$

$$K[\tau] = I, \quad t_0 \leq t \leq \tau$$

must satisfy the property

$$\det\ K[t] \neq 0 \ \ \forall t \in [t_0, \tau] \tag{13.3}$$

We further denote $K[t] = K(t, \tau, M(\cdot))$ with the function $M(\cdot)$ given. Consider the equation

$$\dot{s}(t) = -s(t)(A(t) - L(t)), \quad t_0 \leq t \leq \tau, \tag{13.4}$$

where $L(\cdot)$ is any matrix function from the space $\Re^n[t_0, \tau]$. By analogy with the function $S(t, \xi)$ the matrix solution to (13.4) (in the first variable t) will be denoted as $S_L(t, \xi)$ ($S_L(\xi, \xi) = I$, $S(t, \xi) = S_L(t, \xi)$ for $L(t) \equiv 0$).

It is not difficult to verify that for every $M(\cdot) \in \Re_0^n[t_0, \tau]$ there exists an $L(\cdot) \in \Re^n[t_0, \tau]$ such that for all $t \in [t_0, \tau]$

$$K[t] = K(t, \tau, M(\cdot)) = S_L(t, \tau) \tag{13.5}$$

Indeed, we may set $L(t) = K^{-1}[t]M(t)$. Then the differential equation for $K[t]$ is the same as the one for $S_L[t] = S_L(t, \tau)$, namely

$$\begin{aligned} \dot{K}[t] &= -K[t](A(t) - L(t)), \\ \dot{S}_L[t] &= -S_L[t](A(t) - L(t)) \end{aligned} \tag{13.6}$$

with

$$K[\tau] = S_L[\tau] = S_L(\tau, \tau) = I \tag{13.7}$$

Hence the solutions $K[t]$ and $S_L[t]$ to (13.6), (13.7) must be identically equal in $t \in [t_0, \tau]$.

From (13.2), (13.5), (13.6) it now follows (for $M(\cdot) \in \Re_0^n[t_0, \tau]$)

$$R(\tau, M(\cdot)) = S_L(t_0, \tau)X_0 + \int_{t_0}^{\tau} S_L(\xi, \tau)(P(\xi) + L(\xi)Y(\xi))d\xi \tag{13.8}$$

Observe that the right –hand part of (13.8) is the crossection $Z(\tau, t_0, X_0, L(\cdot))$ at instant τ of the set $Z(\cdot, \tau, t_0, X_0, L(\cdot))$ of all solutions to the differential inclusion

$$\dot{z} \in (A(t) - L(t))z + P(t) + L(t)Y(t) \tag{13.9}$$

$$z(t_0) \in X_0, \quad t_0 \le t \le \tau,$$

Taking into account relations (11.6) and the obvious inclusion

$$\begin{aligned} &\cap \{ \, R(\tau, M(\cdot)) \mid M(\cdot) \in \Re^n[t_0, \tau] \, \} \subseteq \\ &\subseteq \cap \{ \, R(\tau, M(\cdot)) \mid M(\cdot) \in \Re_0^n[t_0, \tau] \, \} \end{aligned}$$

we again arrive (now due to a different manner of reasoning) at the inclusion of Theorem 10.1

$$X[\tau] \subseteq \bigcap \{ \, Z(\tau, t_0, X_0, L(\cdot)) \mid L(\cdot) \in \Re^n[t_0, \tau] \, \} \tag{13.10}$$

However the main point is that the inclusion (13.10) actually turns to be an equality. The respective result is given by

Theorem 13.1. *The following equality is true*

$$X[\tau] = \bigcap \{ \, Z(\tau, t_0, X_0, L(\cdot)) \mid L(\cdot) \in \Re^n[t_0, \tau] \, \} \tag{13.11}$$

for any $\tau \in [t_0, t_1]$.

The proof of this theorem is a rather cumbersome procedure which will have to be divided into several parts.

In the first step observe that without loss of generality one may further assume $A(t) \equiv 0$ $(t_0 \leq t \leq \tau)$ provided $P(t)$ and $Y(t)$ are time dependent.

Due to the state transformation $\tilde{x} = x - x_*[t]$ with $x_*[t]$ being viable trajectory of the system, one may also suppose that the sets X_0, $P(t)$, $Y(t)$ $(t_0 \leq t \leq \tau)$ all contain the origin (the existence of at least one viable trajectory $x_*[t]$ was assumed in Section 3). Therefore, from now on, one may take all the support functions $\rho(l \mid X_0)$, $\rho(l \mid P(t))$, $\rho(l \mid Y(t))$ to be nonnegative.

Before passing to the proof of Theorem 13.1 , we will establish two auxiliary results, beginning with the formula for the support function $\rho(l \mid X[\tau])$ (11.4) under assumption $A(t) \equiv 0$, namely

$$\rho(l \mid X[\tau]) = \inf\{ \ \Phi_\tau(l, \lambda(\cdot)) \ \mid \ \lambda(\cdot) \in C^n_\infty[t_0, \tau] \ \} =$$

$$= \inf\{ \ \Phi_\tau(l, \lambda(\cdot)) \ \mid \ \lambda(\cdot) \in \mathcal{M}^n[t_0, \tau] \ \} \tag{13.12}$$

where

$$\Phi_\tau(l, \lambda(\cdot)) = \rho(l - \int_{t_0}^{\tau} \lambda(\xi) d\xi \mid X_0) + \tag{13.13}$$

$$+ \int_{t_0}^{\tau} \rho(l - \int_{\xi}^{\tau} \lambda(s) ds \mid P(\xi)) d\xi + \int_{t_0}^{\tau} \rho(\lambda(s) \mid Y(s)) ds$$

Denoting

$$g(s) = l - \int_{s}^{\tau} \lambda(\xi) d\xi$$

we may rewrite (13.12), (13.13) as

$$\rho(l \mid X[\tau]) = \inf\{ \ \Psi((g(\cdot)) \ \mid \ g(\cdot) \in C^n_\infty[t_0, \tau] \ , g(\tau) = l \ \} =$$

$$= \inf\{ \ \Psi(g(\cdot)) \ \mid \ g(\cdot) \in \mathcal{M}^n[t_0, \tau], \ g(\tau) = l \ \} \tag{13.14}$$

where

$$\Psi(g(\cdot)) = \rho(g(t_0) \mid X_0) + \int_{t_0}^{\tau} (\rho(g(\xi) \mid P(\xi)) +$$

$$+ \rho(\dot{g}(\xi) \mid Y(\xi))) d\xi \tag{13.15}$$

For any $l \in R^n$ denote $C^n_{\infty,l}$ $[t_0, \tau]$ as the subclass of $C^n_\infty[t_0, \tau]$ for each element of which there exists a function $M(\cdot) \in \Re^n_\infty[t_0, \tau]$ such that $g(t) = M'(t) l$ for any $t \in [t_0, \tau]$ and

$$M(\tau) = I, \quad \det M(t) \neq 0 \ \ \forall t \in [t_0, \tau]$$

The next lemma is the crucial point in the proof of the Theorem 13.1.

Lemma 13.1. *Suppose the vector $l \in R^n$ is such that its coordinates $l_i \neq 0$ for all $i = 1, \dots, n$. Then the following equality is true*

$$\rho(l \mid X[\tau]) = \inf\{ \ \Psi(g(\cdot)) \ \mid \ g(\cdot) \in C^{n}_{\infty,l} \ [t_0, \tau] \ \}. \tag{13.16}$$

Proof. For the proof of this property we will distinguish the cases of n being even or odd.

Suppose $n = 2$. Then for calculating the infimum in (13.14) it suffices to restrict ourselves to the class of functions

$$\mathcal{G} = \{ \ g(\cdot) = \{g_1(\cdot), g_2(\cdot)\} : \ g(\cdot) \in \mathcal{M}^2[t_0, \tau], \ g(\tau) = l,$$

$$(g(t), g(t)) \neq 0 \ \ \forall t \in [t_0, \tau] \ \}$$

Indeed, for any $g(\cdot) \in \mathcal{M}^2[t_0, \tau]$ with $g(\tau) = l$ it is possible to construct a sequence $g_\epsilon(\cdot) \in \mathcal{M}^2[t_0, \tau]$, $g_\epsilon(\tau) = l$ $(\epsilon \to 0)$ for which

$$\lim_{\epsilon \to +0} \Psi(g_\epsilon(\cdot)) = \Psi(g(\cdot)) \tag{13.17}$$

and

$$(g_\epsilon(t), g_\epsilon(t)) \neq 0 \ \ \forall t \in [t_0, \tau]$$

For example, assume

$$g_\epsilon(t) = \{g_1(t), g_{\epsilon,2}(t)\}$$

where

$$g_{\epsilon,2}(t) = l_2 g_2(t + \epsilon)/g_2(\tau + \epsilon)$$

Since it is assumed that $g(\tau) = l$, $l_1 \neq 0$, $l_2 \neq 0$, the function $g_\epsilon(\cdot)$ is well–defined for minor values of $\epsilon > 0$ (i.e., $g_\epsilon(\cdot) \in \mathcal{M}^2[t_0, \tau]$, $g(\tau) = l$). Since the number of zeros of the polynomials $g_1(\cdot)$, $g_2(\cdot)$ is finite, we may select a "shift" $\epsilon = \epsilon^0$ in $g_2(t + \epsilon)$ so that the zeros of $g_1(t)$ and $g_2(t + \epsilon)$ will not coincide for all $\epsilon \in (0, \epsilon^0]$.

For each $t \in [t_0, \tau]$ we now have

$$g_\epsilon(t) \to g(t), \ \ \dot{g}_\epsilon(t) \to \dot{g}(t), \ \ \epsilon \to 0$$

The sequences $\{g_\epsilon(t)\}$, $\{\dot{g}_\epsilon(t)\}$ are equibounded in t for $\epsilon \in (0, \epsilon^0]$. Therefore (13.17) is true.

It is now possible to demonstrate that any function $g(\cdot) \in \mathcal{M}^2[t_0, \tau]$ with $g(\tau) = l$, $(g(t), g(t)) \neq 0 \ \ \forall t \in [t_0, \tau]$, may be presented in the form $g(\cdot) = M'(\cdot)l$ where $\det M(t) \neq 0 \ \forall t \in [t_0, \tau]$, $M(\tau) = I$.

In fact, let

$$M(t) = \begin{pmatrix} g_2(t)l_2^{-1} & (g_2(t) - l_2l_1^{-1}g_1(t))l_1^{-1} \\ (g_1(t) - l_1l_2^{-1}g_2(t))l_2^{-1} & g_1(t)l_1^{-1} \end{pmatrix} \tag{13.18}$$

Then, obviously $g(\cdot) = M'(\cdot)l$, $M(\tau) = I$ and

$$\det M(t) = g_1^2(t)l_1^{-2} + g_2^2(t)l_2^{-2} - g_1(t)g_2(t)l_1^{-1}l_2^{-1} =$$

$$= (g_1^2(t)l_1^{-2} + g_2^2(t)l_2^{-2} + (g_1(t)l_1^{-1} - g_2(t)l_2^{-1})^2)/2 > 0$$

for all $t \in [t_0, \tau]$.

Further on, let us assume that R^n is of even dimension: $n = 2k$. Following the scheme for $n = 2$, it is then possible to check that it suffices to calculate the infimum in (13.14) over such functions $g(\cdot) = \{g_1(\cdot), \ldots, g_{2k}(\cdot)\}$, $g(\tau) = l$, that

$$g_{2i-1}^2(t) + g_{2i}^2(t) > 0 \;\; \forall t \in [t_0, \tau], \;\; i = 1, \ldots, k$$

Any function $g(\cdot)$ of the given type may be presented in the form $g(t) = M'(t)l$ where the $(2k) \times (2k)$–matrix is block–diagonal:

$$M(t) = \begin{pmatrix} M_1(t) & \ldots & 0 \\ \vdots & \ddots & \vdots \\ 0 & \ldots & M_k(t) \end{pmatrix}$$

and each of the matrices $M_i(t)$ is 2×2–dimensional and may be calculated due to (13.18) where in the place of $g_1(t)$, $g_2(t)$ one should substitute $g_{2i-1}(t)$, $g_{2i}(t)$.

The function $M(\cdot) \in \Re_\infty^n[t_0, \tau]$, $M(\tau) = I$ and we have

$$\det M(t) = \prod_{i=1}^{k} \det M_i(t) > 0 \;\; \forall t \in [t_0, \tau]$$

Assume now that n is odd, $n = 2k + 1$.Then we again may calculate the infimum in (13.14) over the class of functions

$$\mathcal{G} = \{ \; g(\cdot) = \{g_1(\cdot), \ldots, g_{2k}(\cdot), g_{2k+1}(\cdot)\} : \; g(\cdot) \in \mathcal{M}^n[t_0, \tau], \; g(\tau) = l,$$

$$g_{2i-1}^2(t) + g_{2i}^2(t) > 0 \;\; \forall t \in [t_0, \tau], \; i = 1, \ldots, k \; \}$$

Each of such functions may be presented in the form $g(t) = M'(t)l$ where

$$M(t) = \begin{pmatrix} M_1(t) & \ldots & 0 & m(t) \\ \vdots & \ddots & \vdots & \vdots \\ 0 & \ldots & M_k(t) & 0 \\ 0 & \ldots & 0 & 1 \end{pmatrix},$$

$$m(t) = (g_{2k+1}(t) - l_{2k+1})/l_1$$

Here $M_i(t)$ is determined similar to (13.18) where $g_1(t)$, $g_2(t)$ are to be substituted by $g_{2i-1}(t)$, $g_{2i}(t)$. Obviously, $M(\tau) = I$, $M(\cdot) \in \Re^n_\infty[t_0, \tau]$ and

$$\det M(t) = \prod_{i=1}^{k} \det M_i(t) > 0 \ \ \forall t \in [t_0, \tau].$$

To finalize the proof of Lemma 13.1 we now have to treat the case $n = 1$. Obviously it suffices to prove (13.16) only for $l = \pm 1$. We formally take $l = 1$, as the other case is similar.

For $n = l = 1$ the class $C^n_{\infty,l}\ [t_0, \tau]$ may be substituted by all positive functions $m(\cdot) \in C^1_\infty\ [t_0, \tau]$, $m(\tau) = 1$.

In view of (13.14), (13.15) we need to demonstrate that

$$\begin{aligned} &\inf \{ \ \Psi(m(\cdot)) \ \mid \ m(\cdot) \in C^1_\infty \ [t_0, \tau] \ , m(\tau) = 1 \ \} = \\ &= \inf \{ \ \Psi(m(\cdot)) \ \mid \ m(\cdot) \in C^1_\infty \ [t_0, \tau] \ , m(\tau) = 1, \\ &\qquad m(t) > 0 \ \ \forall t \in [t_0, \tau] \ \} \end{aligned} \tag{13.19}$$

Let us fix any $m(\cdot) \in C^1_\infty\ [t_0, \tau]$, $m(\tau) = 1$. With t decreasing from the value τ, denote τ^* to be the first instant when $m(t)$ turns to zero. Therefore $m(\tau^*) = 0$, $m(\tau) = 1$ and $m(t) > 0$ for all $t \in (\tau^*, \tau]$.

Denote

$$m^*(t) = \begin{cases} m(t) & \text{for } \tau^* < t \leq \tau \\ 0 & \text{for } t_0 \leq t \leq \tau^* \end{cases}$$

Due to the assumption

$$0 \in X_0 \bigcap Y(t) \bigcap P(t) \ \ \forall t \in [t_0, \tau]$$

one observes

$$\rho(m(t_0) \mid X_0) + \int_{t_0}^{\tau^*} (\rho(m(t) \mid P(t)) + \rho(\dot{m}(t) \mid Y(t)))dt \geq \rho(m^*(t_0) \mid X_0)$$

Hence

$$\begin{aligned} \Psi(m(\cdot)) &= \rho(m(t_0) \mid X_0) + \int_{t_0}^{\tau} \rho(m(t) \mid P(t))dt + \\ &+ \int_{t_0}^{\tau} \rho(\dot{m}(t) \mid Y(t))dt \geq \rho(m^*(t_0) \mid X_0) + \\ &+ \int_{\tau^*}^{\tau} \rho(m^*(t) \mid P(t))dt + \int_{\tau^*}^{\tau} \rho(\dot{m}^*(t) \mid Y(t))dt = \Psi(m^*(\cdot)) \end{aligned} \tag{13.20}$$

Denote $m_\epsilon(t) = (m^*(t) + \epsilon)(1 + \epsilon)^{-1}$ for every $\epsilon > 0$. Then obviously we have

$$m_\epsilon(t) > 0 \ \ \forall t \in [t_0, \tau], \ \ m_\epsilon(\tau) = 1$$

It is not difficult to demonstrate that

$$\lim_{\epsilon \to +0} \Psi(m_\epsilon(\cdot)) = \Psi(m^*(\cdot)) \tag{13.21}$$

Finally, we may always find a function $m^\epsilon(t) \in C^1_\infty[t_0, \tau]$ such that

$$m^\epsilon(t) > 0 \ \ \forall t \in [t_0, \tau], \ \ m^\epsilon(\tau) = 1 \tag{13.22}$$

and

$$\lim_{\epsilon \to +0} \max_{t_0 \le t \le \tau} \| m^\epsilon(t) - m_\epsilon(t) \| = 0,$$

$$\lim_{\epsilon \to +0} \text{mes} \{ \ t \in [t_0, \tau] : \dot{m}^\epsilon(t) \neq \dot{m}_\epsilon(t) \ \} = 0$$

where mes A is the Lebesgue measure of set $A \subseteq [t_0, \tau]$.

The equiboundedness of the sequences $\{m^\epsilon(t)\}$, $\{\dot{m}^\epsilon(t)\}$ yield an equality similar to (13.21)

$$\lim_{\epsilon \to +0} \Psi(m^\epsilon(\cdot)) = \Psi(m^*(\cdot)) \tag{13.23}$$

From relations (13.19), (13.20), (13.22), (13.23) we conclude

$$\inf \{ \ \Psi(m(\cdot)) \ \mid \ m(\cdot) \in C^1_\infty \ [t_0, \tau] \ , m(\tau) = 1 \ \} \geq$$

$$\geq \inf \{ \Psi(m(\cdot)) \mid m(\cdot) \in C^1_\infty[t_0, \tau] \ , m(\tau) = 1, m(t) > 0 \ \forall t \in [t_0, \tau] \}$$

Since the opposite inequality is obvious, the equality (13.19) is thus established. Hence the lemma is finally proved.

The class $C^n_{\infty, l} \ [t_0, \tau]$ in (13.16) may well be substituted by $\Re^n_0[t_0, \tau]$. Lemma 13.1 may therefore be reformulated as follows.

Lemma 13.1'. *For any* $l \in \Lambda = \{l : \ l_i \neq 0, \ i = 1, \dots, n\}$ *one has*

$$\rho(l \mid X[\tau]) = \inf \{ \ \Psi(M'(\cdot)l) \ \mid \ M(\cdot) \in \Re^n_0 \ [t_0, \tau] \ \}. \tag{13.24}$$

Following the suggestions that led to (13.10) we derive

Corollary 13.1. *The relation (13.24) is equivalent to*

$$\rho(l \mid X[\tau]) = \inf \{ \ \rho(l \mid Z(\tau, t_0, X_0, L(\cdot)) \ \mid \ L(\cdot) \in \Re^n_\infty \ [t_0, \tau] \ \} \tag{13.25}$$

Next we need a technical result related to the theory of convex functions.

Lemma 13.2. *Assume $\{X_\alpha\}$, $\alpha \in \Omega$, is a variety of convex compact sets with*

$$X = \bigcap \{ X_\alpha \mid \alpha \in \Omega \} \neq \emptyset$$

Then

$$\rho(l \mid X) = (co\ f)(l) = f^{**}(l)$$

where

$$f(l) = \inf \{\rho(l \mid X_\alpha) \ \mid \ \alpha \in \Omega \}.$$

We are now able to prove Theorem 13.1.

Proof of Theorem 13.1. Applying Lemma 13.2 to $Z_L[\tau] = Z(\tau, t_0, X_0, L(\cdot))$ with $L(\cdot)$ as a parameter we get

$$\rho(l \mid \bigcap \{ Z_L[\tau] \mid \ L(\cdot) \in \Re^n_\infty \ [t_0, \tau] \ \}) = (\text{co } h)(l) \qquad (13.26)$$

where

$$h(l) = \inf \{ \ \rho(l \mid Z_L[\tau] \ \mid \ L(\cdot) \in \Re^n_\infty \ [t_0, \tau] \ \}$$

and

$$h(l) = \rho(l \mid X[\tau]) \ \ \forall l \in \Lambda = \{l \in R^n : l_i \neq 0, \ i = 1 \ldots, n\} \qquad (13.27)$$

Relations (13.21)-(13.27) lead to the equality

$$X[\tau] = \bigcap \{ \ Z_L[\tau] \mid \ L \in \Re^n_\infty \ [t_0, \tau] \ \} \qquad (13.28)$$

Indeed, since always

$$X[\tau] \subseteq Z_L[\tau], \ \ L \in \Re^n_\infty \ [t_0, \tau], \qquad (13.29)$$

assume that there exists a point $x^* \notin X[\tau]$ such that

$$x^* \in \bigcap \{ \ Z_L[\tau] \mid \ L \in \Re^n_\infty \ [t_0, \tau] \ \}$$

Then there must exist a vector l that ensures

$$(l, x^*) > \rho(l \mid X[\tau])$$

Since $X[\tau] \in \text{conv}R^n$ one may assume $l \in \Lambda$. Hence

$$\rho(l \mid \bigcap \{ \ Z_L[\tau] \mid L \in \Re^n_\infty[t_0, \tau]\}) > \rho(l \mid X[\tau])$$

However this contradicts with (13.26), (13.27).

Equality (13.28) is thus true and in view of (13.10), (13.29) Theorem 13.1 is now fully proved. Moreover we have established

Corollary 13.2. *Assume that in (11.1) the matrix function* $A(\cdot) \in \Re^n_k[t_0, \tau]$. *Then*

$$X[\tau] = \bigcap \{ \ Z(\tau, t_0, X_0, L(\cdot)) \ \mid \ L(\cdot) \in \Re^n_k[t_0, \tau] \ \}$$

for any $\tau \in [t_0, t_1]$.

Having figured out the solutions to the linear–convex viability problem, we shall move further by introducing an alternative matrix technique that will prove, as we shall see in the sequel, to be closely connected with stochastic techniques.

14 Viability Tubes: The Linear – Quadratic Approximation

Let us return to the problem of finding the viability tube $X[t] = X(t, t_0, X_0)$ for the system

$$\dot{x} \in A(t)x + P(t), \tag{14.1}$$

$$G(t)x \in Y(t), \tag{14.2}$$

$$x(t_0) \in X_0, \tag{14.3}$$

where the multivalued maps $P(t)$, $Y(t)$ and the set X_0 are the same as in Section 3.

The support function for the convex compact set $X[t]$ may be calculated directly similarly to the relations (11.3)-(11.5), [38]. This gives

$$\rho(l \mid X[t]) = \inf\{ \ \Psi_t(l, \lambda(\cdot)) \mid \lambda(\cdot) \in L_2^m[t_0, \tau] \ \}, \tag{14.4}$$

where

$$\Psi_t(l, \lambda(\cdot)) = \rho(S'(t_0, t)l - \int_{t_0}^{t} S'(t_0, \tau)G'(\tau)\lambda(\tau)d\tau \mid X_0) + \int_{t_0}^{t} \rho(S'(\tau, t)l -$$

$$- \int_{\tau}^{t} S'(\tau, s)G'(s)\lambda(s)ds \mid P(\tau))d\tau + \int_{t_0}^{t} \rho(\lambda(\tau) \mid Y(\tau))d\tau$$

We shall rewrite the system (14.1)-(14.3) in the form

$$\dot{x} = A(t)x + v(t),$$

$$G(t)x = w(t), \tag{14.5}$$

$$x(t_0) = x_0,$$

where the measurable functions $v(t)$, $w(t)$ and the vector x_0 satisfy the inclusions

$$v(t) \in P(t), \quad w(t) \in Y(t), \quad x_0 \in X_0 \tag{14.6}$$

The viability tube $X[t]$ may then be described as the multivalued map $X(t, t_0, X_0)$ generated by the bundle of all the trajectories $x(t, t_0, x_0)$ of system (14.5) that are consistent with the constraints (14.6).

The calculation of $X[t]$ for a given instant $t = \tau$ will now run along the following scheme.

Let us fix a triplet $k^*(\cdot) = \{v^*(\cdot), w^*(\cdot), x_0^*\}$ where $v^*(\cdot)$, $w^*(\cdot)$, x_0^* ($t \in [t_0, \tau]$) satisfy the constraint (14.6):

$$k^*(\cdot) \in \{P(\cdot) \times Y(\cdot) \times X_0\} \tag{14.7}$$

Instead of handling (14.5), (14.6) we shall now consider a "perturbed" system, which is

$$\begin{aligned} \dot{z} &= A(t)z + v^*(t) + \eta(t), \\ G(t)z &= w^*(t) + \xi(t), \\ x(t_0) &= x_0^* + \zeta^0, \quad t_0 \le t \le \tau \end{aligned} \tag{14.8}$$

The elements $\{ \zeta^0, \eta(\cdot), \xi(\cdot) \}$ represent the unknown disturbances bounded jointly by the quadratic inequality

$$\zeta^{0\prime} M \zeta^0 + \int_{t_0}^{\tau} \eta'(t)R(t)\eta(t)dt + \int_{t_0}^{\tau} \xi'(t)H(t)\xi(t)dt \le \mu^2 \tag{14.9}$$

where

$$\{M, R(\cdot), H(\cdot)\} \in \Im$$

and symbol $\Im$ stands for the product space

$$\Im = \Re_+^n \times \Re_+^n[t_0, \tau] \times \Re_+^m[t_0, \tau]$$

with $\Re_+^r[t_0, \tau]$ denoting the class of all $r \times r$-matrix functions $M(\cdot) \in \Re^n[t_0, \tau]$ whose values $M(t)$ are symmetric and positively definite.

For every fixed $k^*(\cdot)$, $\Lambda = \{M, R(\cdot), H(\cdot)\}$ and μ denote $Z[\tau] = Z(\tau, k^*(\cdot), \Lambda, \mu)$ to be the set of all the states $z(\tau)$ of the system (14.8) that are consistent with the constraint (14.9). It is well known that $Z(\tau, k^*(\cdot), \Lambda, \mu)$ is an ellipsoid and that its center $z_0[\tau] = z_0(\tau, k^*(\cdot), \Lambda)$ does not depend on μ [38]. This can be observed without difficulty by direct calculation which also indicate that $z_0[s]$ satisfies the linear differential equation

$$\dot{z} = (A(s) - \Sigma(s)G'(s)H(s)G(s))z +$$

$$+\Sigma(s)G'(s)H(s)w^*(s)+v^*(s), \tag{14.10}$$
$$z(t_o)=x_0^*, \quad t_0\le s\le\tau,$$

where $\Lambda=\{M,R(\cdot),H(\cdot)\}\in\Im$ and $\Sigma(\cdot)$ is the matrix solution to the Riccati equation

$$\dot\Sigma=A(s)\Sigma+\Sigma A'(s)-$$
$$-\Sigma G'(s)H(s)G(s)\Sigma+R^{-1}(s), \tag{14.11}$$
$$\Sigma(t_0)=M^{-1}, \quad t_0\le s\le\tau.$$

Let us now introduce the set

$$Z_0(\tau,\Lambda)=\bigcup\{\ z_0(\tau,k(\cdot),\Lambda)\mid k(\cdot)\in\{P(\cdot)\times Y(\cdot)\times X_0\}\}$$

which is the union of the centers $z_0[\tau]$ over all the triplets $k(\cdot)$ of (14.17).

The set $Z_0(\tau,\Lambda)$ is convex and compact, being the attainability domain for the system (14.10) with constraints (14.6). Our next step will be to find the support function for $Z_0(\tau,\Lambda)$, which will therefore give a complete description of this set.

We need however to set still some more notations. Let l be an arbitrary vector in R^n. Consider the following class of functions $p[\cdot]=p(\cdot;l,\Lambda)$ ($\Lambda=\{M,R(\cdot),H(\cdot)\}\in L_2^m[t_0,\tau]$):

$$p(\cdot;l,\Lambda)=(\mathcal{L}_\Lambda^{-1}\circ\mathcal{D}_\Lambda)l \tag{14.12}$$

where the linear operators $\mathcal{D}_\Lambda: R^n\to L_2^m[t_0,\tau]$ and $\mathcal{L}_\Lambda: L_2^m[t_0,\tau]\to L_2^m[t_0,\tau]$ are defined by

$$(\mathcal{D}_\Lambda b)(t)=G(t)(S(t_0,t)M^{-1}S'(t_0,\tau)+ \tag{14.13}$$
$$+\int_{t_0}^t S(s,t)R^{-1}(s)S'(s,\tau)ds)b, b\in R^n, \ t_0\le t\le\tau,$$
$$\mathcal{L}_\Lambda\lambda(\cdot)=(K_1+K_2)\lambda(\cdot), \quad \lambda(\cdot)\in L_2^m[t_0,\tau] \tag{14.14}$$
$$(K_1\lambda(\cdot))(t)=\int_{t_0}^\tau K(t,s)\lambda(s)ds,$$

$$K(t,s)=G(t)(S(t_0,t)M^{-1}S'(t_0,s)+\int_{t_0}^{\min\{t,s\}}S(\sigma,t)\times$$
$$\times R^{-1}(\sigma)S'(\sigma,s)d\sigma)G'(s), \tag{14.15}$$
$$(K_2\lambda(\cdot))(t)=H^{-1}(t)\lambda(t), \quad t_0\le t\le\tau. \tag{14.16}$$

Here $\mathcal{L}_\Lambda$ is the nondegenerate Fredholm operator of the second kind, so that the functions $\{p(\cdot;l,\Lambda)\}$ are well defined for all $l\in R^n$ and $\Lambda\in\Im$ [29].

Theorem 14.1. *The following equality is true*

$$\rho(l \mid Z_0(\tau, \Lambda)) = \Psi_\tau(l, p(\cdot; l, \Lambda)) \tag{14.17}$$

for all $l \in R^n$ and $\Lambda \in \Im$ where functions Ψ_τ, p are defined by relations (14.4) and (14.11)-(14.15) respectively.

Comparing the equality (14.16) with (14.4) we immediately come to the assertion.

Corollary 14.1. *The viability tube $X[\tau]$ may be estimated from above as follows*

$$X[\tau] \subseteq \bigcap \{ Z_0(\tau, \Lambda) \mid \Lambda \in \Im \} \tag{14.18}$$

As in the previous Section the objective is to indicate that the inclusion (14.17) actually becomes an equality for respective set–valued maps.

The last result may be demonstrated by using the approximation ideas and the smoothing techniques in the fashion of the proof of Lemma 13.1.

Lemma 14.1. *Suppose the $m \times n$–matrix $G(t)$ is of full rank:*

$$r(G(t)) = m \textit{ for any } t \in [t_0, \tau].$$

Then for every $l \in R^n$ the following equalities are true

$$\rho(l \mid X[\tau]) = \inf\{ \Psi_\tau(l, \lambda(\cdot)) \mid \lambda(\cdot) \in L_2^m[t_0, \tau] \} =$$

$$= co \inf\{ \Psi_\tau(l, p(\cdot; l, \Lambda)) \mid \Lambda \in \Im \}. \tag{14.19}$$

Combining Theorem 14.1, Lemma 14.1 and taking into account Lemma 13.2, we obtain

Theorem 14.2. *Let $r(G(t)) = m$ for every $t \in [t_0, \tau]$. Then*

$$X[\tau] = \bigcap \{ Z_0(\tau, \Lambda) \mid \Lambda \in \Im \}. \tag{14.20}$$

Theorem 14.2 gives a precise description of $X[\tau]$ through the solutions $Z_0(\tau, \Lambda)$ of the linear-quadratic problem (14.8)-(14.9) by allowing to vary the matrix parameters $\Lambda = \{M, R(\cdot), H(\cdot)\}$ in the joint integral constraint (14.9).

To finalize this Section we shall present differential relations that describe the set–valued estimators $Z_0(\tau, \Lambda)$. In view of Theorem 14.2 they will also determine the set $X[\tau]$.

Theorem 14.3. *The set $Z_0(\tau, \Lambda)$ is a τ-crossection of the solution tube (the integral funnel) to the following differential inclusion*

$$\dot{z}(t) \in (A(t) - \Sigma(t)G'(t)H(t)G(t))z(t) +$$

$$+ \Sigma(t)G'(t)H(t)Y(t) + P(t), \qquad (14.21)$$

$$z(t_0) \in X_0, \quad t_0 \leq t \leq \tau,$$

where $\Lambda = \{M, R(\cdot), H(\cdot)\} \in \Im$ and $\Sigma(\cdot)$ is the matrix solution to the Riccati equation (14.11).

Remark 14.1. We may observe that the sets $Z_0(\tau, \Lambda)$ are the attainability domains for the system (14.10), $k^*(\cdot) \in \{P(\cdot) \times Y(\cdot) \times X_0\}$, or, what is the same, to the differential inclusion (14.21).

Remark 14.2. In comparison with Theorems 10.4, 13.1, the last result (together with Theorem 14.2) provides a special structure of the matrix functions $L(\cdot)$ from relations (10.8), (13.11). Namely, here we may set

$$L(t) = \Sigma(t)G'(t)H(t), \quad t_0 \leq t \leq \tau,$$

where $\Sigma(\cdot)$ is defined by (14.11) with $\Lambda = \{M, R(\cdot), H(\cdot)\}$ varying within the set $\Im$.

In conclusion it is worthy to mention that several techniques based on the consideration of auxiliary uncertain systems under quadratic integral constraints have been discussed in this context in [32, 39, 54].

One of these is the guaranteed state estimation problem mentioned in Section 3. Apart from the deterministic techniques of this section the problem may be also solved through a stochastic scheme which ends up, however, with results of a similar type.

15 Guaranteed State Estimation: The Deterministic and the Stochastic Filtering Approximations

Suppose that the system (14.1)-(14.3) is specified as follows

$$\dot{x} \in A(t)x + P(t), \qquad (15.1)$$

$$y(t) \in G(t)x + Q(t), \qquad (15.2)$$

$$x(t_0) \in X_0, \qquad (15.3)$$

where (15.2) is the *measurement (observation) equation* and the continuous multifunction $Q(t)$ ($Q : T = [t_0, t_1] \to \text{conv}R^n$) reflects the restriction on the unknown but bounded noise w in the observations as indicated in (3.9).

Given the measurement $y = y^*(t)$, $t \in [t_0, \tau]$ the guaranteed state estimation problem is to specify at a given time-instant τ the set $X[\tau]$ of all states $x[\tau]$ of system (15.1) that are consistent with the equations (15.1)-(15.3) when $y(t) \equiv y^*(t)$. In other words, one is to find the crossections at time $t = \tau$ of the viability tube $X[t]$ for (15.1)-(15.3), $y(t) \equiv y^*(t)$. In the state estimation context the set $X[\tau]$ is known as the *informational domain*, or *consistency domain* [10, 33, 37, 38, 70]. This set depends on the measurement $y_\tau(\sigma) = y(\tau + \sigma)$, $t_0 - \tau \le \sigma \le 0$, namely,

$$X[\tau] = X[\tau, y_\tau(\cdot)]$$

for the linear system under consideration $X[\tau, y_\tau(\cdot)] \in \text{conv}R^n$.

The problem of finding $X[\tau]$ is further propagated into one of describing the evolution of $X[\tau] = X[\tau, y_\tau(\cdot)]$ in time. The evolution equation for $X[\tau]$ would therefore be the "guaranteed filtering" equations for the system (15.1)-(15.3) with unknown but bounded uncertainties (see also (3.5), (3.9)).

Needless to say that the evolution equations of type (6.1) and (8.2) or the equations (14.20), (14.11), together with relations (14.19), serve to be the solution to this problem (provided, of course, that $Y(t) = y(t) - Q(t)$ as indicated above and that $Y(t)$ satisfies the respective assumptions).

It is well–known, however, that a conventional stochastic filtering technique is given by the equations of the "Kalman filter" [27, 38] which turn to solve the stochastic filtering problem for linear systems with Gaussian noise. Our next question thereby will be to see, whether the equations of the Kalman filter could be also used to describe the informational domain $X[\tau, y_\tau(\cdot)]$ for the guaranteed estimation problem of the above.

This question is justified by the fact that on the one hand, the tube $X[t] = X[t, y_\tau(\cdot)]$ may be described through the linear–quadratic approximations of Section 14, while on the other — by the well established connections between the Kalman filtering equations and the solutions to the linear–quadratic problem of control.

Using the solutions of the previous Section, let us fix a triplet $k(\cdot) = k^*(\cdot) = \{v^*(\cdot), w^*(\cdot), x_0^*\}$ with $k^*(\cdot) \in \{P(\cdot) \times (y(\cdot) - Q(\cdot)) \times X_0\}$ and consider the stochastic differential equations

$$dz = (A(t)z + v^*(t))dt + \sigma(t)d\xi \tag{15.4}$$

$$dq = (G(t)z + w^*(t))dt + \sigma_1(t)d\eta \tag{15.5}$$

$$z(0) = x_0^* + \zeta, \quad q(0) = 0 \tag{15.6}$$

where ξ, η are standard normalized Brownian motions with continuous diffusion matrices $\sigma(t)$, $\sigma_1(t)$ and det $(\sigma(t)\sigma'(t)) \neq 0$ for all $t \in T$, ζ is a Gaussian vector with zero mean and variance $M^* = \sigma_0 \sigma_0{}'$.

Denoting $\sigma(t)\sigma'(t) = R^*(t)$, $\sigma_1(t)\sigma_1{}'(t) = H^*(t)$ and treating $q = q(t)$ as the available measurement we may find the equations for the minimum variance estimate

$$z^*(t) = E(z(t) \mid q(s),\ t_0 \leq s \leq t)$$

(the respective "Kalman filter").

These are

$$dz^*(t) = (A(t) - \Sigma(t)G'(t)H^{*-1}(t)G(t))z^*(t)dt +$$

$$+ \Sigma(t)G'(t)w^*(t)dt + \Sigma(t)G'(t)dq(t) + v^*(t)dt, \ z^*(t_0) = x_0^*, \tag{15.7}$$

$$\dot{\Sigma}(t) = A(t)\Sigma(t) + \Sigma(t)A'(t) -$$

$$- \Sigma(t)G'(t)H^{*-1}(t)G(t)\Sigma(t) + R^*(t), \ \Sigma(t_0) = M^* \tag{15.8}$$

It is obvious that the estimate $z^*(t)$ depends on the triplets $k^*(\cdot)$ and $\Lambda^* = \{M^*,\ R^*(\cdot),\ H^*(\cdot)\} \in \Im$.

Let us consider the set

$$Z^*(t) = Z^*(t, \Lambda^*) = \bigcup \{\ z^*(t) \mid k^*(\cdot) \in \{P(\cdot) \times Y(\cdot) \times X_0\}\}$$

which , with a given realization $q(t)$, is the attainability domain for the equation (15.7).

Theorem 15.1. *Assume the equalities*

$$M^* = M^{-1}, \quad R^*(t) \equiv R^{-1}(t), \quad H^*(t) \equiv H^{-1}(t) \tag{15.9}$$

to be true and $Y(t) = y(t) - Q(t)$, $t \in T$. *Also assume*

$$q(t) \equiv \int_{t_0}^{t} y(\tau)d\tau \tag{15.10}$$

Then the sets $Z_0(t, \Lambda)$ *of Section 14 and* $Z^*(t, \Lambda^*)$ *coincide, namely,*

$$Z_0(t, \Lambda) \equiv Z^*(t, \Lambda^*), \quad t \in T, \quad \Lambda = \{M, R(\cdot), H(\cdot)\}.$$

Corollary 15.1. *Under the assumptions of Theorem 15.1 the following equality is true*

$$X[\tau] = \bigcap \{Z^*(\tau, \Lambda^*) \mid \Lambda^* \in \Im\} \tag{15.11}$$

The proof of Theorem 15.1 follows from the fact that under the assumptions of this Theorem equation (15.7), with $k^*(\cdot) \in \{P(\cdot) \times Y(\cdot) \times X_0\}$, and the differential inclusion (14.29) have the same attainability domains. Corollary 15.1 then follows from Theorem 14.2.

Remark 15.1. The last results describe a clear connection between the solutions to the linear–quadratic Gaussian filtering problem (the Kalman filter) and the solutions to the deterministic guaranteed state estimation problems for uncertain systems with unknown but bounded "noise" in the *nonquadratic case* of the instantaneous constraints on the unknowns.

It should be clear that the theorems discussed in Sections 6-9 were proved under Assumptions B1 or B2, while those of Sections 10-15 allowed $Y(t)$ to be only continuous in time t. We shall now formulate a technique of singular perturbations for differential inclusions which will eventually allow to relax the requirements on $Y(\cdot)$ and to accept a state constraint, when $Y(t)$ is only measurable in T.

A nonlinear filtering version of these results could be pursued through a combination of the reasoning of Section 10 and of paper [7].

16 The Singular Perturbation Techniques

In this Section the finite (nondifferential) inclusion (3.8) or (3.11) for the state constraint will be substituted by a differential inclusion with a parameter by the derivative of some additional state space variable (with a *singular perturbation* [9, 28, 71, 72]). This will finally allow to reduce the solution to a scheme similar to the parametrization approach of Sections 10-15.

Consider again the system of inclusions

$$\dot{x}(t) \in F(t, x(t)), \tag{16.1}$$

$$0 \in G(t, x(t)), \tag{16.2}$$

$$x(t_0) \in X_0, \quad t_0 \le t \le t_1 \tag{16.3}$$

The multivalued maps $F : [t_0, t_1] \times R^n \to \text{conv}R^n$ and $G : [t_0, t_1] \times R^n \to \text{conv}R^m$ are now supposed to be measurable in t and continuous in x , $X_0 \in \text{comp}R^n$.

We assume also that the maps F, G satisfy the following conditions

$$\begin{array}{l} \| F(t,x) \| \le g(t)(1+ \| x \|), \\ \| G(t,x) \| \le g(t)(1+ \| x \|) \end{array} \tag{16.4}$$

for some function $g(t) > 0$ integrable on $[t_0, t_1]$ (symbol $\| A \|$ further denotes the norm

$$\| A \| = \sup\{ \| a \| : a \in A \},$$

for any set $A \subseteq R^n$).

The last additional condition on F, G ensures all the solutions of (16.1) (and also of other equation introduced further, see (16.5)) to be defined on the whole interval $[t_0, t_1]$. Here, of course, it is possible to require other types of such extendability conditions [11] instead of (14.4).

As before, we shall study the tube $X(\cdot; \tau, t_0, X_0)$ of trajectories (now taken with respect to state constraints (16.2)) and its crossection $X[\tau] = X(\tau, t_0, X_0)$.

We still keep the assumption on the existence of at least one trajectory viable on $[t_0, t_1]$.

For a fixed $\tau \in [t_0, t_1]$ and for some matrix function $B(\cdot) \in \Re^m[t_0, \tau]$ consider the following auxiliary system of differential inclusions in $R^n \times R^m$:

$$\begin{cases} \dot{x} & \in & F(t, x), \\ \epsilon \dot{y} & \in & B(t)y + G(t, x), \end{cases} \tag{16.5}$$

$$x(t_0) \in X_0, \;\; y(t_0) \in Y_0, \;\; t_0 \le t \le \tau,$$

where $Y_0 \in \mathrm{comp} R^m$.

Denote $z = \{x, y\} \in R^n \times R^m$ and $z[t] = z(t; t_0, z_0, \tau, \epsilon)$ to be the solution to (16.5) that starts at a point $z[t_0] = z_0 = \{x_0, y_0\}$. Also denote $Z(\cdot; t_0, Z_0, \tau, \epsilon)$ as the tube of all such solutions $z[\cdot]$ with $z[t_0] = z_0 \in Z_0 = X_0 \times Y_0$. Let $Z(\tau; t_0, Z_0, \epsilon)$ be the crossection at τ of the assembly $Z(\cdot; t_0, Z_0, \tau, \epsilon)$.

The next result may be easily derived from the definitions of $Z(\cdot; t_0, Z_0, \tau, \epsilon)$, $Z(\tau;\ t_0,\ Z_0, \epsilon)$.

Theorem 16.1. *Let the instant $\tau \in [t_0, t_1]$. Then*

(i) *for every $\epsilon > 0$ and any $B(\cdot) \in \Re^m[t_0, \tau]$ and with $Y_0 \in compR^m$, $0 \in Y_0$, we have*

$$X(\cdot, \tau, t_0, X_0) \times \{0\} \subseteq Z(\cdot; t_0, Z_0, \tau, \epsilon), \tag{16.6}$$

(ii) *for $B(t) \equiv 0$ $(t \in [t_0, \tau])$ and any $Y_0 \in compR^m$ the following equality is true*

$$X(\cdot, \tau, t_0, X_0) \times Y_0 = \bigcap \{ Z(\cdot; t_0, Z_0, \tau, \epsilon) \mid \epsilon > 0 \}. \tag{16.7}$$

From Theorem 16.1 we immediately obtain

Corollary 16.1. *Let $B(\cdot) \in \Re^m[t_0, \tau]$ and $Y_0 \in compR^m$. Then the following estimate for $X[\tau] = X(\tau, t_0, X_0)$ is true*

$$X[\tau] \subseteq \pi_x \left(\bigcap \{ \, Z(\tau; t_0, Z_0, \epsilon) \mid \epsilon > 0 \, \} \right) \tag{16.8}$$

if either $B(t) \equiv 0$ ($t \in [t_0, \tau]$) or $0 \in Y_0$.

The notation $\pi_x W$ is used here for the projection of the set $W \subseteq R^n \times R^m$ into the space R^n of x–variables.

For a system of linear differential inclusions with "linear" state constraints it is possible however to establish an exact equality in (16.8).

To discussed this issue we first introduce a new class of perturbations to the nonlinear system (16.1)-(16.3), namely of type (16.5) but with a time–dependent matrix perturbation $L(\cdot) \in \Re^m[t_0, \tau]$ instead of the scalar $\epsilon > 0$:

$$\begin{cases} \dot{x} & \in \; F(t, x), \\ L(t)\dot{y} & \in \; B(t)y + G(t, x), \end{cases} \tag{16.9}$$

$$z_0 = \{x_0, y_0\} \in Z_0, \quad t_0 \le t \le \tau$$

Denote the class of all continuous invertible matrix functions $L(\cdot) \in \Re^m[t_0, \tau]$ as $\Re^m_*[t_0, \tau]$. Symbol $Z(\cdot; \tau, t_0, Z_0, L)$ will now stand for the solution tube to the system (16.9) with $Z(\tau; t_0, Z_0, L) = Z(\tau; \tau, t_0, Z_0, L)$ where $Z_0 \in$ comp $(R^n \times R^m)$.

The following analogy of Theorem 10.3 is then true.

Theorem 16.2. *Let $\tau \in [t_0, t_1]$, $B(\cdot) \in \Re^m[t_0, \tau]$, $Y_0 \in compR^m$. Then the following inclusions are fulfilled*

(i) *for every $\epsilon > 0$ and any $B(\cdot) \in \Re^m[t_0, \tau]$ and with $Y_0 \in compR^m$, $0 \in Y_0$, we have*

$$X(\cdot, \tau, t_0, X_0) \times \{0\} \subseteq \bigcap \{ Z(\cdot; t_0, Z_0, \tau, L) \mid L \in \Re^m_*[t_0, \tau] \, \} \tag{16.10}$$

if $0 \in Y_0$,

(ii)

$$X(\cdot, \tau, t_0, X_0) \times Y_0 = \bigcap \{ Z(\cdot; t_0, Z_0, \tau, L) \mid L \in \Re^m_*[t_0, \tau] \, \} \tag{16.11}$$

if $B(t) \equiv 0$, $t \in [t_0, \tau]$.

Corollary 16.2. *Under either of the assumptions of Theorem 16.2 one has*

$$\begin{aligned} X[\tau] \;&\subseteq\; \pi_x \left(\bigcap \{ Z(\tau; t_0, X_0, Y_0, L(\cdot)) \mid L(\cdot) \in \Re^m_*[t_0, \tau] \} \right) \subseteq \\ &\subseteq\; \pi_x \left(\bigcap \{ \, Z(\tau; t_0, X_0, Y_0, \epsilon) \mid \epsilon > 0 \, \} \right). \end{aligned}$$

Consider now the maps F, G to be 'inear in x, namely

$$\begin{aligned} \dot{x}(t) &\in A(t)x(t) + P(t), \\ x(t_0) &\in X_0, \quad t_0 \leq t \leq t_1, \end{aligned} \tag{16.12}$$

$$G(t)x(t) \in Q(t), \quad t_0 \leq t \leq \tau, \tag{16.13}$$

Here the $n \times n$– and $m \times n$–matrices $A(t)$, $G(t)$ are taken measurable and bounded on $[t_0, t_1]$ and the set–valued maps $P : [t_0, t_1] \to \mathrm{conv}R^n$ and $Q : [t_0, t_1] \to \mathrm{conv}R^m$ are measurable and bounded.

Further on we consider either the parametrized system of type (16.5), namely

$$\left\{ \begin{aligned} \dot{x} &\in A(t)x + P(t), \\ \epsilon\dot{y} &\in -G(t)x + Q(t), \end{aligned} \right. \tag{16.14}$$

$$x(t_0) \in X_0, \; y(t_0) \in Y_0, \quad t_0 \leq t \leq \tau,$$

with any $\epsilon > 0$, or the system of type (16.9), which is

$$\left\{ \begin{aligned} \dot{x} &\in A(t)x + P(t), \\ L(t)\dot{y} &\in -G(t)x + Q(t), \end{aligned} \right. \tag{16.15}$$

$$x(t_0) \in X_0, \; y(t_0) \in Y_0, \quad t_0 \leq t \leq \tau.$$

with $L(\cdot) \in \Re_n^*[t_0, \tau]$ and $Z_0 = X_0 \times Y_0 \in \mathrm{conv}(R^n \times R^m)$.

We use the same notation for the solution tubes to (16.14) and (16.15) and their crossections as for (16.5), (16.9).

A slight modification of Theorem 16.1 yields

Theorem 16.3. *Assume*

$$X_0 \subseteq \pi_x Z_0. \tag{16.16}$$

Then for every trajectory $x[\cdot] \in X(\cdot, \tau, t_0, X_0)$ of (16.12), (16.13) there exists a vector $y_0 \in R^m$ such that

$$\{x[t_0], y_0\} \in Z_0$$

and for every $\tau \in [t_0, t_1]$

$$z(\tau) = \{x[\tau], y_0\} \in Z(\tau; t_0, Z_0, \epsilon)$$

for all $\epsilon > 0$.

Corollary 16.3. *Assume relation (16.16) to be true. Then for every $\tau \in [t_0, t_1]$ one has*

$$X[\tau] = X(\tau, t_0, X_0) \subseteq \pi_x \left(\bigcap \{ Z(\tau; t_0, Z_{0,\epsilon} \mid \epsilon > 0) \} \right). \qquad (16.17)$$

The following analogy of Theorem 16.2 is also true.

Theorem 16.4. *Assume relation (16.16) to be true. Then for every trajectory $x[\cdot] \in X(\cdot, \tau, t_0, X_0)$ of (16.12), (16.13) there exists a vector $y_0 \in R^m$ such that*

$$\{x[t_0], y_0\} \in Z_0$$

and for every $\tau \in [t_0, t_1]$

$$z(\tau) = \{x[\tau], y_0\} \in Z(\tau; t_0, Z_0, L),$$

whatever is the function $L \in \Re_*^m[t_0, \tau]$.

Corollary 16.4. *Under assumption (16.14) for every $\tau \in [t_0, t_1]$ we have*

$$X[\tau] \subseteq \pi_x \left(\bigcap \{ Z(\tau; t_0, Z_0, L) \mid L \in \Re_*^m[t_0, \tau] \} \right). \qquad (16.18)$$

The principal result of this Section is now formulated as follows.

Theorem 16.5. *Suppose*

$$X_0 \subseteq \pi_x Z_0. \qquad (16.19)$$

Then for every $\tau \in [t_0, \tau]$ one has

$$\pi_x \left(\bigcap \{ Z(\tau; t_0, Z_0, L) \mid L \in \Re_*^m[t_0, \tau] \} \right) \subseteq X[\tau]. \qquad (16.20)$$

The proof of this Theorem more or less follows the schemes of Sections 13-14.

From Corollary 16.3 and Theorem 16.5 we may now obtain the exact description of the set $X[\tau] = X(\tau, t_0, X_0)$ through the solutions of a singularly perturbed differential inclusion without state constraints. Namely we finally come to

Theorem 16.6. *Under the assumption*

$$\pi_x Z_0 = X_0$$

the following formula is true for any $\tau \in [t_0, t_1]$

$$X[\tau] = \pi_x \left(\bigcap \{ \; Z(\tau; t_0, Z_0, L) \mid \; L \in \Re_*^m[t_0, \tau] \; \} \right).$$

It is clear that the conditions of smoothness or continuity of $Y(t)$ in time are not required in this setting.

17 Control Synthesis Under Viability Constraints

The techniques of this paper allow an approach to the problem of control synthesis. We shall treat this problem for the linear-convex case in the context of the terminal control problem under viability constraints. The solution though will be a nonlinear strategy. Returning to the system (8.1) and taking $A(t) \equiv 0$, without loss of generality, we come to the equations

$$\dot{x} = u + f(t), \quad u \in \mathcal{P}(t), \tag{17.1}$$

$$x \in Y(t), \tag{17.2}$$

$$x^0 = x(t_0) \in X^0, \tag{17.3}$$

where $f(t)$ is a given function measurable in t and $\mathcal{P}(t)$, $Y(t)$ satisfy the assumptions of Section 7.

Further on we shall consider

The Problem of Control Synthesis: Given a terminal set $\mathcal{M} \in$ conv R^n, the problem of control synthesis will consist in specifying a solvability set $W^0[\tau] = W(\tau, t_1, \mathcal{M})$ and a set-valued feedback control strategy $u = \mathcal{U}(t,x)$, $\mathcal{U}(\cdot,\cdot) \in \mathcal{U}$, such that all the solutions to the differential inclusion

$$\dot{x} \in \mathcal{U}(t,x) \; + \; f(t), \tag{17.4}$$

that start from a given position $\{\tau, x_\tau\}$,

$$x_\tau = x[\tau], \quad x[\tau] \in W(\tau, t_1, \mathcal{M}), \quad \tau \in [t_0, t_1],$$

would satisfy the inclusion (17.2), $\tau \le t \le t_1$, and $x(t_1) \in \mathcal{M}$.

Here $\mathcal{U}$ is the variety of all set - valued functions $\mathcal{U}(\cdot,\cdot)$ with convex compact values $\mathcal{U}(t,x) \subseteq \mathcal{P}(t)$, that are measurable in t and upper semicontinuous in x, ensuring thus the solvability and extendability of the solutions to (17.4) for any $x^0 \in X^0 \in$ conv R^n.

The given problem is nonredundant provided $W[\tau] = W(\tau, t_1, \mathcal{M}) \neq \emptyset$, where $W[\tau]$ is the largest set of all states from which the solution to the problem does exist at all. We thus define a multifunction $W[t]$, $\tau \leq t \leq t_1$ with values in conv R^n.

It is not difficult to observe that if the system (17.1), (17.2), $x(t_1) \in \mathcal{M}$, would be integrated in backward time, then $W[\tau]$ would be similar to its attainability domain. It will therefore satisfy an evolution equation similar to (8.2) but taken in backward time.

From the results of Sections 6-8 we come, in view of the last remark, to the following

Lemma 17.1. *The multifunction $W[t]$ satisfies the following evolution equation*

$$\lim_{\sigma \to 0} \sigma^{-1} h\left(W[t-\sigma], (W[t] - \sigma \mathcal{P}(t) - f(t)) \cap Y(t-\sigma)\right) = 0, \qquad (17.5)$$

$$W[t_1] = \mathcal{M}, \quad t_0 \leq t \leq t_1.$$

The tube $W[t]$ will now allow to introduce a scheme for finding the synthesizing control strategy $\mathcal{U}(t, x)$.

Given $W[t]$, $t \in [\tau, t_1]$, let

$$\frac{d}{dt} h^+(x(t), W[t]) \mid_{(17.1)} \qquad (17.6)$$

denote the derivative in time $t+0$ of the distance $d(x, W[t]) = h^+(x, W[t])$ due to equation (17.1).

Define

$$\mathcal{U}^*(t, x) = \left\{ u \ : \ \frac{d}{dt} h^+(x(t), W[t]) \mid_{(17.1)} \ \leq \ 0 \right\}$$

We will now have to prove that the set-valued strategy $\mathcal{U}(t, x) \neq \emptyset$ for all $\{t, x\}$ and that it solves the problem of control synthesis.

Calculating $h^+(x, W[t])$, we have

$$h^+(x, W[t]) = \max\{ \ (l, x) - \rho(l \mid W[t]) \mid \parallel l \parallel \leq 1 \ \}$$

so that the maximizer $l^0 = l^0_W(t, x) \neq 0$ yields

$$h^+(x, W[t]) = (l^0, x) - \rho(l^0 \mid W[t])$$

if $h^+(x, W[t]) > 0$ (otherwise $l^0 = 0$ and $h^+(x, W[t]) = 0$).

To calculate the derivative (17.6) we need to know the partial derivative of $\rho(l \mid W[t])$ in time $t+0$, which is

$$\frac{\partial}{\partial(t+0)}\rho(l^0 \mid W[t]) = -\frac{\partial}{\partial(s-0)}\rho(l^0 \mid W[-s]), \quad s=-t.$$

Here the right-hand part can be calculated following Section 9, that is similarly to $\partial\rho(l \mid X[t])/\partial(t+0)$, but in backward time.

Since $W[t] \subseteq Y(t)$, this gives

$$\frac{\partial\rho(l^0 \mid W[t])}{\partial(t-0)} = \begin{cases} \rho(-l^0 \mid \mathcal{P}(t)) - (l^0, f(t)), \\ \qquad\qquad\qquad\qquad \text{if } \rho(l^0 \mid W[t]\ ) < \rho(l^0 \mid Y(t)); \\ \min\{\rho(-l^0 \mid \mathcal{P}(t)) - (l^0, f(t)),\ -\partial/\partial t\ (\rho(l^0 \mid Y(t)))\}, \\ \qquad\qquad\qquad\qquad \text{if } \rho(l^0 \mid W[t]\ ) = \rho(l^0 \mid Y(t)),\ )), \end{cases}$$

and further on,

$$\frac{d}{dt}h^+(x(t), W[t]) = (l^0, u + f(t)) - \frac{\partial\rho(l^0 \mid W[t])}{\partial(t+0)} =$$

$$= \begin{cases} -(-l^0, u) + \rho(-l^0) \mid \mathcal{P}(t)) \\ \qquad\qquad \text{or} \\ l^0(u + f(t)) - \partial\rho(l^0 \mid Y(t))/\partial t \end{cases}$$

The last relation indicates that

$$\mathcal{U}^0(t,x) = \{\ u:\ (-l^0, u) = \rho(-l^0 \mid \mathcal{P}(t))\ \} \neq \emptyset,$$

and therefore $\mathcal{U}^*(t,x) \neq \emptyset$, since

$$\mathcal{U}^0(t,x) \subseteq \mathcal{U}^*(t,x).$$

Thus we come to

Lemma 17.2. *The strategy* $\mathcal{U}^*(\cdot,\cdot) \in \mathcal{U}$.

The necessary properties that ensure $\mathcal{U}^*(\cdot,\cdot) \in \mathcal{U}$ may be checked directly.

A standard type of reasoning yields the following

Theorem 17.1. *The strategy* $u = \mathcal{U}^*(t,x)$ *solves the problem of control synthesis for any position* $\{t,x\}$ *that satisfies the inclusion* $x \in W[t]$.

The theorem follows from

Lemma 17.3. *For any solution* $x[t] = x(t,\tau,x_\tau)$ *to the inclusion*

$$\dot{x} \in \mathcal{U}^*(t,x) + f(t) \tag{17.7}$$

$$x[\tau] = x_\tau \in W[\tau], \ \tau \le t \le t_1,$$

one has

$$x[t] \in W[t], \ \tau \le t \le t_1,$$

and therefore

$$x[t] \in Y(t), \ \tau \le t \le t_1,$$

$$x[t_1] \in \mathcal{M}.$$

Indeed, if we suppose $x[\tau] \in W[\tau]$ and $h^+(x[t^*], W[t]) > 0$ for some $t^* > \tau$, then there exists a point $t^{**} \in (\tau, t^*]$ where

$$dh^+(x[t^{**}], \ W[t^{**}])/dt > 0$$

in contradiction with the definition of $\mathcal{U}^*(t,x)$.

The theory of trajectory tubes and the results given particularly in this section allow to give a further description of the synthesized solution to (17.4), $\mathcal{U}(t,x) \equiv \mathcal{U}^*(t,x)$. Namely, it is now worthy to describe the tube of all solutions to (17.7), $x[\tau] \in X_\tau$. The main point is that this tube may be described without the knowledge of the strategy $\mathcal{U}^*(t,x)$ itself but only on the basis of the information given just for the original Problem of Control Synthesis.

Together with (17.5) consider the following evolution equation

$$\lim_{\sigma \to 0} \sigma^{-1} h(\Gamma(t+\sigma), \ (\Gamma(t) + \sigma \mathcal{P}(t)) \cap W(t+\sigma)) = 0, \tag{17.8}$$

$$\Gamma(\tau) = X_\tau, \ X_\tau \in \mathrm{conv} R^n,$$

$$X_\tau \subseteq Y(\tau), \tau \le t \le t_1.$$

The theory given in this and the previous sections allows to prove the following assertions.

Theorem 17.2. *The solution tube* $\Gamma[t] = \Gamma(t, \tau, X_\tau)$ *to the synthesized system (17.7),* $X[\tau] = X_\tau$ *under the viability constraint (17.2) is the tube of all solutions to the evolution equation (17.8).*

Corollary 17.1. *The tube of all the synthesized trajectories from a given point* $x_\tau \in W[\tau]$ *may be defined as the tube of all solutions to the differential inclusion*

$$\dot{x} \in \mathcal{P}(t) + f(t), \ x(\tau) = x_\tau,$$

that satisfy

$$x(t) \in W(t), \ \tau \le t \le t_1$$

and are therefore viable relative to $W(t)$, where $W[t]$ is the solution to (17.5).

The synthesized trajectory tube is therefore a set-valued solution to the "two-set boundary value problem " (17.5), (17.8) for the respective evolution equations.

The calculation of this solution obviously does not require the knowledge of the strategy $u = \mathcal{U}^*(t, x)$ itself.

18 Conclusion

This paper is an overview of the results of the authors on solution tubes for differential inclusions with state constraints (the "viable trajectory tubes" in terms of J.- P. Aubin). The emphasis of this work is to treat the problem through an evolution theory for respective generalized dynamic systems with set-valued trajectories, on the one hand, and to indicate, on the other, that the solutions to these equations could be described through constructive relations that further allow an algorithmization .

The motivations and applications for this theory come from problems in evolutionary dynamics,in modelling, estimation and control for systems with unknown but bounded uncertainty as indicated in this paper. A research in applicability of the present techniques indicates effective tools based on set-valued approximations (particularly, an ellipsoidal calculus) and on parallelization of the solution algorithms. The latter techniques are, however, already beyond the scope of this paper.

References

[1] J.- P. Aubin, *Viability Theory*, Birkhauser, Boston, 1991.

[2] ———, *A survey of viability theory*, SIAM J. Control Optim., 28 (1990), pp.749-788.

[3] J.- P. Aubin and A. Cellina, *Differential Inclusions*, Heidelberg, Springer-Verlag, 1984.

[4] J.- P. Aubin and I. Ekeland, *Applied Nonlinear Analysis*, Academic Press, New York, 1984.

[5] J.- P. Aubin and H. Francowska, *Set-valued Analysis*, Birkhauser, 1990.

[6] A. V. Balakrishnan, *Applied Functional Analysis*, Springer - Verlag, Heidelberg, 1976.

[7] J. S. Baras, A. Bensoussan and M. R. James, *Dynamic observers and asymptotic limits of recursive filters: special cases*, SIAM J. Appl. Math., 48 (1988), pp.11-47.

[8] E. A. Barbashin, *On the theory of generalized dynamic systems*, Uchen. Zap. Moscow Univ., Matematika, 135 (1949), pp.110-133.

[9] D. J. Bell, *Singular problems in optimal control - a survey*, Int. J. Control, 21 (1975), pp.319-331.

[10] D. P. Bertsekas and I. B. Rhodes, *Recursive state estimation for a set-membership description of uncertainty*, IEEE Trans. Aut. Contr., AC-16 (1971), pp.117-128.

[11] V. I. Blagodatskih and A. F. Filippov, *Differential inclusions and optimal control*, Trudy Mat. Inst. Akad. Nauk SSSR, 169 (1985), pp.194-252.

[12] C. Castaing and M. Valadier, *Convex Analysis and Measurable Multifunctions*, Lect. Notes in Math., 580 (1977).

[13] F. L. Chernousko, *State estimation for dynamic systems. The method of ellipsoids*, Nauka, Moscow, 1988.

[14] F. H. Clarke, *Optimization and Nonsmooth Analysis*, Wiley - Interscience, New-York, 1983.

[15] V. F. Dem'yanov, *Minimax: Directional Differentiation*, Leningrad University Press, 1974.

[16] V. F. Dem'yanov, C. Lemarechal and J. Zowe, *Approximation to a set -valued mapping,I: a proposal*, Appl. Math. Optim., 14 (1986), pp.203-214.

[17] V. F. Dem'yanov and A. M. Rubinov, *Quasidifferential calculus*, Optimization Software,Inc. Publications Division, New York, 1986.

[18] A. L. Dontchev, *Perturbations, Approximations and Sensitivity Analysis of Optimal Control Systems*, Lect. Notes in Control and Inform. Sciences, 52, Springer - Verlag, 1986.

[19] A. L. Dontchev and V. M. Veliov, *Singular perturbation in Mayer's problems for linear systems*, SIAM J. Control Optim., 21 (1983), pp.566-581.

[20] I. Ekeland and R. Temam, *Analyse Convexe at Problemes Variationnelles*, Dunod, Paris, 1974.

[21] A. F. Filippov, *Differential Equations with Discontinuous Right-hand Side*, Nauka, Moscow, 1985.

[22] T. F. Filippova, *A note on the evolution property of the assembly of viable solution to a differential inclusion*, Computers Math. Applic., 25 (1993), pp.115-121.

[23] ———, *On the modified maximum principle in the estimation problems for uncertain systems*, Laxenburg, IIASA Working Paper WP-92-032, 1992.

[24] T. F. Filippova, A. B. Kurzhanski, K. Sugimoto and I. Valyi, *Ellipsoidal calculus, singular perturbations and the state estimation problems for uncertain systems*, Laxenburg, IIASA Working Paper WP-92-51, 1992.

[25] H. G. Guseinov, A. I. Subbotin and V. N. Ushakov, *Derivatives for multivalued mappings with applications to game- theoretical problems of control*, Problems of Control and Inform. Theory, 14, 3 (1985), pp.155-167.

[26] A. D. Joffe and V. M. Tikhomirov, *The Theory of Extremal problems*, Nauka, Moscow, 1979.

[27] R. Kalman and R. Bucy, *New results in linear filtering and prediction theory*, Trans. AMSE, 83, D (1961).

[28] P. Kokotovic, A. Bensoussan and G. Blankenship , eds., *Singular Perturbations and Asymptotic Analysis in Control Systems*, Lect. Notes in Contr. and Inform. Sci., 90, Springer-Verlag, 1986.

[29] A. N. Kolmogorov and S. V. Fomin, *Elements of Function Theory and Functional Analysis*, Nauka, Moscow, 1968.

[30] V. A. Komarov, *The equation for attainability sets of differential inclusions in the problem with state constraints*, Trudy Steklov Matem. Instit. Akad. Nauk SSSR, 185 (1988), pp.116-125.

[31] ———, *The estimates of attainability sets of differential inclusions*, Mat. Zametki, 37 (1985), pp.916-925.

[32] A. S. Koscheev and A. B. Kurzhanski, *On adaptive estimation of multistage systems under uncertaity*, Isvestia AN SSSR, Teh. Kibernetika, 2 (1983), pp.72-93.

[33] N. N. Krasovskii, *The Theory of Control of Motion*, Nauka, Moscow, 1968.

[34] ———, *The Control of a Dynamic System*, Nauka, Moscow, 1986.

[35] ———, *Game-theoretic Problems on the Encounter of Motions*, Nauka, Moscow, 1970.

[36] N. N. Krasovskii and A. I. Subbotin, *Positional Differential Games*, Springer-Verlag, 1988.

[37] A. B. Kurzhanski, *On minmax control and estimation strategies under incomplete information*, Problems of Contr. Inform. Theory, 4 (1975), pp. 205-218.

[38] ———, *Control and Observation under Conditions of Uncertainty*, Nauka, Moscow, 1977.

[39] ———, *Dynamic control system estimation under uncertainty conditions, part* I, Problems of Control and Information Theory, 9 (1980), pp.395-406; *part* II, Problems of Control and Information Theory, 10 (1981), pp.33-42.

[40] ———, *On evolution equations in estimation problems for systems with uncertainty*, Laxenburg, IIASA Working Paper WP-82-49, 1982.

[41] ———, *Evolution equations for problems of control and estimation of uncertain systems*, in Proceedings of the Int. Congress of Mathematicians, Warszawa, 1983, pp.1381-1402.

[42] ———, *On the analytical description of the set of viable trajectories of a differential system*, Doklady Akad. Nauk SSSR, 287 (1986), pp.1047-1050.

[43] ———, *Identification - a theory of guaranteed estimates*, in From Data to Model, J. C. Willems, ed., Springer - Verlag, 1989.

[44] A. B. Kurzhanski and T. F. Filippova, *On the description of the set of viable trajectories of a differential inclusion*, Doklady AN SSSR, 289 (1986), pp.38-41.

[45] ———, *On the description of the set of viable trajectories of a control system*, Different. Uravn., 23 (1987), pp.1303-1315.

[46] ———, *On the set-valued calculus in problems of viability and control for dynamic processes: the evolution equation*, Les Annales de l'Institut Henri Poincare, Analyse non-lineaire, 1989, pp.339-363.

[47] ———, *Dynamics of the set of viable trajectories to a differential inclusion: the evolution equation*, Problems of Contr. Inform. Theory, 17 (1988), pp.137-144.

[48] ———, *Perturbation techniques for viability and control*, in System Modelling and Optimization, P. Kall, ed., Lect. Notes in Control and Inform. Sciences, 180, Springer - Verlag, 1992.

[49] ———, *On the method of singular perturbations for differential inclusions*, Doklady Akad. Nauk SSSR, 321 (1991), pp.454-459.

[50] ———, *Differential inclusions with state constraints. The singular perturbation method*, Trudy Steklov Matem. Inst. Ross. Akad. Nauk, to appear.

[51] A. B. Kurzhanski and O. I. Nikonov, *On the control strategy synthesis problem. Evolution equations and set-valued integration*, Doklady Akad. Nauk SSSR, 311 (1990), pp.788-793.

[52] ———, *Funnel equations and multivalued integration problems for control synthesis*, in Perspectives in Control Theory, Progress in Systems and Control Theory, Vol.2, B. Jakubczyk, K. Malanowski and W. Respondek, eds, Birkhauser, Boston, 1990.

[53] A. B. Kurzhanski, B. N. Pschenichnyi and V. G. Pokotilo, *Optimal inputs for guaranteed identification*, Laxenburg, IIASA Working Paper WP-89-108, 1989.

[54] A. B. Kurshanski and I. F. Sivergina, *On noninvertible evolutionary systems: guaranteed estimates and the regularization problem*, Doklady Akad. Nauk SSSR, 314 (1990), pp.292-296.

[55] A. B. Kurzhanski, K. Sugimoto and I. Valyi, *Guaranteed state estimation for dynamic systems: ellipsoidal techniques*, International Journal of Adaptive Control and Signal Processing, to appear.

[56] A. B. Kurzhanski and I. Valyi, *Set-valued solutions to control problems and their approximations*, in Analysis and Optimization of systems, A. Bensoussan and J. L. Lions, eds, Lect. Notes in Contr. and Inform. Sc., 111, Springer-Verlag, 1988.

[57] ——— , *Ellipsoidal techniques for dynamic systems: the problems of control synthesis*, Dynamics and Control, 1 (1991), pp.357-378.

[58] ——— , *Ellipsoidal techniques for dynamic systems: control synthesis for uncertain systems*, Dynamics and Control, 2 (1992).

[59] P. J. Laurent, *Approximation and Optimization*, Hermann, Paris, 1972.

[60] C. Lemarechal and J. Zowe, *Approximation to a multi-valued mapping: existence, uniqueness, characterization*, Math. Institut, Universitat Bayreuth, 1987, Report N 5.

[61] M. S. Nikol'skii, *On the approximation of the attainability domain of differential inclusions*, Vestn. Mosc. Univ., Ser. Vitchisl. Mat. i Kibern., 4 (1987), pp.31-34.

[62] A. I. Panasyuk and V. I. Panasyuk, *On an equation resulting from a differential inclusion*, Matem. Zametki, 27 (1980), pp.429-445.

[63] ——— , *Asymptotic Magistral Optimization of Control Systems*, Nauka i Teknika, Minsk, 1986.

[64] R. T. Rockafellar, *Convex Analysis*, Princeton University Press, 1970.

[65] ——— , *State constraints in convex problems of Bolza*, SIAM J. Control, 10 (1972), pp.691-715.

[66] ——— , *Augmented Lagrange multiplier functions and duality in nonconvex programming*, SIAM J. Contr., 12 (1974), pp.268-285.

[67] ——— , *Augmented Lagrangians and applications of the proximal point algorithm in nonconvex programming*, Math. Oper. Res., 1 (1976), pp.97-116.

[68] E. Roxin, *On the generalized dynamical systems defined by contigent equations*, J. of Diff. Equations, 1 (1965), pp.188-205.

[69] F. Schlaepfer and F. Schweppe, *Continuous time state estimation under disturbance bounded by convex sets*, IEEE Trans. Autom. Contr., AC-17 (1972), pp.197-206.

[70] F. Schweppe, *Uncertain Dynamic Systems*, Prentice Hall Inc., Englewood Cliffs, New Jersey, 1973.

[71] A. N. Tikhonov, *On the dependence of the solutions of differential equations on small parameter*, Matem. Sbornik, 22 (1948), pp.198-204.

[72] ———, *Systems of differential equations containing a small parameter multiplying the derivative*, Matem. Sbornik, 31 (1952), pp.575-586.

[73] A. A. Tolstonogov, *Differential Inclusions in Banach Space*, Nauka, Novosibirsk, 1986.

[74] I. Valyi, *Ellipsoidal approximations in problems of control*, in Modelling and adaptive control, C. Birnes, A. B. Kurzhanski, eds, Lect.Notes in Contr. and Inform.Sciences, 105, Springer-Verlag, 1988.

[75] ———, *Ellipsoidal Techniques for Dynamic Systems*, Candidate's Thesis, Hungarian Academy of Sciences, Budapest, Hungary, 1992.

[76] V. M. Veliov, *Second order discrete approximations to strongly convex differential inclusions*, Systems and Control Letters, 13 (1989), pp.263-269.

[77] ———, *Approximations to differential inclusions by discrete inclusions*, Laxenburg, IIASA Working Paper WP-89-017, 1989.

[78] R. B. Vinter, *A characterization of the reachable set for nonlinear control systems*, SIAM J. Contr. Optim., 18 (1980), pp.599-610.

[79] R. B. Vinter and P. Wolenski, *Hamilton - Jakobi theory for optimal control problems with data measurable in time*, SIAM J. Control Optim., 28 (1990), pp.1404-1419.

[80] H. S. Witsenhausen, *Set of possible states of linear systems given perturbed observations*, IEEE Trans. Autom. Control, AC-13 (1968), pp.556-558.

[81] P. R. Wolenski, *The exponential formula for the reachable set of a Lipschitz differential inclusion*, SIAM J. Contr. Optim., 28 (1990), pp.1148-1161.

A.B.Kurzhanski
Moscow State University,
Fac. of Comput. Math. & Cybernetics (VMK)
Moscow 119899, Russia

T.F.Filippova
Institute of Mathematics and Mechanics,
Russian Academy of Sciences,
S.Kovalevsky st. 16
Ekaterinburg 620219, Russia

A Theory of Generalized Solutions to First–Order PDEs with an Emphasis on Differential Games

A.I. Subbotin

Key words and phrases: minimax solutions of first order PDE, viscosity solutions, differential games, differential inclusions, weakly invariant sets, viable trajectories.

1 Introduction

It is well known that in many applications connected with boundary value problems and Cauchy problems for Hamilton-Jacobi equations and other types of partial differential equations (PDE) of the first order there exist no classical nonlocal solutions. On the other hand, there exist nonsmooth functions that are crucial for the considered problems (e.g., optimal value functions in control problems, alsoknown as Bellman functions). These functions satisfy the considered equations at each point of differentiability. Thus, there arises a necessity to introduce a notion of generalized solution and to develop the theory and methods for constructing these solutions. Investigations of various approaches to the definition of generalized solutions are dealt with in many papers (see, e.g., the articles [13,28] and the book [32], which contain reviews and bibliography of investigations of 1950s–70s among which we should name the results of S.N.Kruzhkov for the Hamilton-Jacobi equation whose Hamiltonian is convex with respect to the impulse variable).

In the beginning of the 1980's M.G.Crandall and P.-L.Lions introduced the notion of viscosity solution. The first papers [12,13] by these authors were followed by many publications, in which various types of equations and boundary value problems were considered, theorems of the existence and uniqueness of viscosity solutions were proved, and also some applications of the viscosity solutions theory to control problems and differential games were studied.

In the articles [40,41,43,45] and the book [42] another approach was presented. It originates from the theory of positional differential games worked out by N.N.Krasovskiĭ and his colleagues. It is known (see, e.g., [23–26]) that the optimality principle for the value function of a differential game can be formulated in the form of two properties called u-stability and v-stability. The u-stability (resp., v-stability) property of the value function is the weak invariance of epigraph (resp., hypograph) of this function with respect to some family of differential inclusions. The u-stability and v-stability properties can be formulated in different ways, in particular, in the form of inequalities for directional derivatives. These inequalities were introduced in the articles [40,45] in order to define generalized (minimax) solutions of Bellman-Isaacs equation. The proposed approach can applied be to a broad range of boundary value and Cauchy problems for various types of first order PDE (i.e., not only to Bellman-Isaacs equations which give origin to the term "minimax solution"). The present article deals mostly with a definition that can be viewed as a generalization of the classical method of characteristics. In this definition the characteristic system is replaced by a family of "characteristic" differential inclusions. The minimax solution graph is weakly invariant with respect to these differential inclusions.

Investigations of weakly invariant sets[1] form a large section of the theory of differential equations and inclusions and to a great extent are connected with applications to the control theory and the theory of differential games (see, e.g., [1,10,20,21,29,34,39]). These results can facilitate the development of the proposed approach, which in its turn will probably promote new investigations in this field.

The definitions of minimax and viscosity solutions have different form, but their equivalence can be proved (see, e.g., [33,46]). A direct proof of this equivalence given in [42,44] is based on the possibility to separate a convex compact set from the epigraph of the first variation of a lower semicontinuous function, nonconvex in general, (see Lemma 5.1 below). Here one can see the duality between the definition of viscosity solutions in the form of inequalities, in which subdifferentials are used , on the one hand, and on the other hand, the definition of minimax solutions, where the inequalities for directional derivatives are used. This is one of the many aspects of the duality typical to the theory of generalized solutions of first order PDE that confirms the remark of L.Young [48]: "...Actually, Hamiltonians are inseparably intertwined with the notion of convexity and particularly with duality of convex figures".

[1]the property of weak invariance is also called the viability property

Let us outline the contents of the article. In Sect. 2 we recall the facts of the classical method of characteristics which later in Sect. 3 lead to the definition of characteristic differential inclusions.

In Sect. 3 we give the definition of minimax solutions of the first order PDE

$$F(x, u(x), Du(x)) = 0, \quad x \in G \subset R^n$$

and show that it is compatible with the notion of classical solutions.

The notions of upper and lower solutions are essential for the investigations of generalized (viscosity and minimax) solutions of first order PDE. In Sect. 4 we give definitions according to which the epigraph (resp., hypograph) of an upper (resp., lower) solution is weakly invariant with respect to characteristic differential inclusions.

Sect. 5 contains some facts of nonsmooth analysis used in the next section to define the minimax solution in the infinitesimal form.

In Sect. 6 various forms of the definition of minimax solutions are considered and their equivalence is proved, the equivalence of minimax and viscosity solutions is proved also.

In Sect. 7 the Cauchy problem for the Hamilton-Jacobi equation

$$\partial u/\partial t + H(t, x, u, D_x u) = 0, \quad 0 \leq t \leq \theta, \ x \in R^n,$$

$$u(\theta, x) = \sigma(x), \quad x \in R^n$$

is considered. We give sufficient conditions for the existence and uniqueness of a minimax solution of this problem and prove the uniqueness of the solution.

As it is mentioned above, the origination of the proposed approach and methods of investigating minimax solutions is closely connected with the constructions of the theory of positional differential games. To illustrate this fact, we show in Sect. 8 that the value function coincides with the minimax solution of the Bellman-Isaacs equation.

In the theory of differential games much attention is given to the methods of reducing numerical construction of the value function to solving auxiliary problems of program control (see, e.g., [23–26]). These results can be used to find generalized solutions of Cauchy problems for Hamilton-Jacobi equations. In Sect. 9 we consider constructions which originate from the method of program iterations developed in the theory of differential games (see, e.g., [5]).

To simplify the considerations connected with the proposed approach and to eliminate minor details from the construction of minimax solutions it is assumed that in the considered first order PDEs the functions $F(x, z, p)$ and $H(t, x, z, p)$ satisfy Lipschitz conditions in x and p. These assumptions can be weakened. In Sect. 10 conditions taken from the article [11] are given. Under these conditions (which do not imply that the Hamiltonian is Lipschitz continuous) the minimax solutions of Hamilton-Jacobi equations are defined, and theorems on the existence and uniqueness of minimax solutions to the Cauchy problem are formulated.

2 The Method of Characteristics

Recall the statement of the Cauchy problem considered in the classical theory of first order PDEs. Let an $(n-1)$-dimensional manifold C_0 be given in the $(n+1)$-dimensional space with variables (x, z) being defined by the parametric equalities

$$x = \xi_0(s), \quad z = \zeta_0(s), \quad s \in R^{n-1},$$

where $s \mapsto \xi_0(s) : R^{n-1} \to R^n$ and $s \mapsto \zeta_0(s) : R^{n-1} \to R$ are continuously differentiable functions. To simplify the presentation we assume that the whole manifold C_0 is a one-to-one projected on Euclidian space R^{n-1}.

The Cauchy problem is formulated in the following way: find a continuously differentiable function $x \mapsto u(x)$ defined on some open domain $X \subset R^n$ such that

$$F(x, u(x), Du(x)) = 0, \quad x \in X, \tag{2.1}$$

$$\{(x, u(x)) : x \in X\} \supset C_0 := \{(\xi_0(s), \zeta_0(s)) : s \in R^{n-1}\}. \tag{2.2}$$

Here $Du := (\partial u/\partial x_1, \ldots, \partial u/\partial x_n)$, the real function $(x, z, p) \mapsto F(x, z, p)$ is defined on $R^n \times R \times R^n$.

Let us briefly outline the method of characteristics. This method is considered in details in many books (see, e.g., [9,36]). Dealing here with facts concerned with the method of characteristics we assume that the derivatives $D_xF := (\partial F/\partial x_1, \ldots, \partial F/\partial x_n)$, $D_zF := \partial F/\partial z$, $D_pF := (\partial F/\partial p_1, \ldots, \partial F/\partial p_n)$ exist and are continuous on $R^n \times R \times R^n$.

According to one of the main results of the classical theory of first order PDEs, the solution of the Cauchy problem (2.1), (2.2) can be reduced to solving the characteristic system of ordinary differential equations

$$\begin{aligned} \dot{x} &= D_pF(x, z, p) \\ \dot{p} &= -(D_xF(x, z, p) + F_z(x, z, p)p) \\ \dot{z} &= \langle p, D_pF(x, z, p)\rangle \end{aligned} \tag{2.3}$$

Here $\dot{x} := \mathrm{d}x/\mathrm{d}t$, $\dot{p} := \mathrm{d}p/\mathrm{d}t$, $\dot{z} := \mathrm{d}z/\mathrm{d}t$, the symbol $\langle p, f\rangle$ denotes the scalar product of vectors p and f.

Note that the function F is an integral of system (2.3), i.e., its values are constant $F(x(t), z(t), p(t)) = const$ along any solution $t \mapsto (x(t), z(t), p(t))$ of system (2.3). The solutions along which the function F is equal to zero are employed in this method. Thus the third equation in system (2.3) can be written as

$$\dot{z} = \langle \dot{x}, p \rangle - F(x, z, p) \tag{2.4}$$

(this form of the equation will be needed below to define the generalized solution).

It is assumed that for any parameter $s \in R^{n-1}$ there exists a unique vector $p = \pi_0(s)$ that satisfies the system of n equations

$$\begin{aligned} &F(\xi_0(s), \zeta_0(s), \pi_0(s)) = 0, \\ &\frac{\partial \zeta_0(s)}{\partial s_i} - \left\langle \frac{\partial \xi_0(s)}{\partial s_i}, \pi_0(s) \right\rangle = 0, \quad i = 1, \ldots, n-1. \end{aligned} \tag{2.5}$$

Besides that the function $s \mapsto \pi_0(s) : R^{n-1} \to R^n$ should be sufficiently smooth. We consider solutions $(x(t), z(t), p(t))$ of system (2.3) which satisfy the condition

$$(x(0), z(0), p(0)) = (\xi_0(s), \zeta_0(s), \pi_0(s)).$$

Let us denote by $(\xi(s,t), \zeta(s,t), \pi(s,t))$ the solutions, that depend on the parameter $s \in R^{n-1}$. The mapping

$$(s,t) \mapsto (\xi(s,t), \zeta(s,t), \pi(s,t))$$

is considered on some open domain $D \subset R^n$ which contains the hyperplane $\{(s, 0) : s \in R^{n-1}\}$.

According to the method of characteristics, a solution of problem (2.1), (2.2) is given by parametric equalities

$$x = \xi(s,t), \quad u = \zeta(s,t).$$

The assumptions imposed in the method of characteristics imply that the functions ξ and ζ are twice continuously differentiable on the domain D, the mapping $(s,t) \mapsto \xi(s,t) : D \to X$ is one-to-one, and the inclusion

$$C_0 \subset \{(\xi(s,t), \zeta(s,t)) : (s,t) \in D\}$$

is true. Here $X := \{\xi(s,t) : (s,t) \in D\}$. The unknown solution of problem (2.1), (2.2) can be defined as an explicit function of x by the equality

$$u(x) = \zeta(\sigma(x), \tau(x)), \quad x \in X,$$

where $x \mapsto (\sigma(x), \tau(x)) : X \to D$ is the reverse mapping to ξ.

Note that

$$\operatorname{gr} u = \bigcup_{s \in R^{n-1}} \operatorname{gr}(\xi(s,\cdot), \zeta(s,\cdot)).$$

Here $\operatorname{gr} u := \{(x, u(x)) : x \in X\}$ is the graph of the function u, $\operatorname{gr}(\xi(s,\cdot), \zeta(s,\cdot)) := \{(\xi(s,t), \zeta(s,t)) : t \in T(s)\}$, $T(s) := \{t \in R : (s,t) \in D\}$. One can say that the graph of the function u is formed by the parameterized family of graphs of characteristics $(\xi(s,\cdot), \zeta(s,\cdot))$. Usually, the characteristics are understood to be functions $t \mapsto \xi(s,t)$. Deviating somewhat from the established term we treat the characteristics as functions $t \mapsto (\xi(s,t), \zeta(s,t))$.

It is well known that in many applications connected with the Cauchy problem and boundary value problems for first order PDE the classical solutions do not exist. As a rule, the method of characteristics is not applicable to these problems because the mapping ξ is not one-to-one (the characteristics intersect). There are different approaches to define and methods to investigate the generalized solutions. The constructions proposed in Sect. 3 for defining generalized (minimax) solution can be considered as a reduction and relaxation of the classical method of characteristics.

3 The Generalized Characteristics

The notion of weak invariance of a set with respect to a differential inclusion is used in the definition of minimax solution of first order PDE. Let us recall this notion.

A set $W \subset R^m$ is called weakly invariant with respect to differential inclusion

$$\dot{y}(t) \in E(y(t)) \subset R^m, \tag{3.1}$$

if for any point $y_0 \in W$ there exists a solution to the differential inclusion (3.1) such that $y(0) = y_0$ and $y(t) \in W$ for all $t \geq 0$. In this case it is said also that the set W satisfies the viability property. Weakly invariant sets and their connections with control problems were investigated by many authors (see, e.g., the papers [1,10,20,21,30,31,34,39] and the bibliography therein).

Consider a first order PDE

$$F(x, u(x), Du(x)) = 0, \quad x \in G \subset R^n, \tag{3.2}$$

where G is an open domain in R^n. Let us indicate the main assumptions on the function F under which we will define the minimax solutions of equation (3.2).

The function $(x, z, p) \mapsto F(x, z, p)$ is assumed to be continuous, non-increasing in z, i.e.

$$F(x, z_1, p) \geq F(x, z_2, p) \;\; \forall\, (x, z_1, z_2, p) \in G \times R \times R \times R^n, \;\; z_1 \leq z_2, \quad (3.3)$$

and satisfing the conditions

$$\begin{gathered} |F(x, z, 0)| \leq (1 + \|x\| + |z|)\mu, \\ |F(x, z, p) - F(x, z, q)| \leq \|p - q\|\rho(x) \\ \forall (x, z, p, q) \in G \times R \times R^n \times R^n, \end{gathered} \quad (3.4)$$

where $\rho(x) := (1 + \|x\|)\mu$, μ is a positive number.

The above-mentioned assumptions simplify the definition of the minimax solution. Without them the construction of these solutions involves additional approximating elements (cf. Sect. 10).

Let us now define the minimax solution. Consider a system that consists of a differential inequality and differential equation

$$\|\dot{x}\| \leq \rho(x), \quad \dot{z} = \langle \dot{x}, p \rangle - F(x, z, p). \quad (3.5)$$

Here p is a vector in R^n. Note that the differential equation for the variable z coincides with equation (2.4). System (3.5) can be written in the form of a differential (characteristic) inclusion

$$(\dot{x}(t), \dot{z}(t)) \in E(x(t), z(t), p), \quad (3.6)$$

where

$$E(x, z, p) := \{(f, g) \in R^n \times R : \|f\| \leq \rho(x),\; g = \langle f, p \rangle - F(x, z, p)\} \quad (3.7)$$

$$(x, z, p) \in G \times R \times R^n.$$

Since the multivalued mapping E is not defined for $x \notin G$, we use the following notion of solution to the inclusion (3.6). Let $R^+ := [0, \infty[$ and

$$(x(\cdot), z(\cdot)) : R^+ \to R^n \times R$$

be a continuous function, $(x_T(\cdot), z_T(\cdot))$ be the restriction of this function to $T := [0, \tau] \subset R^+$. We understand a solution of differential inclusion (3.6) to be a continuous function $(x(\cdot), z(\cdot))$ that satisfies the following condition: if $\{(x(t), y(t)) : t \in T\} \subset G$, then the function $(x_T(\cdot), z_T(\cdot))$ is absolutely continuous and for almost all $t \in T$ the inclusion (3.6) is true. By $\mathcal{S}(x_0, z_0, p)$ we denote the set of solutions to the differential inclusion (3.6) which satisfies the initial condition $(x(0), z(0)) = (x_0, z_0)$.

The symbol $\mathrm{Inv}_{(3.6)}(p)$ denotes the family of closed sets $W \subset \mathrm{cl}\, G \times R$ weakly invariant with respect to the differential inclusion (3.6) for a fixed $p \in R^n$.

Definition 3.1. *A minimax solution of equation* (3.2) *is a continuous function* $u : \mathrm{cl}\, G \to R$ *such that*

$$\mathrm{gr}\, u \in \mathrm{Inv}_{(3.6)}(p), \quad \forall p \in R^n. \tag{3.8}$$

Here $\mathrm{gr}\, u := \{(x, u(x)) : x \in \mathrm{cl}\, G\}$ is the graph of function u, $\mathrm{cl}\, G := G \cup \partial G$ is the closure of set G, ∂G is the boundary of G.

Thus, a continuous function u is called a minimax solution of equation (3.2), if for any $p \in R^n$ and $x_0 \in \mathrm{cl}\, G$ there exists a solution (generalized characteristic) of the differential inclusion (3.6) such that $x(0) = x_0, \quad z(t) = u(x(t))$ for all $t \geq 0$. One can also say that a minimax solution is a function u whose graph contains viable trajectories of the characteristic differential inclusion (3.6) for any parameter $p \in R^n$.

Let us now demonstrate that the definition of a minimax solution is compatible with the notion of classical solution. Note that

$$E(x, z, p) \cap E(x, z, q) \neq \emptyset \quad \forall (x, z, p, q) \in G \times R \times R^n \times R^n. \tag{3.9}$$

Really, for $p \neq q$ this intersection contains an element (f_*, g_*) of the form

$$f_* := \frac{[F(x, z, p) - F(x, z, q)](p - q)}{\|p - q\|^2},$$

$$g_* := \langle f_*, p\rangle - F(x, z, p) = \langle f_*, q\rangle - F(x, z, q).$$

Let u be a classical solution of equation (3.2), i.e., the function u is continuously differentiable and satisfies equation (3.2). Let us show that this function is a minimax solution of this equation. Take arbitrary $(x_0, z_0) \in \mathrm{gr}\, u$, $p \in R^n$. Using (3.9), it is not difficult to check that there exists a solution $(x(t), z(t))$ to the differential inclusion

$$(\dot{x}, \dot{z}) \in E(x, u(x), Du(x)) \cap E(x, u(x), p), \tag{3.10}$$

that satisfies the initial condition $(x(0), z(0)) = (x_0, z_0)$. It follows from (3.7) and the equality $F(x, u(x), Du(x)) = 0$ that

$$\dot{z}(t) = \langle Du(x(t)), \dot{x}(t)\rangle.$$

Taking into account the initial condition $z(0) = u(x(0))$, we obtain that $z(t) = u(x(t))$. According to (3.10), $(x(t), z(t))$ is a solution of the differential inclusion (3.6). Thus, condition (3.8) is valid for a classical solution u.

The reverse is also true: if a minimax solution of equation (3.2) is differentiable then it is a classical solution of this equation. Really, let u be a minimax solution and let the function u be differentiable at a point $x_0 \in G$. Suppose $z_0 = u(x_0)$, $p = Du(x_0)$. According to (3.8), there exists a solution $(x(t), z(t))$ of the differential inclusion (3.6) such that

$$u(x(\delta)) = z(\delta) = u(x_0) + \langle p, x(\delta) - x_0 \rangle - F(x_0, u(x_0), p)\delta + \alpha_1(\delta)\delta,$$

where $\alpha_1(\delta) \to 0$ as $\delta \downarrow 0$. On the other hand, the differentiability of function u at point x_0 and the equality $p = Du(x_0)$ imply

$$u(x(\delta)) = z(\delta) = u(x_0) + \langle p, x(\delta) - x_0 \rangle + \alpha_2(\delta)\delta, \quad \alpha_2(\delta) \to 0 \text{ as } \delta \downarrow 0.$$

Thus, $F(x_0, u(x_0), Du(x_0)) = \alpha_1(\delta) - \alpha_2(\delta)$. Taking the limit as $\delta \downarrow 0$, we obtain the desired equality $F(x_0, u(x_0), Du(x_0)) = 0$.

Note that in the proof of this equality we have actually assumed that the function u is differentiable only at point x_0. It is thus shown that a minimax solution satisfies equation (3.2) at every point of differentiability.

4 Upper and Lower Minimax Solutions

The notions of upper and lower minimax solutions are essentially used in investigations of generalized (viscosity and minimax) solutions. These notions can be defined in different forms. In the present paper we use the following definition.

Definition 4.1. *An upper (resp., a lower) minimax solution of equation* (3.2) *is a lower (resp., an upper) semicontinuous function* $u : \mathrm{cl}\, G \mapsto R$ *which satisfies condition* (4.1) *(resp., condition* (4.2)*)*

$$\mathrm{epi}\, u \in \mathrm{Inv}_{(3.6)}(p), \quad \forall p \in R^n, \tag{4.1}$$

$$\mathrm{hypo}\, u \in \mathrm{Inv}_{(3.6)}(p), \quad \forall p \in R^n, \tag{4.2}$$

where $\mathrm{epi}\, u := \{(x, z) \in \mathrm{cl}\, G \times R : z \geq u(x)\}$ *and* $\mathrm{hypo}\, u := \{(x, z) \in \mathrm{cl}\, G \times R : z \leq u(x)\}$ *are respectively the epigraph and the hypograph of function* u.

Thus, a lower semicontinuous function u is called an upper minimax solution, if for every $p \in R^n$, $x_0 \in \mathrm{cl}\, G$ and $z_0 \geq u(x_0)$ there exists a solution (a viable trajectory) $(x(t), z(t))$ of the differential inclusion (3.6) such that $(x(0), z(0)) = (x_0, z_0)$, $z(t) \geq u(x(t))$ for all $t \geq 0$. A similar comment is true for a lower solution.

Proposition 4.1. *For a continuous function $u : \mathrm{cl}\, G \mapsto R$ the following equivalence holds* [(4.1), (4.2)] $\Leftrightarrow$ (3.8).

Proof. The implication (3.8) $\Rightarrow$ [(4.1), (4.2)] is obvious. Let us prove the implication [(4.1), (4.2)] $\Rightarrow$ (3.8). Let u be a continuous function that satisfies conditions (4.1) and (4.2). It is clear that

$$\mathrm{epi}\, u \cap D(x_0, u(x_0), p, \tau) \neq \emptyset, \quad \mathrm{hypo}\, u \cap D(x_0, u(x_0), p, \tau) \neq \emptyset \tag{4.3}$$

for any $(x_0, p, \tau) \in R^n \times R^n \times R^+$. Here

$$D(x_0, z_0, p, \tau) := \{(x(\tau), z(\tau)) : (x(\cdot), z(\cdot)) \in S(x_0, z_0, p)\}$$

is the attainability set for the differential inclusion (3.6).

In the theory of differential inclusions it is known (see, e.g., [15]) that the attainability set $D(x_0, z_0, p, \tau)$ is connected. We therefore obtain

$$\mathrm{gr}\, u \cap D(x_0, u(x_0), p, \tau) \neq \emptyset \tag{4.4}$$

$(x_0, z_0) \in \mathrm{gr}\, u$, $(p, \tau) \in R^n \times R^+$. Using this property, we will demonstrate that the function u satisfies condition (3.8).

Let $x(\cdot) : R^+ \to R^n$ be a continuous function. Define

$$\tau^0(x(\cdot)) := \begin{cases} \infty & \text{if } x(t) \in G \text{ for all } t \geq 0, \\ \min\{t \geq 0 : x(t) \notin G\} & \text{otherwise.} \end{cases}$$

Choose an arbitrary $\delta > 0$. It follows from (4.4) that for any $(x_0, z_0) \in \mathrm{gr}\, u$ and $p \in R^n$ there exists an element $(x_\delta(\cdot), z_\delta(\cdot)) \in S(x_0, z_0, p)$ such that

$$(x_\delta(i\delta), z_\delta(i\delta)) \in \mathrm{gr}\, u, \quad i = 0, 1, 2, \ldots, \tag{4.5}$$

$$(x_\delta(t), z_\delta(t)) = (x_\delta(\tau_\delta), z_\delta(\tau_\delta)), \quad t \geq \tau_\delta := \tau^0(x_\delta(\cdot)).$$

One can choose a sequence $\{\delta_k, x_{\delta_k}(\cdot), z_{\delta_k}(\cdot)\}$ such that it converges as $k \to \infty$ to a limit $(0, x(\cdot), z(\cdot))$. The convergence $(x_{\delta_k}(\cdot), z_{\delta_k}(\cdot)) \to (x(\cdot), z(\cdot))$ means that

$$\lim_{k\to\infty} \max_{0\leq t\leq\theta} [\|x_{\delta_k}(t) - x(t)\| + |z_{\delta_k}(t) - z(t)|] = 0, \quad \forall \theta > 0.$$

Note that $(x(\cdot), z(\cdot)) \in S(x_0, z_0, p)$. It follows from (4.5) that for the limit trajectory the condition $(x(t), z(t)) \in \mathrm{gr}\, u$ is true for $t \geq 0$. □

The minimax solution can thus be defined as a function that simultaneously is an upper solution and a lower one.

Let us note that the set $\mathcal{S}(x_0, z_0, p)$ (the family of solutions of differential inclusion (3.6)) has the following property: let $(x_*(\cdot)), z_*(\cdot)) \in \mathcal{S}(x_0, \zeta_*, p)$ and $\zeta^* \geq \zeta_*$, then $(x^*(\cdot)), z^*(\cdot)) \in \mathcal{S}(x_0, \zeta^*, p)$ exists such that $x^*(t) = x_*(t)$, $z^*(t) \geq z_*(t)$. This property follows from condition (3.3). Taking into account the mentioned property, we obtain that condition (4.1) is equivalent to the following: for any point $(x_0, z_0) \in \operatorname{gr} u$ and any vector $p \in R^n$ there exists a solution $(x(\cdot), z(\cdot)) \in \mathcal{S}(x_0, z_0, p)$ such that $z(t) \geq u(x(t))$ for all $t \geq 0$. Thus, instead of the initial points $(x_0, z_0) \in \operatorname{epi} u$ it suffices to deal with the initial points $(x_0, z_0) \in \operatorname{gr} u$. A similar remark is true for a lower solution.

5 Some Facts from Nonsmooth Analysis

In Sect. 6 the upper and lower solutions will be defined in an infinitesimal form. The techniques of nonsmooth analysis are used in this definitions. Let us recall the necessary facts from this field.

Let W be a nonempty set in R^m. The value

$$\operatorname{dist}(y; W) := \min_{w \in W} \|y - w\|$$

is called the distance from point y to the set W. The set

$$T(w; W) := \{h \in R^m : \liminf_{\delta \downarrow 0} \frac{\operatorname{dist}(w + \delta h; W)}{\delta} = 0\} \tag{5.1}$$

is called the contingent cone to the set W at the point w (it is also called the Bouligand cone, the upper tangent cone, some other terms for this set are used too). Note that $T(w; W)$ is a closed cone.

Let $u : \operatorname{cl} G \to R$ be a function (recall that G is an open domain in R^n). We use the following notations for the lower and upper Dini semiderivatives of function u at a point $x_0 \in G$ in direction f

$$d^- u(x_0; f) := \sup_{\varepsilon > 0} \inf_{(\delta, h) \in \Delta} [u(x_0 + \delta h) - u(x_0)]/\delta, \tag{5.2}$$
$$d^+ u(x_0; f) := \inf_{\varepsilon > 0} \sup_{(\delta, h) \in \Delta} [u(x_0 + \delta h) - u(x_0)]/\delta,$$

where $\Delta := \{(\delta, h) : \delta \in (0, \varepsilon),\ \|f - h\| < \varepsilon,\ x_0 + \delta h \in G\}$.

Let $T((x, u(x)); \operatorname{epi} u)$ and $T((x, u(x)); \operatorname{hypo} u)$ be respectively the contingent cones to the epigraph and the hypograph of function u at point $(x, u(x))$. The following relations hold (see, e.g., [2,16])

$$d^- u(x_0; f) = \inf\{g \in R : (f, g) \in T((x, u(x)); \mathrm{epi}\, u)\}, \qquad (5.3)$$
$$d^+ u(x_0; f) = \sup\{g \in R : (f, g) \in T((x, u(x)); \mathrm{hypo}\, u)\}.$$

Let us note that the definitions of epigraph, hypograph, and contingent cone imply the following relations

$$T((x, z_*; \mathrm{epi}\, u) \subset T((x, z^*; \mathrm{epi}\, u), \quad T((x, z_*; \mathrm{hypo}\, u) \supset T((x, z^*; \mathrm{hypo}\, u), \quad \forall z_* \leq z^*. \qquad (5.4)$$

Below we also use the contingent cone to the graph of function u at point $(x, u(x)) \in G \times R$. It is denoted and defined in the following way

$$T(u)(x) := T((x, u(x)); \mathrm{gr}\, u) := \qquad (5.5)$$

$$:= \{(f, g) \in R^n \times R : \liminf_{\delta \downarrow 0} \frac{\mathrm{dist}((x + \delta f, u(x) + \delta g); \mathrm{gr}\, u)}{\delta} = 0\}.$$

Consider some relations between the contingent cone $T(u)(x)$ and the derivatives $d^- u(x; f)$, $d^+ u(x; f)$.

$$[\, -\infty < d^- u(x; f) < \infty \,] \Rightarrow [\, (f, d^- u(x; f)) \in T(u)(x) \,] \qquad (5.6)$$

$$[\, -\infty < d^+ u(x; f) < \infty \,] \Rightarrow [\, (f, d^+ u(x; f)) \in T(u)(x) \,] \qquad (5.7)$$

$$[\, (f, g) \in T(u)(x) \,] \Rightarrow [\, d^- u(x; f) \leq g \leq d^+ u(x; f) \,] \qquad (5.8)$$

$$[\, u \text{ l.s.c.}, d^- u(x; f) = -\infty \,] \Rightarrow [\, (0, -r) \in T(u)(x), \ \forall r \geq 0 \,] \qquad (5.9)$$

$$[\, u \text{ u.s.c.}, d^+ u(x; f) = \infty \,] \Rightarrow [\, (0, r) \in T(u)(x), \ \forall r \geq 0 \,] \qquad (5.10)$$

Relations (5.6) – (5.8) follow directly from definitions (5.2) and (5.5).

Let us prove relation (5.10). The condition $d^+ u(x; f) = \infty$ means that there exist sequences $\{x_k\}$ and $\{\delta_k\}$ such that

$$(x_k - x)/\delta_k \to f, \quad \delta_k \downarrow 0, \quad d_k/\delta_k \to \infty, \text{ as } k \to \infty,$$

where $d_k := u(x_k) - u(x)$. Without a loss of generality we can assume that $d_k > 0$ for all $k = 1, 2, \ldots$ and that there exists $\lim_{k \to \infty} d_k \in [0, \infty]$. The upper semicontinuity of function u implies that $\lim d_k = 0$. Since $(x_k - x)\delta_k^{-1} \to f$ and $d_k/\delta_k \to \infty$ we have $(x_k - x)d_k^{-1} \to 0$ as $k \to \infty$. Besides that, $(u(x_k) - u(x))d_k^{-1} = 1$ for $k = 1, 2, \ldots$. Therefore we have $(0, 1) \in T(u)(x)$. The set $T(u)(x)$ is a cone, consequently $(0, r) \in T(u)(x)$ for all $r \geq 0$. □

Recall the definitions of subdifferential and superdifferential. Let

$$\partial_p^* u(x) := \sup\{\langle p, f \rangle - d^- u(x; f) : f \in R^n\},$$

$$\partial_{*p}u(x) := \inf\{\langle p, f\rangle - d^+u(x; f) : f \in R^n\},$$

$$D^-u(x) := \{p \in R^n : \partial_p^* u(x) \le 0\}, \quad D^+u(x) := \{p \in R^n : \partial_{*p}u(x) \ge 0\}. \tag{5.11}$$

The set $D^-u(x)$ (resp., $D^+u(x)$) is called a subdifferential (resp., a superdifferential) of function u at point $x \in G$. The elements of $D^-u(x)$ (resp., $D^+u(x)$) are called subgradients (resp., supergradients).

The following lemma holds.

Lemma 5.1. *Let v be a lower semicontinuous function defined on an open set $Y \subset R^m$. Let H be a convex compact set in R^m, and let $y_0 \in Y$. Assume that the inequality holds*

$$d^-v(y_0; h) > 0, \ \forall\, h \in H. \tag{5.12}$$

Then for any $\varepsilon > 0$ there exist a point $y_\varepsilon \in Y$ and a subgradient $s_\varepsilon \in D^-v(y_\varepsilon)$ such that

$$\|y_0 - y_\varepsilon\| < \varepsilon, \quad \langle s_\varepsilon, h\rangle > 0, \ \ \forall\, h \in H. \tag{5.13}$$

The proof is given in [42,44]. Consider a geometric interpretation of this lemma. Let

$$H^-(y) := \{h \in R^m : d^-v(y; h) \le 0\}.$$

Note that the function $h \mapsto d^-v(y; h)$ is positively homogeneous and lower semicontinuous. Therefore $H^-(y)$ is a closed cone. Condition (5.12) means that $H \cap H^-(y_0) = \emptyset$. However the cone $H^-(y_0)$ can be nonconvex. Thus a hyperplane that separates the convex compactum H and the cone $H^-(y_0)$ does not necessary exists. According to Lemma 5.1 in any neighbourhood of point y_0 there exists a point y_ε in which the function v is subdifferentiable, i.e. $D^-v(y_\varepsilon) \neq \emptyset$. Moreover, there exists a subgradient $s_\varepsilon \in D^-v(y_\varepsilon)$ such that the hyperplane $\{h \in R^m : \langle s_\varepsilon, h\rangle = 0\}$ separates the convex compact set H from the cone $H^-(y_\varepsilon)$. Indeed, (5.11) implies that $\langle s_\varepsilon, h\rangle \le 0$, $\forall\, h \in H^-(y_\varepsilon)$. On the other hand, it follows from (5.13) that $\langle s_\varepsilon, h\rangle > 0$, $\forall\, h \in H$.

6 The Equivalence of Minimax and Viscosity Solutions

Let us recall the definition of viscosity solution introduced by M.G. Crandall and P.-L. Lions [12,13].

Definition 6.1 *A viscosity supersolution (resp., subsolution) of equation* (3.2) *is a lower (resp., an upper) semicontinuous function* $u : \mathrm{cl}\, G \to R$ *that satisfies the following condition: if the difference* $u(x) - \varphi(x)$ *takes its maximal (resp., minimal) value at a point* $x_0 \in G$ *and the function* φ *is differentiable at this point then inequality* (6.1) *(resp., inequality* (6.2)*) holds*

$$F(x_0, u(x_0), D\varphi(x_0)) \leq 0, \tag{6.1}$$

$$F(x_0, u(x_0), D\varphi(x_0)) \geq 0. \tag{6.2}$$

A viscosity solution is a function that is simultaneously a supersolution and subsolution.

Note that the inequality signs in (6.1) and (6.2) are opposite to those in the corresponding inequalities in the definition given in articles [12,13]. The point is that in the mentioned articles the functions $z \to F(x, z, p)$ are nondecreasing. In the constructions of the present paper nonincreasing functions $z \to F(x, z, p)$ are more convenient.

It is known that conditions (6.1) and (6.2) are equivalent to conditions (6.3) and (6.4) respectively

$$F(x, u(x), p) \leq 0, \quad \forall x \in G, \ \forall p \in D^- u(x), \tag{6.3}$$

$$F(x, u(x), p) \geq 0, \quad \forall x \in G, \ \forall p \in D^+ u(x). \tag{6.4}$$

In the sequel the equivalence of minimax and viscosity solutions will be proved. Let us consider several other constructions which can be used to define generalized (minimax and viscosity) solutions and let us prove their equivalence. The constructions presented below enable us to look at one and the same notion from different perspectives and to broader the possibilities to investigate generalized solutions of the first order PDE. Consider the following conditions.

$$\inf\{d^- u(x; f) - \langle p, f\rangle + F(x, u(x), p) : \|f\| \leq \rho(x)\} \leq 0, \quad \forall x \in G, \ \forall p \in R^n, \tag{6.5}$$

$$\sup\{d^+ u(x; f) - \langle p, f\rangle + F(x, u(x), p) : \|f\| \leq \rho(x)\} \geq 0, \quad \forall x \in G, \ \forall p \in R^n, \tag{6.6}$$

$$\inf\{d^- u(x; f) - \langle p, f\rangle + F(x, u(x), p) : f \in R^n\} \leq 0, \quad \forall x \in G, \ \forall p \in R^n, \tag{6.7}$$

$$\sup\{d^+ u(x; f) - \langle p, f\rangle + F(x, u(x), p) : f \in R^n\} \geq 0, \quad \forall x \in G, \ \forall p \in R^n, \tag{6.8}$$

$$T(u)(x) \cap E^0(x, z, p) \neq \varnothing, \quad \forall (x, z) \in \mathrm{gr}\, u, \ \forall p \in R^n. \tag{6.9}$$

In the last relation we use the notation

$$E^0(x, z, p) := \{(f, g) \in R^n \times R : g = \langle f, p\rangle - F(x, z, p)\}. \tag{6.10}$$

It is clear that in condition (6.9) the contingent cone $T(u)(x)$ can be substituted by its convex hull $\mathrm{co}T(u)(x)$.

The following theorem shows the equivalence of some conditions which can be used to define upper, lower, minimax, and viscosity solutions of equation (3.2).

Theorem 6.1. *For a lower semicontinuous function $u : \mathrm{cl}\, G \to R$ the following conditions are equivalent*

$$(4.1) \Leftrightarrow (6.3) \Leftrightarrow (6.5) \Leftrightarrow (6.7).$$

For an upper semicontinuous function u the following conditions are equivalent

$$(4.2) \Leftrightarrow (6.4) \Leftrightarrow (6.6) \Leftrightarrow (6.8).$$

For a continuous function u the following conditions are equivalent

$$(3.8) \Leftrightarrow [\,(4.1),\ (4.2)\,] \Leftrightarrow [\,(6.3),\ (6.4)\,] \Leftrightarrow$$

$$\Leftrightarrow [\,(6.5),\ (6.6)\,] \Leftrightarrow [\,(6.7),\ (6.8)\,] \Leftrightarrow (6.9).$$

Note that the named conditions are equivalent if they are valid for all points x from an open set G. If the conditions are considered for a fixed point x then the relations given in Theorem 6.1 can be violated.

Let us prove the theorem. Equivalence (3.8) $\Leftrightarrow$ [(4.1), (4.2)] is proved in Proposition 4.1. The implications (6.5) $\Rightarrow$ (6.7) and (6.6) $\Rightarrow$ (6.8) are obvious.

Let us prove the implication (4.1) $\Rightarrow$ (6.5). According to the theorem of Haddad [21] condition (4.1) is equivalent to the condition

$$T((x,z);\mathrm{epi}\, u) \cap E(x,z,p) \neq \emptyset \quad \forall (x,z) \in \partial \mathrm{epi}\, u, \quad \forall p \in R^n. \quad (6.11)$$

Note that $(x, u(x)) \in \partial\mathrm{epi}\, u$. Let $(f,g) \in T((x,z);\mathrm{epi}\, u) \cap E(x,z,p)$. By (5.3) $g \geq d^- u(x;f)$. According to (3.7) we have $g = \langle f,p\rangle - F(x,z,p)$, $\|f\| \leq \rho(x)$. And so, $d^- u(x;f) - \langle p,f\rangle + F(x,u(x),p) \leq 0$. □

Let us prove the implication (6.5) $\Rightarrow$ (6.11) $\Leftrightarrow$ (4.1). Let $(x,z) \in \partial\mathrm{epi}\, u$.

Case 1. Assume that $g(f) := d^- u(x;f) > -\infty$ for all $f \in R^n$. According to (6.5) and (3.3) there exists an $f \in R^n$ such that

$$g(f) \leq \langle f,p\rangle - F(x,u(x),p) \leq g_* := \langle f,p\rangle - F(x,z,p).$$

Since (5.3) holds and the cone $T((x, u(x)); \operatorname{epi} u)$ is closed we have $(f, g(f)) \in T((x, u(x)); \operatorname{epi} u)$. The lower semicontinuity of function u implies $z \geq u(x)$. It thus follows from (5.4) that

$$(f, g(f)) \in T((x, u(x)); \operatorname{epi} u) \subset T((x, z); \operatorname{epi} u).$$

Note that

$$[\ (f, g) \in T((x, z); \operatorname{epi} u),\ g_* \geq g\ ] \Rightarrow (f, g_*) \in T((x, z); \operatorname{epi} u). \quad (6.12)$$

Consequently $(f, g_*) \in T((x, z); \operatorname{epi} u) \cap E(x, z, p)$.

Case 2. Assume that there exists an $f \in R^n$ such that $d^- u(x; f) = -\infty$. According to (5.9) we have

$$(0, r) \in T((x, u(x)); \operatorname{gr} u) \subset T((x, u(x)); \operatorname{epi} u) \subset T((x, z); \operatorname{epi} u), \quad \forall r \leq 0.$$

Taking (6.12) into account, we obtain

$$(0, r) \in T((x, z); \operatorname{epi} u), \quad \forall r \in R.$$

Therefore $(0, F(x, z, p)) \in T((x, z); \operatorname{epi} u) \cap E(x, z, p)$.

Implication (6.5) $\Rightarrow$ (6.11) is thus proved. Due to the theorem of Haddad [21] condition (4.1) is equivalent to condition (6.11). We thus get the implication (6.5) $\Rightarrow$ (4.1). □

Similarly we also get the equivalence (4.2) $\Leftrightarrow$ (6.6).

Let us prove the implication (6.5) $\Rightarrow$ (6.3). Let $x \in G$ and $p \in D^- u(x)$. According to (6.5), there exists an $f \in R^n$ such that $d^- u(x; f) - \langle p, f \rangle + F(x, u(x), p) \leq 0$. By (5.11) we have $\langle p, f \rangle - d^- u(x; f) \leq 0$. Combining these inequalities, we come to (6.3). □

Let us prove the implication (6.3) $\Rightarrow$ (6.5). Let a l.s.c. function u satisfy condition (6.3) for all $x \in G$. To prove (6.5), let us assume the contrary: there exist $x_0 \in G$ and $q \in R^n$ such that

$$\min\{d^- u(x_0; f) - \langle q, f \rangle + F(x_0, u(x_0), p) : \|f\| \leq \rho(x_0)\} > 0. \quad (6.13)$$

Since the mapping $f \mapsto \partial_f^- u(x_0)$ is lower semicontinuous there exists a $\beta > 0$ such that

$$\min\{d^- u(x_0; f) - \langle q, f \rangle + F(x_0, u(x_0), p) : \|f\| \leq \rho(x_0) + \beta\} > \beta. \quad (6.14)$$

Fix a number $r > 0$ such that $B(x_0, r) := \{x \in R^n : \|x - x_0\| \leq r\} \subset G$. Let

$$z_0 := u(x_0), \quad a := \inf_x u(x), \quad b := \sup_x u(x) \quad \text{for } x \in B(x_0, r).$$

The function $z \mapsto F(x_0, z, q)$ is continuous, therefore a number $\delta > 0$ exists such that

$$|F(x_0, z, q) - F(x_0, z_0, q)| < \beta/3 \quad \forall z \in [z_0 - \delta, z_0 + \delta].$$

Define

$$\lambda := \sup \left\{ \frac{|F(x_0, z, q) - F(x_0, z_0, q)|}{|z - z_0|} : z \in [a, b],\ |z - z_0| \geq \delta \right\}.$$

The definitions of the values δ and λ imply the estimate

$$|F(x_0, z, q) - F(x_0, z_0, q)| \leq \lambda |z - z_0| + \beta/3 \quad \forall z \in [a, b]. \tag{6.15}$$

Let

$$c := F(x_0, u(x_0), q) + \lambda u(x_0) - \beta, \tag{6.16}$$

$$Y := \{y = (t, x) \in R \times R^n : |t| < 1/\lambda,\ \|x - x_0\| < r\}.$$

Consider the function

$$Y \ni y \mapsto v(y) = v(t, x) := (1 - \lambda t)u(x) - \langle q, x\rangle + ct. \tag{6.17}$$

Note that the function v is lower semicontinuous on Y, and the equality

$$d^- v(y; h) = (1 - \lambda t)d^- u(x; f) - \langle q, f\rangle + \alpha(c - \lambda u(x)), \tag{6.18}$$

holds, where $h := (\alpha, f) \in R \times R^n$. According to (5.11) we have

$$(\sigma, p) \in D^- v(t, x) \Rightarrow [\, \sigma = c - \lambda u(x),\ s := (p + q)(1 - \lambda t)^{-1} \in D^- u(x)\,]. \tag{6.19}$$

Let $y_0 := (0, x_0)$, $H := \{(1, f) \in R \times R^n : \|f\| \leq \rho(x_0) + \beta\}$. Condition (6.14) implies that the function v of the form (6.17) satisfies condition (5.12). Using Lemma 5.1 and relation (6.19), we obtain that for every $\varepsilon > 0$ there exist a pair $y_\varepsilon = (t_\varepsilon, x_\varepsilon) \in Y$ and a subgradient $s_\varepsilon \in D^- u(x_\varepsilon)$ such that

$$\|x_\varepsilon - x_0\| < \varepsilon, \quad |t_\varepsilon| < \varepsilon,$$

$$\min_f [\langle s_\varepsilon, f\rangle - \nu_\varepsilon \langle q, f\rangle + \nu_\varepsilon(c - \lambda u(x_\varepsilon))] > 0 \quad \text{for } \|f\| \leq \rho(x_0) + \beta, \tag{6.20}$$

where $\nu_\varepsilon := (1 - \lambda t_\varepsilon)^{-1}$. Note that $\nu_\varepsilon \to 1$ as $\varepsilon \downarrow 0$. The lower semicontinuity of function u, continuity of function F, and estimate (6.15) imply the inequalities

$$\lambda(u(x_\varepsilon) - u(x_0)) \geq \lambda |u(x_\varepsilon) - u(x_0)| - \gamma_\varepsilon^{(1)},$$

$$F(x_\varepsilon,u(x_\varepsilon),q)\geq F(x_0,u(x_\varepsilon),q)-\gamma_\varepsilon^{(2)}\geq$$
$$F(x_0,u(x_0),q)-\lambda(u(x_\varepsilon)-u(x_0))-\beta/3-\gamma_\varepsilon, \qquad (6.21)$$

where $\gamma_\varepsilon=\gamma_\varepsilon^{(1)}+\gamma_\varepsilon^{(2)}\to 0$ as $\varepsilon\downarrow 0$. For a sufficiently small ε, according to (6.16),(6.21), and (6.20) we obtain

$$F(x_\varepsilon,u(x_\varepsilon),\nu_\varepsilon q)-\beta/3\geq \nu_\varepsilon[F(x_\varepsilon,u(x_\varepsilon),q)+\gamma_\varepsilon-2\beta/3]\geq \nu_\varepsilon(c-\lambda u(x_\varepsilon)),$$

$$\langle s_\varepsilon,f\rangle-\nu_\varepsilon\langle q,f\rangle+F(x_\varepsilon,u(x_\varepsilon),\nu_\varepsilon q)-\beta/3>0 \quad \text{for } \|f\|\leq\rho(x_\varepsilon).$$

Condition (3.9) implies that there exists a vector $f\in R^n$ such that

$$\langle s_\varepsilon,f\rangle-\nu_\varepsilon\langle q,f\rangle+F(x_\varepsilon,u(x_\varepsilon),\nu_\varepsilon q)=F(x_\varepsilon,u(x_\varepsilon),s_\varepsilon), \quad \|f\|\leq\rho(x_\varepsilon).$$

Thus for a sufficiently small ε we have $F(x_\varepsilon,u(x_\varepsilon),s_\varepsilon)>\beta/3>0$, where $x_\varepsilon\in G$, $s_\varepsilon\in D^-u(x_\varepsilon)$. This is in contradiction with (6.3). The obtained contradiction proves the implication (6.3) $\Rightarrow$ (6.5). □

In a similar way we obtain the equivalence (6.4) $\Leftrightarrow$ (6.6).

The implication (6.7) $\Rightarrow$ (6.5) is proved by contradiction. Assume that inequality (6.13) holds. Then, as it shown in the proof of implication (6.3) $\Rightarrow$ (6.5), the inequality $F(x_\varepsilon,u(x_\varepsilon),s_\varepsilon)>0$ is valid, where $x_\varepsilon\in G$, $s_\varepsilon\in D^-u(x_\varepsilon)$. According to (5.11) the inequality $d^-u(x_\varepsilon;f)-\langle s_\varepsilon,f\rangle\geq 0$ holds for all $f\in R^n$. Therefore

$$\inf\{d^-u(x_\varepsilon;f)-\langle s_\varepsilon,f\rangle+F(x_\varepsilon,s_\varepsilon,u(x_\varepsilon)):f\in R^n\}>0$$

This contradicts with condition (6.7). □

The implication (6.8) $\Rightarrow$ (6.6) is obtained in a similar way.

The proof of the equivalence (6.9) $\Leftrightarrow$ [(6.7), (6.8)].

Choose an arbitrary $(x,z)\in \operatorname{gr} u$ and $p\in R^n$. Let us introduce the notation $L(f):=\langle p,f\rangle-F(x,z,p)$. Assume that condition (6.9) is true, i.e., a vector $f\in R^n$ exists such that $(f,L(f))\in T(u)(x)$. According to (5.8) we have the estimate $L(f)\geq d^-u(x;f)$, thus $d^-u(x;f)-\langle p,f\rangle+F(x,z,p)\leq 0$. Therefore the implication (6.9) $\Rightarrow$ (6.7) is proved. The implication (6.9) $\Rightarrow$ (6.8) is obtained similarly.

Let us now prove the implication [(6.5), (6.6)] $\Rightarrow$ (6.9). Denote

$$E^-:=\{(f,g)\in R^n\times R: g\leq L(f),\ \|f\|\leq\rho(x)\},$$

$$E^+:=\{(f,g)\in R^n\times R: g\geq L(f),\ \|f\|\leq\rho(x)\}.$$

The implications

$$(6.5)\Rightarrow [\,T(u)(x)\cap E^-\,]\neq\emptyset, \quad (6.6)\Rightarrow [\,T(u)(x)\cap E^+\,]\neq\emptyset. \quad (6.22)$$

are valid. Let us prove the first of them. First assume that $d^- u(x; f) > -\infty$ for all $f \in R^n$. According to (6.5), there exists an $f_* \in R^n$ such that $g_* := d^- u(x : f_*) \leq L(f_*) < \infty$, $\|f_*\| \leq \rho(x)$. Employing (5.6) we obtain that $(f_*, g_*) \in T(u)(x) \cap E^-$. Now consider the other case. Suppose there exists an $f \in R^n$ such that $d^- u(x; f) = -\infty$. Let $f_* = 0$ and $g_* = -|F(x, u(x), p)|$. Using (5.9) we obtain $(f_*, g_*) \in T(u)(x) \cap E^-$. The first implication in (6.22) is proved. The second implication in (6.22) can be proved similarly.

Thus, if conditions (6.5), (6.6) are true then $(f_*, g_*) \in T(u)(x)$ and there exists a pair $(f^*, g^*) \in T(u)(x)$ such that $g_* \leq L(f_*)$ and $g^* \geq L(f^*)$. Note that the function $(f, g) \mapsto g - L(f)$ is continuous, and that $T(u)(x)$ is a connected set. Therefore, $T(u)(x)$ contains an element (f, g) such that $g = L(f)$, i.e., $T(u)(x) \cap E^0(x, z, p) \neq \emptyset$. □

Theorem 6.1 is proved.

7 The Cauchy Problem for the Hamilton-Jacobi Equation with a Lipschitz Hamiltonian

In the framework of the proposed approach the existence and the uniqueness of a minimax solution for a broad class of boundary value problems and Cauchy problems can be proved. As an example let us consider an existence and a uniqueness theorem for the Cauchy problem for the Hamilton-Jacobi equation. At first we will assume that the Hamiltonian satisfies the Lipschitz condition in the phase and impulse variables. More general restrictions on the Hamiltonian are given in Sect. 10.

Indeed consider the Cauchy problem for the Hamilton-Jacobi equation

$$\partial u/\partial t + H(t, x, u, D_x u) = 0, \quad 0 \leq t \leq \theta, \ x \in R^n, \tag{7.1}$$

$$u(\theta, x) = \sigma(x), \quad x \in R^n. \tag{7.2}$$

Assume that the Hamiltonian $H(t, x, z, p)$ and the function σ satisfy the following conditions.

H1. The Hamiltonian $H(t, x, z, p)$ is continuous on $D :=]0, \theta[\times R^n \times R \times R^n$, the function $z \mapsto H(t, x, z, p)$ is nonincreasing, i.e.,

$$H(t, x, z_1, p) \geq H(t, x, z_2, p), \quad (t, x, z_i, p) \in D, \ z_1 \leq z_2. \tag{7.3}$$

H2. The Lipschitz condition in the variable p is true, namely

$$|H(t, x, z, p^{(1)}) - H(t, x, z, p^{(2)})| \leq \mu(x)\|p^{(1)} - p^{(2)}\|, \quad \forall (t, x, z, p^{(i)}) \in D \tag{7.4}$$

and the inequality

$$|H(t,x,z,0)| \leq (1+\|x\|+|z|)c, \quad \forall\,(t,x,z)\in]0,\theta[\times R^n\times R \tag{7.5}$$

is valid, where $\mu(x) := (1+\|x\|)c$, and c is a positive number.

H3. For any bounded set $B \subset R^n$ there exists a constant $\lambda(B)$ such that

$$|H(t,x,z,p)-H(t,y,z,p)| \leq \lambda(B)\|x-y\|(1+\|p\|), \tag{7.6}$$
$$\forall\,(x,y,t,z,p)\in B\times B\times]0,\theta[\times R\times R^n.$$

H4. The function $x \to \sigma(x) : R^n \to R$ is continuous.

Observe that equation (7.1) is a special case of the following equation of the form (3.2)

$$F(y,u(y),D_y u(y)) = 0, \tag{7.7}$$

where

$$y=(t,x)\in G :=]0,\theta[\times R^n,\ F(y,z,s)=p_0+H(t,x,z,p),$$
$$s=(p_0,p)\in R\times R^n. \tag{7.8}$$

According to the construction given in Sect. 3, consider the characteristic differential inclusion

$$(\dot y,\dot z)\in E(y,z,s).$$

Recall that

$$E(y,z,s)=\{(f,g):\|f\|\leq\rho(y),\ g=\langle f,s\rangle-F(y,z,s)\}. \tag{7.9}$$

The basic property of the multivalued mapping E is that

$$E(y,z,s)\cap E(y,z,l)\neq\emptyset\quad\forall\,(y,z,s,l)\in G\times R\times R^{n+1}\times R^{n+1}. \tag{7.10}$$

In the considered case $f=(h_0,h)\in R\times R^n$. According to (7.8), the equality (7.9) can be written in the form

$$E(t,x,z,s)=\{(h_0,h,g):|h_0|^2+\|h\|^2\leq\rho^2(t,x),$$
$$g=h_0p_0+\langle h,p\rangle-p_0-H(t,x,z,p)\}.$$

Due to the form of equation (7.1), it is convenient to take $h_0=1$.

Thus we obtain for equation (7.1) the characteristic differential inclusion

$$(\dot t,\dot x,\dot z)\in E(t,x,z,p), \tag{7.11}$$

where

$$E(t,x,z,p):=\{(1,h,g)\in R\times R^n\times R:\|h\|\leq r(t,x),g=\langle h,p\rangle-H(t,x,z,p)\}. \tag{7.12}$$

The differential inclusion (7.11) can be written as the following system of differential equations and inequalities

$$\dot{t} = 1, \quad \|\dot{x}\| \leq r(t,x), \quad \dot{z} = \langle \dot{x}, p\rangle - H(t,x,z,p). \tag{7.13}$$

In the equality (7.12) and in the system (7.13) we put $r(t,x) = \mu(x)$, where $\mu(x)$ is the value from the estimate (7.4). The multivalued mapping E defined in this way satisfies the required condition (7.10). Indeed, if $s = (p_0, p) \neq l = (q_0, q)$ then the intersection (7.10) contains an element $(1, h_*, g_*)$ of the form

$$h_* := \frac{[H(t,x,z,p) - H(t,x,z,q)](p-q)}{\|p-q\|^2},$$

$$g_* := \langle h_*, p\rangle - H(t,x,z,p) = \langle h_*, q\rangle - H(t,x,z,q).$$

Let a point $(t_0, x_0, z_0, p) \in D$ be fixed. The symbol $\mathcal{S}(t_0, x_0, z_0, p)$ denotes the set of absolutely continuous functions $(x(\cdot), z(\cdot)) : [t_0, \theta] \to R^n \times R$ that satisfy the condition $x(t_0) = x_0$, $z(t_0) = z_0$ and for almost all $t \in [t_0, \theta]$ satisfy system (7.13) (or, in other words, satisfy the differential inclusion (7.11)). According to the general construction, let us now define the minimax solution of equation (7.1).

Definition 7.1. *A lower semicontinuous function $(t,x) \mapsto u(t,x) : [0,\theta] \times R^n \to R$ is called an upper minimax solution of equation* (7.1) *if for any parameter $p \in R^n$ the epigraph* epi u *is weakly invariant with respect to the differential inclusion* (7.11), *i.e., for any $(t_0, x_0, z_0) \in$ epi u any $p \in R^n$ a trajectory $(x(\cdot), z(\cdot)) \in \mathcal{S}(t_0, x_0, z_0, p)$ exists such that*

$$z(t) \geq u(x(t), t) \quad \forall\, t \in [t_0, \theta]. \tag{7.14}$$

Similarly, an upper semicontinuous function $u : [0,\theta] \times R^n \to R$ is called a lower minimax solution of equation (7.1) *if for any parameter $p \in R^n$ the hypograph* hypo u *is weakly invariant with respect to the differential inclusion* (7.11). *A minimax solution of equation* (7.1) *is a continuous function $u : [0,\theta] \times R^n \to R$ whose graph is weakly invariant with respect to the differential inclusion* (7.11) *for any parameter $p \in R^n$.*

A minimax solution of the Cauchy problem (7.1), (7.2) *is a minimax solution of equation* (7.1) *that satisfies equality* (7.2). *Similarly, an upper (resp., a lower) minimax solution of the Cauchy problem* (7.1), (7.2) *is an upper (resp., a lower) minimax solution of equation* (7.1) *that satisfies the equality* (7.2).

It is clear that the minimax solutions of the Hamilton-Jacobi equation (7.1) can be defined in an infinitesimal form with the help of contingent cones, or directional derivatives, or subdifferentials and superdifferentials. Also note that all the results of Sect. 6 imply the equivalence of minimax and viscosity solutions of equation (7.1). The following assertions are true.

Theorem 7.1. *For any upper minimax solution u of problem* (7.1), (7.2) *and any lower minimax solution v of this problem the inequality $u \geq v$ holds.*

Theorem 7.2. *There exist an upper minimax solution u of problem* (7.1), (7.2) *and a lower minimax solution v of this problem such that $u \leq v$.*

Theorem 7.3. *There exists a unique minimax solution u of the Cauchy problem* (7.1), (7.2).

Theorem 7.3 follows from Theorems 7.1 and 7.2. The proof of Theorem 7.1 is given below in the present section. This proof is based on the method of Lyapunov functions.

The proof of Theorem 7.2 can be given according to the following scheme. Consider the lower envelope of upper minimax solutions of problem (7.1), (7.2). The function u defined in this way is shown to satisfy the definition of an upper minimax solution. Then, it is shown that the upper closure of the function u given by the equality

$$u^*(t,x) := \inf_{\varepsilon>0} \sup\{u(\tau,\xi) : \|\xi - x\| \leq \varepsilon,\ |\tau - t| \leq \varepsilon,\ \tau \in [0,\theta]\}, \qquad (t,x) \in \operatorname{cl} G,$$

satisfies the definition of lower minimax solution. It is clear that $v := u^* \geq u$. The approach used here originate from the constructions developed in the theory of positional differential games for proving the theorem on the alternative in the differential game of pursuit-evasion [25]. The complete proof of Theorem 7.2 according to the outlined scheme is given in [41,42,43]. In [41] we consider the Bellman-Isaacs equation that is satisfied (in the generalized sense) by the value function of the differential game. In [42,43] the mentioned construction is given in the case when the Hamiltonian $H(t,x,z,p)$ is independent of the variable z. Also note that the upper (lower) envelopes of the family of the viscosity subsolutions (supersolutions) were considered in [22], where the existence of viscosity solutions to equation (3.2) were investigated. Below, in Sect. 9, another proof of Theorem 7.2 will be given.

Proof of Theorem 7.1.

By contradiction, assume that there exists an upper solution u of problem (7.1), (7.2), a lower solution v of this problem, a point $(t_0, x_0) \in \operatorname{cl} G$, and a number $a > 0$ such that $u(x_0, t_0) < v(x_0, t_0) - a$.

Denote by $\mathcal{X}(t_0, x_0)$ the family of absolutely continuous functions $x(\cdot) : [t_0, \theta] \to R^n$ that satisfy the differential inequality $\|\dot{x}(t)\| \leq \mu(x(t))$ and the condition $x(t_0) = x_0$ (see (7.13)). Let λ be the constant in the Lipschitz condition (7.6) with

$$B := \{x(t) : t \in [t_0, \theta],\ x(\cdot) \in \mathcal{X}(t_0, x_0)\}.$$

We use the notation

$$r_\varepsilon(x, y) := \sqrt{\varepsilon^4 + \|x - y\|^2}, \quad \alpha_\varepsilon(t) := (e^{-\lambda t} - \varepsilon)/\varepsilon,$$
$$w_\varepsilon(t, x, y) := \alpha_\varepsilon(t) r_\varepsilon(x, y). \tag{7.15}$$

Consider the following function (the Lyapunov function for system (7.17))

$$L_\varepsilon(t, x, y, \xi, \eta) := w_\varepsilon(t, x, y) + \xi - \eta, \tag{7.16}$$

where $(t, x, y, \xi, \eta) \in [0, \theta] \times R^n \times R^n \times R \times R$. Choose a number $\varepsilon > 0$ such that the inequality $0 < \varepsilon^2 \alpha_\varepsilon(t) < a$ holds for all $t \in [t_0, \theta]$.

Find the derivative of L_ε with respect to the system of differential equations and differential inequalities

$$\|\dot{x}\| \leq \mu(x), \quad \dot{\xi} = \langle \dot{x}, p \rangle - H(x, t, \xi, p), \quad \|\dot{y}\| \leq \mu(y),$$
$$\dot{\eta} = \langle \dot{y}, q \rangle - H(y, t, \eta, q), \tag{7.17}$$
$$p = q = D_y w_\varepsilon(t, x, y) = -D_x w_\varepsilon(t, x, y) = -\alpha_\varepsilon(t)(x - y)/r_\varepsilon(x, y).$$

On the interval $[t_0, \theta]$ we consider the solutions of this system which satisfy the initial condition $x(t_0) = y(t_0) = x_0$, $\xi(t_0) = \xi_0 := u(x_0, t_0)$, $\eta(t_0) = \eta_0 := v(x_0, t_0)$. Note that $(x_0, y_0, \xi_0, \eta_0) \in N$, where

$$N := \{(x, y, \xi, \eta) \in B \times B \times R \times R : \xi \leq \eta\}$$

Assume that the point $(x(t), y(t), \xi(t), \eta(t))$ whose motion is described by system (7.17) does not leave the set N on a time interval $[t_0, \tau] \subset [t_0, \theta]$. Let $L_\varepsilon[t] := L_\varepsilon(t, x(t), y(t), \xi(t), \eta(t))$. Due to (7.3), (7.6) we obtain $\dot{L}_\varepsilon[t] \leq 0$ a.e. $t \in [t_0, \tau]$.

Since $L_\varepsilon[t_0] = \varepsilon^2 a_\varepsilon(t_0) + \xi_0 - \eta_0 < \varepsilon^2 a_\varepsilon(t_0) - a < 0$ and $\dot{L}_\varepsilon[t] \leq 0$ a.e. $t \in [t_0, \tau]$ then the motions of system (7.17) do not leave the set N for all $t \in [t_0, \theta]$.

The definition of an upper minimax solution u and a lower minimax solution v implies that there exists a trajectory $(x_\varepsilon(\cdot), y_\varepsilon(\cdot), \xi_\varepsilon(\cdot), \eta_\varepsilon(\cdot))$ of system (7.17) such that

$$\xi_\varepsilon(\theta) \geq u(x_\varepsilon(\theta), \theta) = \sigma(x_\varepsilon(\theta)), \quad \eta_\varepsilon(\theta) \leq v(y_\varepsilon(\theta), \theta) = \sigma(y_\varepsilon(\theta)). \tag{7.18}$$

The inequality $\dot{L}_\varepsilon[t] \le 0$ is valid for any solution of system (7.17). In particular, it holds for a solution which satisfies inequalities (7.18). Thus, we obtain

$$L_\varepsilon[t_0] = \varepsilon^2 \alpha_\varepsilon(t_0) + u(x_0, t_0) - v(x_0, t_0) \ge L_\varepsilon[\theta] \ge$$

$$\ge \alpha_\varepsilon(\theta) r_\varepsilon(x_\varepsilon(\theta), y_\varepsilon(\theta)) + \sigma(x_\varepsilon(\theta)) - \sigma(y_\varepsilon(\theta)) \ge \sigma(x_\varepsilon(\theta)) - \sigma(y_\varepsilon(\theta)).$$

Note that

$$\varepsilon^2 \alpha_\varepsilon(\theta) \to 0, \quad \alpha_\varepsilon(\theta) \to \infty, \quad \text{as } \varepsilon \to 0.$$

It is also clear that

$$|\sigma(x_\varepsilon(\theta)) - \sigma(y_\varepsilon(\theta))| \le \max_{(x,y) \in B \times B} |\sigma(x) - \sigma(y)| := K < \infty.$$

Thus the inequality

$$\alpha_\varepsilon(\theta) r_\varepsilon(x_\varepsilon(\theta), y_\varepsilon(\theta)) \le \varepsilon^2 \alpha_\varepsilon(t_0) + u(x_0, t_0) - v(x_0, t_0) + K$$

gives

$$\|x_\varepsilon(\theta) - y_\varepsilon(\theta)\| \to 0 \quad \text{as } \varepsilon \to 0.$$

Further on, passing to the limit as $\varepsilon \to 0$ in the inequality

$$\varepsilon^2 \alpha_\varepsilon(t_0) + u(x_0, t_0) - v(x_0, t_0) \ge \sigma(x_\varepsilon(\theta)) - \sigma(y_\varepsilon(\theta)),$$

we obtain the inequality $u(x_0, t_0) \ge v(x_0, t_0)$. The assumption $u(x_0, t_0) < v(x_0, t_0)$ thus yields a contradiction. □

Remark. The function w_ε of the form (7.15) satisfies condition A4 (cf. Sect. 10) introduced in [11]. One can see from the proof of the above that condition H3 may be replaced by A4.

8 Minimax Solutions and the Value Function of the Differential Game

As indicated in the Introduction, the origin of the proposed approach and the methods of investigating minimax solutions are connected with the theory of positional differential games (see, e.g., [23–26]). Let us recall some notions of this theory.

Consider a differential game in which the motion of a controlled system is governed by the ordinary differential equation

$$\dot{x}(t) = h(t, x(t), p(t), q(t)), \quad t_0 \le t \le \theta, \quad x(t_0) = x_0 \in R^n. \qquad (8.1)$$

Here $p(t) \in P$ and $q(t) \in Q$ are the controls of the first and the second players respectively, the sets P and Q are compact. The controls are chosen on the basis of the feedback principle depending on the on line position $(t, x(t))$. The first player tries to guarantee the minimal value of the pay-off functional

$$\gamma(x(\cdot), p(\cdot), q(\cdot)) := \sigma(x(\theta)) - \int_{t_0}^{\theta} g(t, x(t), p(t), q(t))dt. \qquad (8.2)$$

On the contrary, the second player tries to guarantee its maximal value.

We assume that functions h and g are continuous on $[0, \theta] \times R^n \times P \times Q$ and satisfy a Lipschitz condition in x, and that the function σ is continuous. It is also assumed that

$$\begin{aligned} &\min_{p \in P} \max_{q \in Q} [\langle s, h(t, x, p, q)\rangle - g(t, x, p, q)] = \\ &\max_{q \in Q} \min_{p \in P} [\langle s, h(t, x, p, q)\rangle - g(t, x, p, q)] := H(t, x, s), \qquad (8.3) \end{aligned}$$

$$(t, x, s) \in [0, \theta] \times R^n \times R^n.$$

It is known that in the considered case the optimal guaranteed results of the first and second players coincide and that their common value is called the value of the positional differential game. The value of the game depends on the initial position. Thus the value function $(t_0, x_0) \mapsto u(t_0, x_0) : [0, \theta] \times R^n \to R$ can be introduced. It is also known that the value function at the points of differentiability satisfies the Hamilton-Jacobi-Bellman-Isaacs equation

$$\partial u / \partial t + H(t, x, D_x u) = 0. \qquad (8.4)$$

In the theory of positional differential games the notions of u-stable and v-stable functions are introduced. These notions are of great importance for this theory. For the differential game (8.1), (8.2) a function $[0, \theta] \times R^n \ni (t, x) \mapsto u(t, x) \in R$ is called u-stable if the following condition holds: for any $(t_0, x_0) \in [0, \theta] \times R^n$, $\tau \in [t_0, \theta]$ and $q \in Q$ there exists a solution of the differential inclusion

$$(\dot{x}(t), \dot{z}(t)) \in \text{co}\{(h(t, x(t), p, q), g(t, x(t), p, q)) : p \in P\}, \qquad (8.5)$$

$$t_0 \le t \le \theta, \quad x(t_0) = x_0, \quad z(t_0) = 0$$

such that

$$u(\tau, x(\tau)) \le u(t_0, x_0) + z(\tau).$$

It is not difficult to demonstrate that for a lower semicontinuous function u the u-stability property means that for an arbitrary $q \in Q$ the epigraph epi u is weakly invariant with respect to the differential inclusion (8.5) (it satisfies the viability property).

Similarly, the v-stability property means that for an arbitrary $p \in Q$ the hypograph is weakly invariant with respect to the differential inclusion

$$(\dot{x}(t), \dot{z}(t)) \in \text{co}\{(h(t, x(t), p, q), g(t, x(t), p, q)) : q \in Q\}.$$

It is known (see, e.g., [25,26,41]) that there exists a unique continuous function that satisfies the boundary condition $u(\theta, x) = \sigma(x)$ and is both u-stable and v-stable. This is the value function of the differential game of the form (8.1), (8.2).

It can be shown that the definitions of u-stable (v-stable) function and of the upper (lower) solution of equation (8.4) coincide. Thus, the value function of the differential game (8.1), (8.2) coincides with the minimax and/or viscosity solution of the Cauchy problem for equation (8.4) (the terminal condition in this problem is $u(\theta, x) = \sigma(x)$). Various methods can be used in order to prove this fact (see, e.g., [42,46]). In particular, the equivalence of the u-stable (v-stable) function and the upper (lower) solution can be obtained through the constructions of Sect. 6.

Let us note that now there are quite many publications devoted to the connections of the theory of viscosity solutions with the theory of optimal control and differential games. In particular, in a number of publications (see, e.g., [3,4,17,33,46]) it is shown that for different types of control problems and differential games the value function coincides with the viscosity solution of the corresponding Bellman-Isaacs equation. On the other hand, it is known that first order PDEs of sufficiently general form can be considered as a Bellman-Isaacs equation for a specially constructed differential game. These constructions were described and used to prove the existence of viscosity solutions of boundary value problems and Cauchy problems (see, e.g., [14,17,19]). The characteristic differential inclusion (3.6) incorporates the elements of these constructions.

The proof of the equivalence of minimax and viscosity solutions given in Sect. 6 is based on the lemma on separability (Lemma 5.1). Here one can observe the duality which is typical for the theory of generalized solutions of first-order PDE. We should also mention that replacing the third equation of system (2.3) by an equation of the form (2.4) resembles the problem of selecting a "genuine" Hamiltonian for an optimal control problem, cf. [8,38,48].

The approach used in this article originally was developed in the framework of the theory of positional differential games and was used for investigating the generalized solutions of the Bellman-Isaacs equations (see, e.g., [40,45]). Its further development, presented in the book [42], has shown that that this approach can be used for studying of a broad scope of problems not necessarily connected with differential games. Here the constructive basis remains the same: the designing of solutions to the first-order PDEs along the trajectories of ordinary differential inclusions which are considered as generalized characteristics.

In the theory of differential games much attention is given to program (open loop) schemes (auxiliary problems of open loop control), that are used to find the value of the differential game (see, e.g., [23–26]). These results and general numerical methods of the theory of positional differential games can be used to construct generalized solutions of the Cauchy problem for the Hamilton-Jacobi equation. In particular, in the next section we consider constructions that originate from the method of program iterations developed in the theory of positional differential games (see, e.g., [5]). Concluding this section, let us state a result obtained in the article [47], that connects Pontryagin's maximum principle, the classical method of characteristics, and the theory of generalized solutions of first-order PDEs.

Consider the Cauchy problem for the Hamilton-Jacobi equation

$$\partial u/\partial t + H(t,x,D_x u) = 0, \quad 0 \le t \le \theta,\ x \in R^n, \tag{8.6}$$

$$u(\theta,x) = \sigma(x), \quad x \in R^n. \tag{8.7}$$

By

$$(x(\cdot,y), p(\cdot,y), z(\cdot,y)) : [0,\theta] \to R^n \times R^n \times R$$

we denote a solution of the characteristic system

$$\begin{aligned} \dot{x} &= D_p H(t,x,p), \\ \dot{p} &= -D_x H(t,x,p), \\ \dot{z} &= \langle p, D_p H(t,x,p)\rangle - H(t,x,p), \end{aligned}$$

which satisfies the condition

$$x(\theta,y) = y, \quad p(\theta,y) = D\sigma(y), \quad z(\theta,y) = \sigma(y). \tag{8.8}$$

Define the set

$$Y(t,\xi) = \{y \in R^n : x(t,y) = \xi\}, \quad (t,\xi) \in \operatorname{cl} G. \tag{8.9}$$

Firstly, recall the following well-known result (see, e.g., [9,36]): if problem (8.6), (8.7) admits a classical solution u then the set $Y(t,\xi)$ is a singleton, i.e. $Y(t,\xi) = \{y(t,\xi)\}$, and for this solution the formula

$$u(t,\xi) = z(t, y(t,\xi)), \quad (t,\xi) \in \operatorname{cl} G$$

is true.

As shown in [47], the classical characteristics can be used to construct a generalized solution under the following assumptions. Let the function $p \mapsto H(t,x,p)$ be convex for all $(t,x) \in \mathrm{cl}\, G$, and let the function $H(t,x,p)$ be continuously differentiable in t, x, p, and derivatives $D^2_{tx}H$, $D^2_{tp}H$ exist. Then a generalized (minimax and/or viscosity) solution of problem (8.6), (8.7) can be expressed through the characteristics due to the equality

$$u(t,\xi) = \max\{z(t,y) : y \in Y(t,\xi)\}, \quad \forall\,(t,\xi) \in \mathrm{cl}\, G.$$

9 The Operator Equations for Minimax Solutions

The present section gives constructions and results whose origination is connected with the method of program iterations developed in the theory of differential games with the aim of finding the value function (see, e.g., [5,6]).

Let us introduce some notations. By LB we denote the set of locally bounded functions $\mathrm{cl}\, G \ni (t,x) \mapsto u(t,x) \in R$ that satisfy condition (7.2). The set of lower (resp., upper) semicontinuous functions $\mathrm{cl}\, G \ni (t,x) \mapsto u(t,x) \in R$ that satisfy condition (7.2) is denoted by LSC (resp., USC).

Let $u \in LB$, $(t_0,x_0) \in \mathrm{cl}\, G$, $(\zeta,\tau,p) \in R \times [t_0.\theta] \times R^n$. Define

$$\psi_*(u)(t_0,x_0,\zeta,\tau,p) := \inf\{u(\tau,x(\tau)) - z(\tau) : (x(\cdot),z(\cdot)) \in \mathcal{S}(t_0,x_0,\zeta,p)\}, \tag{9.1}$$

$$\psi^*(u)(t_0,x_0,\zeta,\tau,p) := \sup\{u(\tau,x(\tau)) - z(\tau) : (x(\cdot),z(\cdot)) \in \mathcal{S}(t_0,x_0,\zeta,p)\}.$$

It can be shown that functions $\zeta \mapsto \psi_*(u)(t_0,x_0,\zeta,\tau,p)$ and $\zeta \mapsto \psi^*(u)(t_0,x_0,\zeta,\tau,p)$ are continuous and monotone in the following sense

$$\psi_*(u)(t_0,x_0,\zeta+r,\tau,p) + r \le \psi_*(u)(t_0,x_0,\zeta,\tau,p), \tag{9.2}$$

$$\psi^*(u)(t_0,x_0,\zeta+r,\tau,p) + r \le \psi^*(u)(t_0,x_0,\zeta,\tau,p), \quad \forall\, r > 0.$$

Thus the set $\{\zeta \in R : \psi_*(u)(t_0,x_0,\zeta,\tau,p) = 0\}$ consists of a single element. Let

$$\{v_*(u)(t_0,x_0,\tau,p)\} := \{\zeta \in R : \psi_*(u)(t_0,x_0,\zeta,\tau,p) = 0\},$$

$$\Gamma_*(u)(t_0,x_0) := \sup_{t_0 \le \tau \le \theta}\ \sup_{p \in R^n} v_*(u)(t_0,x_0,\tau,p). \tag{9.3}$$

and similarly

$$\{v^*(u)(t_0, x_0, \tau, p)\} := \{\zeta \in R : \psi^*(u)(t_0, x_0, \zeta, \tau, p) = 0\},$$

$$\Gamma^*(u)(t_0, x_0) := \inf_{t_0 \le \tau \le \theta} \inf_{p \in R^n} v^*(u)(t_0, x_0, \tau, p). \tag{9.4}$$

Note that

$$\Gamma_*(u)(t_0, x_0) \ge u(t_0, x_0), \quad \Gamma^*(u)(t_0, x_0) \le u(t_0, x_0), \\ \forall u \in LB, \forall (t_0, x_0) \in \mathrm{cl}\, G. \tag{9.5}$$

The following estimate is true

$$\psi_*(u)(t_0, x_0, \zeta, \tau, p) \le u(\tau, x(\tau)) - z(\tau), \\ \forall (x(\cdot), z(\cdot)) \in \mathcal{S}(t_0, x_0, \zeta, p) \cap \mathcal{S}(t_0, x_0, \zeta, 0).$$

By (9.2) we have

$$v_*(u)(t_0, x_0, \tau, p) \le \psi^*(t_0, x_0, 0, \tau, 0) \le |\psi^*(t_0, x_0, 0, \tau, 0)|,$$

$$\Gamma_*(u)(t_0, x_0) \le \sup_{t_0 \le \tau \le \theta} \sup\{|u(\tau, x(\tau)) - z(\tau)| : (x(\cdot), z(\cdot)) \in \mathcal{S}(t_0, x_0, 0, 0).$$

Similarly, we come to a lower estimate for the operator Γ^*. Taking (9.5) into account, we see that

$$\Gamma_*(LB) \subset LB, \quad \Gamma^*(LB) \subset LB.$$

Further, it can be shown that

$$\Gamma_*(LB \cap LSC) \subset LB \cap LSC, \quad \Gamma^*(LB \cap USC) \subset LB \cap USC. \tag{9.6}$$

It should be noted that the upper and lower minimax solutions of Cauchy problem (7.1), (7.2) are locally bounded. The weak invariance of epigraph (resp., hypograph) of lower (resp., upper) semicontinuous function u is equivalent to inequality (9.7) (resp., (9.8))

$$\Gamma_*(u)(t_0, x_0) \le u(t_0, x_0), \quad \forall (t_0, x_0) \in \mathrm{cl}\, G, \tag{9.7}$$

$$\Gamma^*(u)(t_0, x_0) \ge u(t_0, x_0), \quad \forall (t_0, x_0) \in \mathrm{cl}\, G. \tag{9.8}$$

Using (9.5), we obtain

Proposition 9.1. *A function $u \in LB \cap LSC$ is an upper minimax solution of the Cauchy problem* (7.1), (7.2) *if and only if this function is a fixed point of the operator Γ_*. Similarly, a function $u \in LB \cap USC$ is a lower minimax solution of the Cauchy problem* (7.1), (7.2) *if and only if this function is a fixed point of the operator Γ^*. A continuous function*

u is a minimax solution of the Cauchy problem (7.1), (7.2) *if and only if this function is a solution of the system of operator equations*

$$\Gamma_*(u) = u, \quad \Gamma^*(u) = u.$$

The operators introduced here can be used to prove Theorem 7.2. Let

$$\psi_-(t_0, x_0, \zeta) := \inf\{\sigma(x(\theta)) - z(\theta) : (x(\cdot), z(\cdot)) \in S(t_0, x_0, \zeta, 0)\}, \quad (9.9)$$

$$\psi_+(t_0, x_0, \zeta) := \sup\{\sigma(x(\theta)) - z(\theta) : (x(\cdot), z(\cdot)) \in S(t_0, x_0, \zeta, 0)\}, \quad (9.10)$$

$$\{u_-(t_0, x_0)\} := \{\zeta \in R : \psi_-(t_0, x_0, \zeta) = 0\}, \quad (9.11)$$

$$\{u_+(t_0, x_0)\} := \{\zeta \in R : \psi_+(t_0, x_0, \zeta) = 0\}. \quad (9.12)$$

It can be shown that u_- is a lower minimax solution, and u_+ is an upper minimax solution. The functions u_- and u_+ are continuous.

Consider the sequence

$$u_1 := \Gamma^*(u_+), \quad u_{k+1} := \Gamma^*(u_k), \quad k = 1, 2, \ldots \quad (9.13)$$

The following result holds

Proposition 9.2. *The sequence $\{u_k\}_1^\infty$ of* (9.13) *has a pointwise convergence to a lower minimax solution u^* of problem* (7.1), (7.2) *on* $\operatorname{cl} G$. *The function u_* given by*

$$u_*(t, x) := \sup_{\varepsilon > 0} \inf\{u_k(\tau, \xi) : \|\xi - x\| \le \varepsilon, \ |\tau - t| \le \varepsilon, \ \tau \in [0, \theta], \ k = 1, 2, \ldots\}, \quad (9.14)$$

is an upper minimax solution of this problem.

In the proof of Proposition 9.2 we use Propositions 9.3 and 9.4 which are given below. Let us introduce the following notation. Assume

$$E(u) := \{(t, x, z) : (t, x) \in \operatorname{cl} G, \ z > u(t, x)\}.$$

By $\mathcal{U}$ we denote the set of functions $u \in LB \cap USC$ that satisfy the condition

$$\psi_*(u)(t_0, x_0, \zeta, \tau, p) < 0, \quad \forall (t_0, x_0, \zeta) \in E(u), \ \forall (\tau, p) \in [t_0, \theta] \times R^n. \quad (9.15)$$

Proposition 9.3. *Let $\mathcal{U}_*$ be a subset of $\mathcal{U}$. The function v given by*

$$v(t, x) := \sup_{\varepsilon > 0} \inf\{u(\tau, \xi) : \|\xi - x\| \le \varepsilon, \ |\tau - t| \le \varepsilon, \ \tau \in [0, \theta], \ u \in \mathcal{U}_*\},$$

$$(t, x) \in \operatorname{cl} G \quad (9.16)$$

is an upper minimax solution of problem (7.1), (7.2).

We omit the proof of Proposition 9.3. Let us only note that it is based on the upper semicontinuity with respect to inclusion of the mapping $(t_0, x_0, z_0) \mapsto \mathcal{S}(t_0, x_0, z_0, p)$.

Remark. Let $u \in \mathcal{U}$. Consider the function

$$\underline{u}(t,x) := \sup_{\varepsilon>0} \inf\{u(\tau,\xi) : \|\xi - x\| \le \varepsilon,\ |\tau - t| \le \varepsilon,\ \tau \in [0,\theta]\},\ (t,x) \in \operatorname{cl} G.$$

The function $\underline{u}$ is called the lower closure of the function u. It follows from Proposition 9.3 that $\underline{u}$ is an upper minimax solution of problem (7.1), (7.2).

In the proof of Proposition 9.2 we essentially use

Proposition 9.4.

$$\Gamma^*(\mathcal{U}) \subset \mathcal{U}. \tag{9.17}$$

Proof of Proposition 9.4. Select an arbitrary $u \in \mathcal{U}$, $r \in [0,\theta]$, $q \in R^n$. Consider a function $v_{r,q} : [0,\theta] \times R^n \mapsto R$ given by

$$v_{r,q}(t_0, x_0) := \begin{cases} v^*(u)(t_0, x_0, r, q), & (t_0, x_0) \in [0, r] \times R^n, \\ u(t_0, x_0), & (t_0, x_0) \in]r, \theta] \times R^n. \end{cases} \tag{9.18}$$

Recall that the value $v^*(u)(t_0, x_0, r, q)$ is defined by (9.4). Note that according to (9.4) we have

$$\Gamma^*(u)(t,x) = \inf\{v_{r,q}(t,x) : r \in [0,\theta],\ q \in R^n\}. \tag{9.19}$$

Let us demonstrate that $v_{r,q} \in \mathcal{U}$. If $r < \theta$ then $v_{r,q}(\theta, x_0) = u(\theta, x_0) = \sigma(x_0)$. If $r = \theta$ then $v_{r,q}(\theta, x_0) = v^*(u)(\theta, x_0, \theta, q) = \sigma(x_0)$. We therefore observe that $v_{r,q}$ satisfies the equality (7.2). Since $u \in LB \cap USC$, it follows from (9.1), (9.2) and (9.18) that $v_{r,q} \in LB \cap USC$.

It only remains to indicate that function $v_{r,q}$ satisfies condition (9.15), i.e., that for any $(t_0, x_0, \zeta) \in E(v_{r,q})$, $\tau \in [t_0, \theta]$ and $p \in R^n$ the inequality

$$\inf\{v_{r,q}(\tau, x(\tau)) - z(\tau) : (x(\cdot), z(\cdot)) \in \mathcal{S}(t_0, x_0, \zeta, p)\} < 0 \tag{9.20}$$

is true. Three cases are possible here. These are: (i) $t_0 \ge r$, (ii) $\tau \le r$, (iii) $t_0 < r < \tau$.

Case (i). Suppose $t_0 > r$. The function u satisfies (9.15). Conditions (9.15) and (9.18) imply inequality (9.20). We now observe that for $t_0 = r$, according to (9.1) and (9.18) we have $\psi^*(r, x_0, \zeta, r, q) = u(r, x_0) - \zeta$ and $v_{r,q}(r, x_0) = u(r, x_0)$. Therefore (9.20) holds for both $t_0 > r$ and $t_0 = r$.

Case (ii). Suppose

$$\mathcal{S}_\tau := \{(x(\cdot), z(\cdot)) \in \mathcal{S}(t_0, x_0, \zeta, q) : x(t) = x_*(t),\ z(t) = z_*(t),\ t \in [t_0, \tau]\},$$

$$\mathcal{S}^\tau := \mathcal{S}(\tau, x_*(\tau), z_*(\tau), q),$$

where $(x_*(\cdot), z_*(\cdot)) \in \mathcal{S}(t_0, x_0, \zeta, p) \cap \mathcal{S}(x_0, t_0, \zeta, q)$. Since $\zeta > v_{r,q}(t_0, x_0)$ then (9.2) implies that $\psi^*(t_0, x_0, \zeta, r, q) < 0$. According to (9.1), we have

$$0 > \psi^*(t_0, x_0, \zeta, r, q) \geq \sup\{u(r, x(r)) - z(r) : (x(\cdot), z(\cdot)) \in \mathcal{S}_\tau\} =$$

$$\sup\{u(r, x(r)) - z(r) : (x(\cdot), z(\cdot)) \in \mathcal{S}^\tau\} = \psi^*(\tau, x_*(\tau), z_*(\tau), r, q).$$

Therefore we obtain from (9.2) and (9.18) that $v_{r,q}(\tau, x_*(\tau)) < z_*(\tau)$. The inequality (9.20) is thus proved for case (ii). Note that for $\tau = r$ we have

$$0 > \psi^*(r, x_*(r), z_*(r), r, q) = u(r, x_*(r)) - z_*(r).$$

Case (iii). Suppose

$$\mathcal{S}_* := \{(x(\cdot), z(\cdot)) \in \mathcal{S}(t_0, x_0, \zeta, p) : x(t) = x_*(t),\ z(t) = z_*(t),\ t \in [t_0, r]\}.$$

The function u satisfies (9.15). Condition (9.15) and inequality $z_*(r) > u(r, x_*(r))$ imply

$$\inf\{v_{r,q}(\tau, x(\tau)) - z(\tau) : (x(\cdot), z(\cdot)) \in \mathcal{S}(t_0, x_0, \zeta, p)\} =$$
$$= \inf\{u(\tau, x(\tau)) - z(\tau) : (x(\cdot), z(\cdot)) \in \mathcal{S}(t_0, x_0, \zeta, p)\} \leq$$
$$\leq \inf\{u(\tau, x(\tau)) - z(\tau) : (x(\cdot), z(\cdot)) \in \mathcal{S}_*\} =$$
$$= \inf\{u(\tau, x(\tau)) - z(\tau) : (x(\cdot), z(\cdot)) \in \mathcal{S}(r, x_*(r), z_*(r), p)\} < 0.$$

We have thus demonstrated that the function $v_{r,q}$ satisfies condition (9.15).

We obtain that $v_{r,q} \in \mathcal{U}$ for all $r \in [0, \theta]$ and $q \in R^n$. According to (9.6) and (9.19)

$$u_0 := \inf\{v_{r,q} : r \in [0, \theta],\ q \in R^n\} \in LB \cap USC.$$

It is not difficult to observe that the lower envelope u_0 satisfies (9.15) as long as the functions $v_{r,q}$ do. Hence, $u_0 := \Gamma^*(u) \in \mathcal{U}$. □

Let us now give an outline of the proof of Proposition 9.2. It should be noted that $u_+ \in \mathcal{U}$ (the proof of this fact coincides on the whole with the proof of Proposition 9.4). It can be demonstrated that $u \geq u_-$ for any function $u \in \mathcal{U}$. Thus, according to (9.17) and (9.5) a sequence of the form (9.13) satisfies the following relations

$$u_k \in \mathcal{U} \subset LB \cap USC, \quad u_k \geq u_-, \quad u_{k+1} \leq u_k.$$

Therefore there exists a pointwise limit

$$u^*(t,x) := \lim_{k\to\infty} u_k(t,x) \quad (t,x) \in \operatorname{cl} G. \tag{9.21}$$

The limit function u^* is upper semicontinuous.

It follows from the above that $\lim_{k\to\infty} \Gamma^*(u_k) = u^*$. Let us now demonstrate that

$$\lim_{k\to\infty} \Gamma^*(u_k) = \Gamma^*(u^*). \tag{9.22}$$

Assume $\zeta_k := u_k(t_0, x_0)$, $\zeta^* := u^*(t_0, x_0)$. Select an arbitrary instant $\tau \in [t_0, \theta]$ and a vector $p \in R^n$. According to the definition of Γ^*, there exists a pair $(x_k(\cdot), z_k(\cdot)) \in \mathcal{S}(t_0, x_0, \zeta_{k+1}, p)$ such that $(\tau, x_k(\tau), z_k(\tau)) \in \operatorname{hypo} u_k$. The sequence $\{(x_k(\cdot), z_k(\cdot))\}_1^\infty$ contains a subsequence that converges to a limit $(x^*(\cdot), z^*(\cdot)) \in \mathcal{S}(t_0, x_0, \zeta^*, p)$. Note that the closed sets $\operatorname{hypo} u_k$ satisfy the relations

$$\operatorname{hypo} u_{k+1} \subset \operatorname{hypo} u_k, \quad \operatorname{hypo} u^* = \bigcap_1^\infty \operatorname{hypo} u_k.$$

Hence, $(\tau, x^*(\tau), z^*(\tau)) \in \operatorname{hypo} u^*$. We thus obtain

$$\psi^*(u^*)(t_0, x_0, \zeta^*, \tau, p) \leq 0, \quad \forall \tau \in [t_0, \theta], \ \forall p \in R^n.$$

Therefore, $\Gamma^*(u^*) \geq \zeta^* = u^*(t_0, x_0)$. Taking (9.5) into account, we observe that $\Gamma^*(u^*) = u^*$ and besides, that $u^* \in LB \cap USC$. The Proposition 9.1 thus implies that u^* is a lower minimax solution of problem (7.1), (7.2).
□

From (9.14), (9.21) it follows that $u_* \leq u^*$. In other words, Proposition 9.2 implies Theorem 7.2. Moreover, by using Theorem 7.1 we obtain $u_* = u^*$. Therefore a sequence of the form given by (9.13) converges from the above to a minimax solution of problem (7.1), (7.2). Similarly, it may be proved that a sequence given by relations

$$u^{(1)} := \Gamma_*(u_-), \quad u^{(k+1)} := \Gamma_*(u^{(k)}), \quad k = 1, 2, \ldots$$

will converge to a minimax solution from below.

10 The Cauchy Problem for the Hamilton-Jacobi Equation With a Continuous Hamiltonian

To simplify the presentation and to eliminate the inessential details from the construction of minimax solutions, we had assumed in the above that the functions $F(x, z, p)$ and $H(t, x, z, p)$ in (3.2) and (7.1) satisfy the Lipschitz condition in x and p. In the sequel we shall indicate some weaker assumptions A0 – A4. These are taken from [11] and do not imply that the Hamiltonian is Lipschitz. Under these assumptions we define the minimax solutions of the Hamilton-Jacobi equation and formulate theorems for the uniqueness and the existence theorems of minimax solutions to the Cauchy problem.

Indeed, consider the Cauchy problem (7.1), (7.2). We assume that the Hamiltonian $H(t, x, z, p)$ and the function σ satisfy the following conditions.

A0. The Hamiltonian $H(t, x, z, p)$ is continuous on $]0, \theta[\times R^n \times R \times R^n$.

A1. The function $z \to H(t, x, z, p)$ is nonincreasing.

A2. There exist a continuously differentiable function $\mu : R^n \to R^+$ and a continuous function $\varphi : R^+ \times R^+ \to R^+$ such that

$$H(t, x, z, p) - H(t, x, z, p + \lambda D\mu(x)) \leq \varphi(\lambda, \|p\|),$$

$$\forall\, (t, x, z, p) \in]0, \theta[\times R^n \times R \times R^n, \;\; \lambda \in [0, 1].$$

It is assumed that $\sup_{x \in R^n} \|D\mu(x)\| < \infty$, that

$$\lim_{\|x\| \to \infty} \mu(x) = \infty;$$

and that the function φ is increasing monotononously in each argument.

A3. There exist a continuously differentiable function $\nu : R^n \to R^+$ and for any number $h > 0$ a constant C_h such that

$$H(t, x, z, p) - H(t, x, z, p + \lambda D\nu(x)) \leq C_h \qquad \forall\, (t, x, z, p) \in]0, \theta[\times R^n \times R \times B_h, \;\; \lambda \in [0, h],$$

$$\nu(x) \geq \|x\| \quad \text{for } \|x\| \text{ sufficiently large}$$

where $B_h := \{p \in R^n : \|p\| \leq h\}$.

A4. For any number $\varepsilon > 0$ there exists a continuous function w_ε : $\mathrm{cl}\,\Gamma \to R^+$ differentiable in $\Gamma := \{(t, x, y) \in]0, \theta[\times R^n \times R^n : \|x - y\| < 1\}$ such that

$$\partial w_\varepsilon / \partial t + H(t, y, z, D_y w(t, x, y)) - H(t, x, z, -D_x w(t, x, y)) \leq 0$$

$$\forall\,(t,x,y)\in\Gamma,\ \ z\in R.$$

It is assumed that

$$\|D_x w_\varepsilon(t,x,y)\|\le \Lambda_\varepsilon,\quad \|D_y w_\varepsilon(t,x,y)\|\le \Lambda_\varepsilon\ \ \forall\,(t,x,y)\in\Gamma,$$

$$w_\varepsilon(t,x,x)\le\varepsilon\ \ \forall\,(t,x)\in R^n\times[0,\theta],$$

$$w_\varepsilon(t,x,y)\ge 1/\varepsilon,\ \ \text{if } \|x-y\|=1,$$

$$\lim_{\varepsilon\downarrow 0}\inf\{w_\varepsilon(\theta,x,y):\|x-y\|\ge r\}=\infty\ \ \text{for all } r\in]0,1].$$

A5. The function $\sigma : R^n \to R$ is uniformly continuous.

Let us define a minimax solution for the considered case. Let $p\in R^n$, ε and r be some positive numbers. Consider the system of differential inequalities

$$\|\dot x(t)\|\le r,\quad |\dot z(t)-\langle p,\dot x(t)\rangle+H(t,x(t),z(t),p)|\le\varepsilon. \tag{10.1}$$

It should be noted that the solutions of system (10.1) are not necessarily extendable up to the instant $t=\theta$. Taking this fact into account, we consider the solutions of the system only until they leave some bounded set. The respective formulation is as follows.

Let N be an open bounded set in $R^n\times R$. Let $(x_0,z_0)\in N$, $t_0\in[0,\theta]$. Denote by $\mathcal{S}_N(t_0,x_0,z_0,p,r,\varepsilon)$ the set of absolutely continuous functions $(x(\cdot),z(\cdot)) : [t_0,\theta]\mapsto R^n\times R$ such that $(x(t_0),z(t_0))=(x_0,z_0)$ and that the following condition is true: if $(x(t),z(t))\in N$ for $t\in T:=[t_0,\tau]$ then the inequalities (10.1) are valid for almost all $t\in T$.

System (10.1) can be written in the form of a differential inclusion

$$(\dot x,\dot z)\in E_N(t,x,z,p,r,\varepsilon), \tag{10.2}$$

where

$$E_N(t,x,z,p,r,\varepsilon):=\{(f,g)\in R^n\times R:\|f\|\le r,\ |g-\langle p,f\rangle+H(t,x,z,p)|\le\varepsilon\}, \tag{10.3}$$
$$t\in[0,\theta],\ (x,z)\in N,\ p\in R^n,\ r>0,\ \varepsilon>0.$$

A set $M\in[0,\theta]\times R^n\times R$ is said to be weakly invariant with respect to the differential inclusion (10.2) provided for any point $(t_0,x_0,z_0)\in M\cap([0,\theta]\times N)$ there exists a solution $(x(\cdot),z(\cdot))\in\mathcal{S}_N(t_0,x_0,z_0,p,r,\varepsilon)$ such that $(t,x(t),z(t))\in M$ for $t\in[t_0,\tau[$ where $\tau:=\min\{t\in[t_0,\theta]:(x(t),z(t))\notin N\}$.

Definition 10.1. *A lower (resp., an upper) semicontituous function* $(t,x)\mapsto u(t,x):[0,\theta]\times R^n\to R$ *is called an upper (resp., a lower)* *minimax solution of equation* (7.1) *if for any open bounded set* $N \subset R^n \times R$, *any* $p \in R^n$, *and any* $\varepsilon > 0$ *there exists a number* $r > 0$ *such that the epigraph* $\operatorname{epi} u$ *(resp., hypograph* $\operatorname{hypo} u$*) is weakly invariant with respect to the differential inclusion* (10.2). *Similarly, a continuous function* $u : [0, \theta] \times R^n \to R$ *is called a minimax solution of equation* (7.1) *if for any open bounded set* $N \subset R^n \times R$, *any* $p \in R^n$, *and any* $\varepsilon > 0$ *there exists a number* $r > 0$ *such that the graph* $\operatorname{gr} u$ *is weakly invariant with respect to the differential inclusion* (10.2). *A minimax solution of the Cauchy problem* (7.1), (7.2) *is a minimax solution of equation* (7.1) *that satisfies the equality* (7.2).

Let us formulate theorems on the existence and uniqueness of a minimax solution to the Cauchy problem (7.1), (7.2). These theorems do not involve the assumption that the Hamiltonian $H(t, x, z, p)$ satisfies the Lipschitz condition in x and p.

Theorem 10.1. *Suppose that assumptions* A0, A1, A2, A4 *and* A5 *are true and that* u *is an upper minimax solution and* v *a lower minimax solution of problem* (7.1), (7.2). *Assume that there exist constants* C_1 *and* C_2 *exist such that*

$$v(t, x) - u(t, x) \leq C_1 \quad \forall\, (t, x) \in [0, \theta] \times R^n, \tag{10.4}$$

and at least one of the following two estimates does hold

$$|u(t, x) - u(y, t)| \leq C_2, \quad |v(t, x) - v(y, t)| \leq C_2, \tag{10.5}$$

$$\forall\, (x, y, t) \in \operatorname{cl} \Gamma := \{(t, x, y) \in [0, \theta] \times R^n \times R^n : \|x - y\| \leq 1\}.$$

Then $u \geq v$.

Theorem 10.2. *Suppose that assumptions* A0, A1, A3 – A5 *are true and that* u *is an upper minimax solution and* v *a lower minimax solution of problem* (7.1), (7.2). *Assume that there exist such constants* C_2 *and* K *that the estimate*

$$v(t, x) \leq u(t, x) + K(1 + \|x\|) \quad \forall\, (t, x) \in [0, \theta] \times R^n$$

is valid and one of estimates (10.5) *is true. Then the estimate* (10.4) *does hold.*

Theorem 10.3. *Suppose that conditions* A0, A1 *are true, and that the following estimate is valid*

$$\sup\left\{ \frac{|H(t, x, z, p) - H(t, x, z, q)| - \varepsilon}{\|p - q\|} : t \in \,]0, \theta[\,, \right.$$

$$(x,z) \in N,\ (p,q) \in R^n \times R^n,\ p \neq q \Big\} < \infty$$

for any bounded set $N \subset R^n \times R$ *and any* $\varepsilon > 0$. *Assume that there exist a continuous upper minimax solution* u_+ *and a continuous lower minimax solution* u_- *such that* $u_- \leq u_+$. *Then there exists a minimax solution of problem* (7.1), (7.2).

The statements of Theorems 10.1 and 10.2 coincide with the corresponding statements for viscosity solutions obtained in [11]. The proof of the theorems for the minimax solutions is based on the method of Lyapunov functions.

REFERENCES

1. Aubin J.-P., A survey of viability theory, *SIAM J. on Control and Optimization* **28** (1990), 749-788.

2. Aubin J.-P. and Ekeland I., *Applied nonlinear analysis*, John Wiley & Sons, Inc., New York, 1984.

3. Bardi M. and Soravia P., Hamilton-Jacobi equations with singular boundary conditions on a free boundary and applications to differential games, *Trans. Amer. Math. Soc.* **325** (1991), 205–229.

4. Barron E.N., Evans L.C., and Jensen R., Viscosity solutions of Isaacs' equations and differential games with Lipschitz controls, *J. Different. Equat.* **53** (1984), 213-233.

5. Chentsov A.G., On a game problem of converging at a given instant of time, *Math. of the USSR Sbornik* **28** (1976), 353-376.

6. Chistyakov S.V., On solutions of game problems of pursuit, *Prikl. Matemat. i Mehanika* **41** (1977), No. 5 (in Russian).

7. Clarke F.H., Generalized gradients and applications, *Trans. Amer. Math. Soc.* **205** (1975), 247-262.

8. Clarke F.H., *Optimization and nonsmooth analysis*, John Wiley & Sons, Inc., New York, 1983.

9. Courant R. and Hilbert D., *Methods of mathematical physics, Vol. I and II*, John Wiley & Sons, New-York, 1953, 1962.

10. Crandall M.G., A generalization of Peano's existence theorem and flow invariance, *Proc. Amer. Math. Soc.* **36** (1972), 151-155.

11. Crandall M.G., Ishii H., and Lions P.-L., Uniqueness of viscosity solutions revisited, *J. Math. Soc. Japan* **39** (1987), 581-596.

12. Crandall M.G. and Lions P.-L., Condition d'unicité pour les solutions généralisées des équations Hamilton-Jacobi du 1er ordre, *C. R. Acad. Sci. Paris Sér. A - B* **292** (1981), 183-186.

13. Crandall M.G. and Lions P.-L., Viscosity solutions of Hamilton-Jacobi equations, *Trans. Amer. Math. Soc.* **277** (1983), 1-42.

14. Crandall M.G. and Lions P.-L., Hamilton-Jacobi equations in infinite dimensions, Part II. Existence of viscosity solutions, *J. Func. Anal.* **65** (1986), 368-405.

15. Davy J.L., Properties of the solution set of generalized differential equation, *Bull. Austral. Math. Soc.* **6** (1972), 379-398.

16. Demyanov V.F. and Rubinov A.M., *Foundations of nonsmooth analysis and quasidifferential calculus*, Nauka, Moscow, 1990 (in Russian).

17. Evans L.C. and Souganidis P.E., Differential games and representation formulas for solutions of Hamilton-Jacobi-Isaacs equations, *Indiana Univ. Math. J.* **33** (1984), 773-797.

18. Fleming W.H., A note on differential games of prescribed durations, *Contributions to the theory of games, Ann. Math. Stud.* **3** (1957), 407-412.

19. Fleming W.H., The Cauchy problem for degenerate parabolic equations, *J. Math. Mech.* **13** (1964), 987-1008.

20. Guseinov H.G., Subbotin A.I., and Ushakov V.N., Derivatives for multivalued mappings with applications to game theoretical problems of control, *Problems of Contr. Inform. Theory* **14** (1985), 155-167.

21. Haddad G., Monotone trajectories of differential inclusions and functional – differential inclusions with memory, *Izrael J. of Math.* **39** (1981), 83-100.

22. Ishii H., Perron's method for Hamilton-Jacobi equations, *Duke math. J.* **55** (1987), 369-384.

23. Krasovskiĭ N.N., *Game - theoretical problems on the encounter of motions*, Nauka, Moscow, 1970 (in Russian).

24. Krasovskiĭ N.N., *Control of dynamical system*, Nauka, Moscow, 1985 (in Russian).

25. Krasovskiĭ N.N. and Subbotin A.I., *Positional differential games*, Nauka, Moscow, 1974 (in Russian).

26. Krasovskiĭ N. N. and Subbotin A. I., *Game - Theoretical Control Problems*, Springer, New-York, 1988.

27. Kruzhkov S.N., Generalized solutions of nonlinear first order equations with several variables, I, *Matemat. Sbornik (N.S.)* **70(112)** (1966), 394-415 (in Russian).

28. Kruzhkov S.N., Generalized solutions of the Hamilton-Jacobi equations of Eikonal type, I, *Math. USSR-Sb.* **27** (1975), 406-446.

29. Kurzhanski A.B., On the analytical description of the set of viable trajectories to a differential system, *Uspehi mat. nauk*, **40** (1985), 183-194 (in Russian).

30. Kurzhanski A.B. and Filippova T.F., On the set-valued calculus in problems of viablility and control for dynamic processes: the evolution equation, *Les Annales de l'Institut Henri Poincare, Analyse non-linear* (1989), 339-363.

31. Ledyaev Yu.S., Criteria for viability of trajectories of nonautonomous differential inclusion and their applications, *Preprint CMR-1583, Centre de Recherches Math., Univ. de Montreal* (1988), 1-22.

32. Lions P.-L., *Generalized solutions of Hamilton-Jacobi equations*, Pitman, Boston, 1982.

33. Lions P.-L. and Souganidis P.E., Differential games, optimal control and directional derivatives of viscosity solutions of Bellman's and Isaacs equations, *SIAM J. on Control and Optimization* **23** (1985), 556-583.

34. Nagumo M., Uber die Laga der integralkurven gewöhnlicher differential Gleichungen, *Proc. Phys. Math. Japan* **24** (1942), 399-414.

35. Osipov Yu.S., On the theory of differential games to systems with distributed parameters, *Dokl. Akad. Nauk SSSR* **223** (1975), 1314-1317 (in Russian).

36. Petrovskiĭ I.G., *Lectures on the theory of ordinary differential equations*, Nauka, Moscow, 1964 (in Russian).

37. Pontryagin L.S., On the theory of differential games, *Uspehi mat. nauk*, **21** (1966), 219-274 (in Russian).

38. Rockafellar R.T., Existence and duality theorems for convex problems of Bolza, *Trans. Amer. Math. Soc.* **159** (1971), 1-40.

39. Sonnevend G., Existence and numerical computation of extremal invariant sets in linear differential games, *Lect. Notes in Control Inf. Sci.* **22** (1981), 251-260.

40. Subbotin A.I., A generalization of the basic equation of the theory of differential games, *Soviet Math. Dokl.* **22** (1980), 358-362.

41. Subbotin A.I., Generalization of the main equation of differential game theory, *J. Optimiz. Theory and Appl.* **43** (1984), 103-133.

42. Subbotin A.I., *Minimax inequalities and Hamilton – Jacobi equations*, Nauka, Moscow, 1990 (in Russian).

43. Subbotin A.I., Existence and uniqueness results for Hamilton-Jacobi equations, *Nonlinear Analysis* **16** (1991), 683-689.

44. Subbotin A.I., On a property of subdifferential, *Matemat. Sbornik* **182** (1991), 1315-1330 (in Russian).

45. Subbotin A.I. and Subbotina N.N., Necessary and sufficient conditions for a piecewise smooth value of a differential game, *Soviet Math. Dokl.* **19** (1978), 1447-1451.

46. Subbotin A.I. and Tarasyev A.M., Stability properties of the value function of a differential game and viscosity solutions of Hamilton-Jacobi equations, *Problems Contr. Inform. Theory* **15** (1986), 451-463.

47. Subbotina N.N., The method of Cauchy characteristics and generalized solutions of Hamilton-Jacobi-Bellman equations, *Dokl. Akad. Nauk SSSR* **320** (1991), 556-561 (in Russian).

48. Young L.C., *Lectures on the calculus of variations and optimal control theory*, Sanders, Philadelphia, 1969.

Institute of Mathematics and Mechanics,
Ural Branch, Russian Acad. Sci.,
S.Kovalevsky st. 16, Ekaterinburg 620219, Russia

Adaptivity and Robustness in Automatic Control Systems

Ya.Z. Tsypkin

Abstract

This paper deals with various approaches to the solution of control problems for dynamic plants under uncertainty conditions. It gives a description of the principles for designing discrete–time adaptive and robust control systems and presents a discussion of their properties and specificities. A broader range of possibilities is given by robustly nominal systems that incorporate, together with a feedback loop from the system output, also a feedback that depends on the bias between the system output and the output of a special nominal model.

Robustly nominal systems allow to eliminate or to diminish substantially the effect of input as well as parametric disturbances that are due to parameter fluctuations within the range of uncertainty.

The principles of constructing robustly nominal systems are described together with the conditions for their realization based on criteria for robust stability.

For improving the quality of such systems we use algorithms for identifying the bias between the system output and the nominal model rather than those of identifying the dynamic plant itself. This simplifies the system structure and broadens the range of applications of robustly nominal systems as compared with traditional adaptive systems.

0 Introduction.

The problem of controlling dynamic plants under uncertainty attracts a large attention of a broad community of specialists in the theory and techniques of automatic control.

The uncertainty in the dynamic plants is basically due to an incomplete knowledge of their structure and parameters and also of the acting external disturbances. The solution of the problem of control under uncertainty is based on a broad application of various robust and adaptive control systems.

The developed principles of constructing adaptive linear control systems are based on using the online information on the state of an incompletely defined plant. This information is then incorporated into the scheme of selecting the parameters of the controller for ensuring the preassigned requirements on the system, which are usually reduced to the optimality in some sense of the control process. These requirements are traditionally formulated in the form of a closed–loop reference model (CLRM), which operates in the absence of disturbances. The optimality criterion is then determined through a functional of the error between the synthesized adaptive system and the CLRM. Once the structure of the plant is known, then in the absence of an external disturbance or under a standard disturbance (of either stochastic or of an "unknown but bounded" nature) one may determine the optimal structure of the controller. This is done through a deterministic [1, 2], stochastic [3, 4] or a game–theoretical, minmax approach [5, 6].

The adaptive system performs a direct or an indirect estimation of the parameters of the plant and uniquely defines related parameter estimates for the controller in its normal operational mode. In these systems the CLRM usually appears in an implicit form, defining the requirements on the system.

It can be observed without difficulty that the mentioned principles of constructing adaptive systems are related to a complete or incomplete cancellation of the system uncertainty in the normal operational mode and to the application of either the classical control system methods for deterministic systems or (under an incomplete cancellation of the uncertainty) of the game–theoretical, minmax methods for control synthesis.

The "traditional" adaptive systems known up to now contain con trollers with tunable parameters or often even with tunable models used in the algorithms for estimating the plant parameters (indirect adaptive systems) or the parameters of the "optimal" controller (direct adaptive systems) [7, 8].

A somewhat different class of adaptive systems that explicitly incorporates the CLRM is based on the principle if directly optimalizing the optimality criterion with respect to the parameters of the controller [9, 10]. If one considers the uncertain dynamic plant as a family of plants that belong to some set, then one may attempt to synthesize a control system with a fixed controller. The fixed controller should ensure the persistence of the qualitative properties of the system, for example of its stability, aperiodicity or should ensure a guaranteed damping of the transient process (a degree of stability). Such systems are said to the robustly stable, robustly aperiodic or robustly modal. Sometimes it is said that robust systems are systems for the control of a set of plants for each of which the appropriate properties are ensured [11, 12]. These properties do not change from plant to plant. However in their quantitative form they may differ considerably. In general, the robust systems guarantee some properties that are not worse than in the least favourable case.

Robust systems are conceptually close to systems of the minmax or game type [13], a particular case of which are systems optimal relative to the H^∞ criterion [14, 15]. The question that arises here is whether one may design robust control systems that would not only ensure the qualitative properties of the system, but would also guarantee, for the whole set of plants considered, some properties specific for optimal systems. All this should be ensured by one nominal plant selected among the given set of plants under consideration.

The robustly nominal or robustly optimal systems of such type should combine the properties of robustness (low sensitivity) with optimality properties (relative to some preassigned requirements).

The present paper gives a positive answer to this question and discusses the principles for constructing such systems.

A broad application of computers in automatic control systems led to specific interest in adaptive discrete–time control systems (DACS), the majority of publications on adaptive and robust control is devoted specifically to these types of systems. The consideration of DACS allows a simple and clear exposure of the specificities and the possibilities of robust control systems and a comparison of these with traditional adaptive and robust systems.

This paper considers structures and properties of linear discrete–time control systems. The necessary information on traditional adaptive systems is given together with a treatment of robustly stable systems based on frequency domain techniques. The main substance of the paper is devoted to robustly nominal systems or robustly optimal control systems with unknown and bounded uncertainty.

The difference between robustly nominal and traditional adaptive systems lies in the absence of an identification procedure aimed at reducing the uncertainty, since the uncertainty that is due to the deviations of the family of plants under consideration from the nominal plant is eliminated or compensated either on the basis of an a priori information or through a procedure of adaptation.

Finally, some further direction for the development of nominal systems are discussed.

1 The Equations of Discrete Systems.

A typical feature that distinguishes modern discrete control systems for dynamic plants is the use of computers. The general block–scheme for such systems is given in fig.1. Its continuous part consists of the dynamic plant and the performing device. Its discrete part is represented by a computer, whose program defines the controller. The connection of these parts is implemented through a digital–analogue (DA) and an analogue–digital (AD) transformers. For designing the equations of the digital system on the basis a discrete Laplace transform and other related transforms [1, 16, 17] one has to pass from continuous to discrete transfer functions. This is equivalent to a passage from differential equations to equations in finite differences or to difference equations.

Consider a dynamic plant, described by linear difference equations

$$Q(q)y(n) = qP_u(q)u(n) + P_f(q)f(n), \tag{1.1}$$

where $y(n)$ is the output variable, $u(n)$ – the control, $f(n)$ – the disturbance, q – the delay operator which is such that $q^m x(n) = x(n-m)$, n is the discrete time, $Q(q)$, $P_u(q)$ and $P_f(q)$ are polynomials of order N, N_1, N_2 respectively. Here $Q(q)$, $P_f(q)$ are monopolynomials, i.e. $Q(0) = P_f(0) = 1$. For the sake of simplicity we denote the delay operator and the independent variable with the same letter q.

The equation of the plant also may be written as

$$y(n) = W_0(q)u(n) + W_f(q)f(n), \tag{1.2}$$

where

$$W_0(q) = q\frac{P_u(q)}{Q(q)} \tag{1.3}$$

is the transfer function of the plant relative to the control and

$$W_f(q) = \frac{P_f(q)}{Q(q)} \tag{1.4}$$

is the transfer function of the plant relative to the disturbance.

The equation of the controller may in general be written as

$$R(q)u(n) = P_r(q)r(n) - P_y(q)y(n), \tag{1.5}$$

where $r(n)$ is a preassignable input and $R(q)$, $P_r(q)$, $P_y(q)$ are polynomials of order N_3, N_4, N_5, or

$$u(n) = W_r(q)r(q) - W_y(q)y(n), \tag{1.6}$$

where

$$W_r(q) = \frac{P_r(q)}{R(q)} \tag{1.7}$$

is a transfer function of the controller in its preassigned part and

$$W_y(q) = \frac{P_y(q)}{R(q)} \tag{1.8}$$

is the transfer function of the feedback loop.

Eliminating the control variable from equations (1.2), (1.6) we come to an equation for a closed loop system

$$y(n) = K(q)r(n) + K_f(q)f(n), \tag{1.9}$$

where

$$K(q) = q\frac{P_u(q)P_r(q)}{G(q)} \tag{1.10}$$

is the transfer function of the closed–loop system in its preassigned part

$$K_f(q) = \frac{R(q)P_f(q)}{G(q)} \tag{1.11}$$

is the transfer function of the closed loop system relative to the disturbance.

In (1.10), (1.11)

$$G(q) = Q(q)r(q) + qP_u(q)P_y(q) \tag{1.12}$$

is the characteristic polynomial for the closed–loop system.

Substituting $K(q)$ of (1.10) and $K_f(q)$ of (1.11) into (1.9) we come to the equation of a closed– loop system in the form

$$G(q)y(n) = qP_u(q)P_r(q)r(n) + R(q)P_f(q)f(n). \tag{1.13}$$

We will say that a polynomial is external if all its zeros lie beyond a unit circle $|q| \leq 1$.

For a stable and phase minimal plant the polynomials $Q(q)$ and $P_u(q)$ will be external. If $Q(q)$ is not an external polynomial, then the plant will be unstable. If a polynomial $P_u(q)$ is not external, then the plant will not be phase minimal. Comparing the transfer functions of the closed system

$K(q)$ (1.10) and $K_f(q)$ (1.11) with the transfer functions $W_0(q)$ (1.3) and $W_f(q)$ (1.4) of the plant itself, we conclude that the presence of a feedback changes the poles of the latter but does not affect its zeros that are determined by the polynomials $P_u(q)$ and $P_f(q)$. Of course one may select the feedback so as to cancel the external multipliers of this polynomial with the respective multipliers of the characteristic polynomial. However without losing robustness one cannot cancel out the nonexternal multipliers [16, 18]. This means that the introduction of a controller allows to stabilize the closed–loop system, but does not allow to get rid of the absence of phase minimality. This option will be considered later.

The block–scheme of the closed–loop system described by equations (1.1), (1.5) or (1.2), (1.6) is given in fig. 2a or in equivalent but simpler form in fig. 2b. The requirement of stability of a DAS (discrete automatic system) is a necessary condition for its workability. The selection of the structure and of the parameters of the controller then allows to assign one or another qualitative property to the system. For the analysis of stability and of the quality of a DAS one may use frequency criteria of stability and of modality ensuring a required pole assignment for the characteristic polynomial. The most acknowledged methods of synthesis, are those connected with the distribution of zeros of the characteristic polynomial for the closed loop system which define the qualitative properties of the system.

Let us introduce a reference model that determines the requirements to the closed loop system. Let us also represent its equation as

$$y^0(n) = K^{\mathrm{ref}}(q)r(n), \tag{1.14}$$

where

$$K^{\mathrm{ref}}(q) = q\frac{H^{\mathrm{ref}}(q)}{G^{\mathrm{ref}}(q)} \tag{1.15}$$

is the transfer function for the referencc modcl. Hcrc G^{ref} and H^{ref} are external monopolynomials. Let us require that the processes in the closed loop system (in the absence of a disturbance $f_n \equiv 0$) and in the reference model would coincide. Following this aim let us equalize the transfer functions of the closed loop system relative respectively to the preassigned part (1.10), (1.12) and that of the reference model (1.15).

$$\frac{P_u(q)P_r(q)}{Q(q)R(q) + qP_u(q)P_y(q)} = \frac{H^{\mathrm{ref}}(q)}{G^{\mathrm{ref}}(q)}\,, \tag{1.16}$$

This leads to the polynomial or diophantine equations [1, 16, 18]

$$Q(q)R(q) + qP_u(q)P_y(q) = T(q)G^{\mathrm{ref}}(q) \tag{1.17}$$

and

$$P_r(q) = T(q)H^{\text{ref}}(q). \tag{1.18}$$

Here $T(q)$ is the external polynomial whose selection defines one or another property of the closed system. The polynomial equations (1.17), (1.18) define the polynomials $R(q)$, $P_r(q)$, $P_y(q)$ of the controller depending on the given requirements $[G^{\text{ref}}(q),\ H^{\text{ref}}(q),\ T(q)]$ and the known polynomials $Q(q)$, $P_u(q)$ and $P_f(q)$. For a nonphase–minimal plant the polynomial equations turn to be more complicated since there arises a necessity to factorize the polynomials $P_u(q)$, $P_f(q)$ [20]. This leads to more complicated structures of the controller.

2 The A Priori Information on the Dynamic Plant and the Disturbances.

For an incompletely defined dynamic plant one usually assumes that it is also described by the equation (1.1)

$$Q(q)y(n) = qP_u(q)u(n) + P_f(q)f(n), \tag{2.1}$$

which now incorporates polynomials $Q(q)$, $P_u(q)$, $P_f(q)$, whose degrees or their upper bounds are taken to be known while the coefficients are unknown. Such an a priori information is usually accepted in the traditional theory of adaptive systems [1–4]. It is often assumed that the polynomials $P_u(q)$ and $P_f(q)$ are external, so that the dynamic plants under consideration are phase minimal relative to both disturbance and control. A somewhat different but perhaps more realistic a priori information on the plant assumes that the coefficients of the polynomial belong to certain preassigned sets. In this case the polynomials $Q(q)$, $P_u(q)$ and $P_f(q)$ may be presented as

$$\begin{aligned} Q(q) &= Q^0(q) + \delta Q(q), \\ P_u(q) &= P_u^0(q) + \delta P_u(q), \\ P_f(q) &= 0 + \delta P_f(q). \end{aligned} \tag{2.2}$$

Here $Q^0(q)$, $P_u^0(q)$ are nominal polynomials with given coefficients and

$$\delta Q(q) \in \mathcal{A},\ \ \delta P_u(q) \in \mathcal{A}_u,\ \ \delta P_f(q) \in \mathcal{A}_f \tag{2.3}$$

are the variations of polynomials that belong to sets $\mathcal{A}$, $\mathcal{A}_u$ and $\mathcal{A}_f$. These sets may be of various form depending on the specific problem. The "size" of these sets defines the degree of uncertainty while its specific form depends on the a priori information on the parameters of the plant. The plant described by equation (2.1) under conditions (2.2) will be further referred to as an interval plant.

An interval plant defines a family of plants rather than a single plant. Due to (2.2) the equation of an interval plant (2.1) may be presented as

$$Q^0(q)y(n) = qP_u^0(q)u(n) + \varphi(n), \tag{2.4}$$

where

$$\varphi(n) = P_f(q)f(n) + q\delta P_u(q)u(n) - \delta Q(q)y(n). \tag{2.5}$$

The block–scheme of an interval plant is given in fig.3. As it can be seen from fig.3, the interval plant may be considered as a certain preselected given nominal plant, subjected to a generalized disturbance $\varphi(n)$. A generalized disturbance depends both on the external disturbance $f(n)$ applied to the system and also on the deviation of the parameters of the plant from the given parameters of the nominal plant. Usually in the theory of control one assumes that the external disturbance is either absent, i.e. $f(n) \equiv 0$, or that it is given in a certain standard form: a jump $F(n) = 1(n)$, a harmonic disturbance $f(n) = a\sin(\omega n + \nu)$, a stationary random independent input $f(n) = \xi(n)$ with a bounded variance. In a number of problems of control theory instead of a typical unit disturbance one may consider classes of disturbances: $|f(n)| < \Delta$ which are bounded disturbances.

Let us represent a generalized disturbance (2.5) in the form

$$\varphi(n) = P_f(q)f(n) + \varphi_p(n), \tag{2.6}$$

where

$$\varphi_p(n) = q\delta P_u(q)u(n) - \delta Q(q)y(n) \tag{2.7}$$

is a parameter perturbation. It depends on both the deviation of the parameter of an interval plant from the nominal plant and also on the $y(n)$ and the control $u(n)$.

Let us say that an external perturbation $f(n)$ is regular, if it may be predicted through previous data, i.e. if there exists a predictor polynomial $D_f(q)$ of degree M_f, such that

$$f(n) = D_f(q)f(n-1),\ n \geq M_f. \tag{2.8}$$

Obviously, a regular perturbation is the solution of a homogeneous difference equation of the form

$$(1 - qD_f(q))f(n) = 0,\ n \geq M_f. \tag{2.9}$$

Discrete time filters with transfer functions $D_f(q)$ and $1 - qD_f(q)$ could be named respectively as predictors and absorbers. Equation (2.9) gives the condition for absorbing the class of external disturbances $f(n)$. The specific representatives of this class are determined by fixing M_f for the initial values $f(m)$, $m = 0, 1, ..., M_f - 1$. The possibility of representing a broad class of continuous functions in the form of solutions of homogeneous linear differential equations was used by Guerrneri [21] and Shannon [22] for investigating the realizability of various functions by a differential analyzer. V.S. Kulebakin developed these possibilities using them for the solution of a large number of problems in electromechanics and automation [23]. Particularly, he introduced the notion of selectively invariant systems. Somewhat later a similar representation of disturbances, named as disturbances of the wave structure type were used and widely propagated by S. Johnson [24, 25]. These approaches are closely connected with the so-called "principle of an internal model" [26-28], see also [29-30].

The representation of grid–type functions $f(n)$ in the form of solutions of homogeneous difference equations was considered in [31] and was used independently in [32].

Suppose that the parametric disturbance $\varphi_p(n)$ is regular. Then

$$\varphi_p(n) = D_p(q)\varphi_p(n-1) \tag{2.10}$$

and therefore

$$(1 - qD_p(q))\varphi_p(n) = 0 \tag{2.11}$$

For a generalized disturbance

$$\varphi(n) = P_f(q)f(n) + \varphi_p(n) \tag{2.12}$$

we have

$$\begin{aligned}(1 - qD_f(q))(1 - qD_p(q))\varphi(n) &= \\ = (1 - qD_p(q))P_f(q)[(1 - qD_f(q))f(n)]&+ \\ +(1 - qD_f(q))[(1 - qD_p(q))\varphi(n)] &= 0,\end{aligned} \tag{2.13}$$

since due to (2.9) and (2.11) the relations in the square brackets turn to be zero.

However,

$$\begin{aligned}(1 - qD_f(q))(1 - qD_p(q)) &= \\ = 1 - q(D_f(q) + D_p(q)) &+ Q^2 D_f(q)D_p(q).\end{aligned} \tag{2.14}$$

This yields, that the predictor polynomial for the generalized disturbance is equal to

$$D(q) = D_f(q) + D_p(q) - qD_f(q)D_p(q), \tag{2.15}$$

so that

$$\varphi(n) = D(q)\varphi(n-1) \tag{2.16}$$

and therefore

$$(1 - qD(q))\varphi(n) = 0. \tag{2.17}$$

If the external disturbance contains an irregular component $\chi(n)$, then instead of (2.8) we will have

$$f(n) = D_f(q)f(n-1) + \chi(n), \tag{2.18}$$

and the homogeneous difference equations (2.9) will be substituted by a nonhomogeneous one, which is

$$(1 - qD_f(q))f(n) = \chi(n). \tag{2.19}$$

In this case, in view of (2.17) we have, instead of (2.13)

$$\begin{aligned}(1 - qD_f(q))(1 - qD_p(q))\varphi(n) = \\ = (1 - qD_p(q))P_f(q)\chi(n).\end{aligned} \tag{2.20}$$

If we represent $P_f(q)$ as

$$P_f(q) = 1 + qP'_f(q), \tag{2.21}$$

then (2.20) may be transformed to the form

$$\varphi(n) = D(q)\varphi(n-1) - D_1(q)\chi(n) + \chi(n), \tag{2.22}$$

where the predictor polynomial $D(q)$ is determined by the expression (2.14) and the polynomial $D_1(q)$ is derived from $D(q)$ by substituting $D_f(q)$ for $-P'_f(q)$, so that

$$D_1(q) = D_p(q) - P'_f(q) + qD_f(q)P'_f(q). \tag{2.23}$$

The irregular component $\chi(n)$ appears when the external disturbance is a sequence of disturbances from one and the same class. We recall that a similar a priori information on the external disturbances of type (2.8), (2.9) and (2.16), (2.17) was used in paper on selectively invariant continuous control systems [23, 39] and on continuous and discrete systems, that are adaptive relative to disturbances [24, 25, 32]. In the classical and modern control theory the external disturbances are unfortunately not in the center of attention. It is usually assumed that they are either absent or that the disturbances are of standard types.

The irregular component also may turn to be a stochastic stationary input $\chi(n) = \xi(n)$, where $\xi(n)$ satisfies the conditions

$$E\{\xi(n)\} = 0, \; E\{\xi(n)\,\xi(m)\} = \begin{cases} \sigma^2, & n = m, \\ 0, & n \neq m, \end{cases} \tag{2.24}$$

where $E\{\cdot\}$ stands for the mathematical expectation. The stochastic disturbance of the autoregressive type (the AR – disturbance) is represented as

$$f(n) = D_f(q)f(n-1) + \xi(n). \tag{2.25}$$

For a generalized disturbance with a stochastic external disturbance the following equation of type (2.22) is true

$$\varphi(n) = D(q)\varphi(n-1) - D_1(q)\xi(n) + \xi(n). \tag{2.26}$$

This disturbance is of the autoregression sliding mean value type (the ARSM disturbance).

From (2.26) it follows that

$$(1 - qD(q))\varphi(n) = (1 - qD_1(q))\xi(n) \tag{2.27}$$

In this case, as in the case of a nonregular component, a full compensation is impossible. We emphasize that it was usually assumed that the polynomials $D(q)$ and $D_1(q)$ are known. In the general case the polynomials $D(q)$ and $D_1(q)$ may not be known, however. Then one comes to the problem of estimating the degrees and coefficients of these polynomials, which in turn leads to the necessity of introducing an adaptive scheme.

3 On the Possibility of Controlling an Interval Plant.

An incompletely known or interval plant determines a set of plants so that the problem of control that arises is for a set of plants.

A natural possibility of solving this problem consists in studying the plant through identification in its operational mode and in eliminating the uncertainty. The refinement of the parameter estimates through on–line identification then allows to determine the related values of the parameters of the controller. Such systems with a tuned controller constitute a broad class of traditional adaptive control systems, which could be treated as asymptotically optimal controllers. The investigation of traditional adaptive control systems and of the adaptive algorithms used in these systems is traced in a large literature [1–4].

Considerably later another opportunity appeared which is based on the possibility of controlling the set of plants with a fixed controller. Such systems form the class of robust control systems that keeps the property of stability, aperiodicity and provides the guaranteed estimates for the degree of oscillation.

Robust control systems thus provide stability and some guaranteed indices of the control process's quality while controlling a whole set of plants. The robust control systems can of course be minimally optimal only. A rather significant number of works is devoted to various kinds of robust control systems, see, for example, the review [11].

Another realization of an interval plant control, that has not been under investigation until recently, is based on studying a generalized disturbance characterizing the difference between the interval plant and a preselected nominal plant, known a priori, rather than the plant itself. The results of this study are used for creating a controller that would eliminate or significantly reduce the generalized disturbance influence on control process for the nominal plant.

Let us further refer to these systems as robustly adaptive nominal plants. It is a better process quality than in traditional adaptive systems and also in robustly stable systems that can be achieved in these systems. This is due to an ability of choicing a nominal plant that belongs to a family defined by an interval plant and has the appropriate properties of phase–minimality, low degree, given distribution of the poles and further on due to an application of an optimal controller for this plant.

In the absence of necessity for studying the generalized disturbance and the adaptation through a priori information we arrive at robustly nominal systems. Inthe absence of parametric disturbances the robustly nominal systems could be treated as merely selectively or adaptively invariant systems [34].

Let us now discuss the capabilities for the construction of controllers for interval plants among those mentioned above . The attention will be mostly devoted to considerably new classes of adaptive systems, namely robustly adaptive nominal systems, while traditional adaptive systems and robustly stable systems will be just briefly discussed as they are covered in a large number of publications. It is the essential information for comparing traditional adaptive systems and robustly stable systems with robustly adaptive nominal systems that will be given here.

4 Traditional Adaptive Systems.

For completely defined adaptive systems the parameters of the controller are determined through polynomial equations (1.17), (1.18). Once the parameters are unknown, it is natural to identify them in a normal operating mode and then to use the parameter estimates in calculating the control through (1.17), (1.18). Such implicit adaptive systems were among the first types. In the absence of disturbances i.e. with $f(n) = 0$ the equation of the plant (1.1) has the form

$$Q(q)y(n) = qP_u(q)u(n). \tag{4.1}$$

Denote $Q^*(Q)$ to be the vector of the actual plant parameters

$$Q^*(q) = \text{coeff}[Q(q) - 1,\ P_u(q)]. \tag{4.2}$$

For estimating these parameters we introduce a tunable model

$$\hat{y}(n, \theta) = [1 - \hat{\theta}(q)]y(n) + q\hat{P}_u(q)u(n) = \theta^T x(n), \tag{4.3}$$

where θ is the vector of parameter estimates that are incorporated in $\hat{\theta}(q)$ and $\hat{P}_u(q)$, and

$$x(n) = (y(n-1), ..., y(n-N), u(n-1), ..., u(n-N_1))^T \tag{4.4}$$

is the vector of observations.

The bias

$$\mathcal{E}(n, \theta) = y(n) - \hat{y}(n, \theta), \tag{4.5}$$

is equal to

$$\mathcal{E}(n, \theta) = \hat{Q}(q)y(n) - q\hat{P}(q)u(n). \tag{4.6}$$

due to (4.3).

The identification algorithms that generate the estimates of the parameters $\theta(n)$ in each step, could be presented in the form of recurrence relations

$$\theta(n) = \theta(n-1) + \Gamma(n)\mathcal{E}(n, \theta(n-1))x(n), \tag{4.7}$$

where $\Gamma(n)$ is the gain matrix. The selection of the gain matrix determines the specific algorithms. Thus, with

$$\Gamma(n) = \gamma I, \tag{4.8}$$

where I is the unit matrix, the algorithm (4.7) corresponds to the gradient algorithm. With

$$\Gamma(n) = [x^T(n)x(n)]^{-1}I \tag{4.9}$$

the relation (4.7) leads to the Kačmaž algorithm. With

$$\Gamma(n) = [x^T(n)\Gamma_0(n-1)x(n)]^{-1}\Gamma_0(n-1), \tag{4.10}$$

where

$$\Gamma_0(n) = \Gamma_0(n-1) - [x^T(n)\Gamma_0(n-1)x(n)]^{-1}\times \\ \times\Gamma_0(n-1)x(n)x^T(n)\Gamma_0(n-1), \tag{4.11}$$

when

$$x^T(n)\Gamma_0(n-1)x(n) \neq 0,$$

and

$$\theta(n) = \theta(n-1) \text{ and } \Gamma_0(n) = \Gamma_0(n-1), \tag{4.12}$$

when

$$x^T(n)\Gamma_0(n-1)x(n) = 0,$$

then the algorithm (4.7) corresponds to the optimal finite–step algorithm.

The given algorithms are aimed at an exact or an approximate solution of systems of linear equations. An overview of various algorithms and their comparison is given in [2].

The plant parameter estimates allow to find $\hat{Q}(q)$, $\hat{P}_u(q)$, and on the basis of the solution of polynomial equations (1.17), (1.18)

$$\hat{Q}(q)\hat{R}(q) + q\hat{P}_u(q)\hat{P}_y(q) = T(q)G^{\text{ref}}(q), \tag{4.13}$$

$$\hat{P}_u(q)\hat{P}_r(q) = T^0(q)H^0(q)$$

We find the parameter estimates for the controller

$$\hat{R}(q)u(n) = \hat{P}_r(q)r(n) - \hat{P}_y(q)y(n). \tag{4.14}$$

Equations (4.1), (4.14) define the basic control loop, while (4.6), (4.7), (4.13) – the adaptation loop. A block–scheme of the adaptive system, which incorporates the tuned model is given in fig. 4.

If, with $n \to \infty$, the parameter estimates, being refined, finally tend to the actual values, the adaptation loop tunes the controller so that the system comes to a required mode.

Under an external stochastic disturbance $\xi(n)$ that satisfies conditions (2.22), the equations of the plant are transformed into

$$Q(q)y(n) = qP_u(q)u(n) + P_f(q)\xi(n). \tag{4.15}$$

Now the vector θ^* of the plant parameters has a larger dimension so that

$$\theta^* = \text{coeff}[Q(q) - 1,\ P_u(q),\ P_f(q) - 1], \tag{4.16}$$

and the equation of the tuned model has the following form [35]

$$\hat{y}(n,\theta) = [1-\hat{Q}(q)]y(n) + q\hat{P}_u(q)u(n) + \\ +[\hat{P}_f(q)-1](y(n)-\hat{y}(n,\theta)), \tag{4.17}$$

which yields that the bias $\mathcal{E}(n,\theta)$ (4.5) is now determined by the equation

$$\mathcal{E}(n,\theta) = \hat{Q}(q)y(n) - q\hat{P}_u(q)u(n) - \\ -[\hat{P}_f(q)-1]\mathcal{E}(n,\theta), \tag{4.18}$$

or

$$\hat{P}_f(q)\mathcal{E}(n,\theta) = \hat{Q}(q)y(n) - q\hat{P}_u(q)u(n). \tag{4.19}$$

The identification algorithms in this stochastic case are similar to (4.7)

$$\theta(n) = \theta(n-1) + \Gamma(n)\mathcal{E}(n,\theta(n-1))x(n), \tag{4.20}$$

Now, however, they incorporate the extended vector of observables

$$x(n) = (y(n-1),...,y(n-N),u(n-1),... \\ ...,u(n-N_1),\mathcal{E}(n-1),...,\mathcal{E}(n-N_2))^T \tag{4.21}$$

and other gain matrices $\Gamma(n)$.

Thus, with

$$\Gamma(n) = \frac{\gamma}{n}I \tag{4.22}$$

algorithm (4.21) corresponds to the stochastic approximation method, [1, 35].

With

$$\Gamma(n) = \left[\sum_{m=1}^{n} x^T(m)x(m)\right]^{-1} \cdot I \tag{4.23}$$

we come to the Goodwin algorithm [1, 35].

If $\Gamma(n)$ satisfies the recurrence relation

$$\Gamma(n) = \Gamma(n-1) - \frac{\Gamma(n-1)x(n)x^T(n)\Gamma(n-1)}{1+x^T(n)\Gamma(n-1)\chi(n)}, \tag{4.24}$$

we arrive at one of the versions of the least–square method, the one due to Paniuška [35].

Optimal identification algorithms that allow the best possible convergence rate are of the form

$$\theta(n) = \theta(n-1) + \Gamma(n)\mathcal{E}(n,\theta(n-1))v(n), \tag{4.25}$$

$$\hat{P}_f(q)v(n) = x(n),$$

$$\Gamma(n) = \Gamma(n-1) - \frac{\Gamma(n-1)v(n)v^T(n)\Gamma(n-1)}{1 + x^T(n)\Gamma(n-1)x(n)}\ .$$

Here $v(n)$ are the sensitivity functions that are equal to the derivatives of $\hat{y}(n, \theta)$ over the vector of the coefficients of the polynomial $\hat{P}_f(q)$. With $\hat{P}_f(q) \equiv 1$ we have $v(n) = x(n)$, and the algorithms (4.25), (4.20) and (4.24) turn into those of the least–square method.

The estimates $\hat{Q}(q)$, $\hat{P}_u(q)$, $P_f(q)$ of the plant parameters on the basis of polynomial equations (1.17), (1.18) with $T(q) = T_1(q)\hat{P}_f(q)$ allow, due to the equations

$$\hat{Q}(q)\hat{R}(q) + q\hat{P}_u(q)\hat{P}_y(q) = T_1(Q)\hat{P}_f(q)G^{\text{ref}}(q),$$

$$\hat{P}_u(q)\hat{P}_2(q) = T_1(q)\hat{P}_f(q)H^{\text{ref}}(q) \tag{4.26}$$

to determine the estimates of the parameters of the controller

$$\hat{R}(q)u(n) = \hat{P}_r(q)r(n) - \hat{P}_y(q)y(n). \tag{4.27}$$

Specifying the polynomial equations (4.26), taking $R(q) = P_u(q)$ and cancelling the polynomials $P_u(q)$, $P_f(q)$, one may derive results already known from respective literature [1–4].

The block–scheme of a tunable model with a stochastic disturbance is given in fig. 5. For the interval dynamic plant the given identification algorithms of the above are supplemented by projectors that ensure such estimates $\theta(n)$ with which the variations $\delta Q(q)$, $\delta P_u(q)$, $\delta P_f(q)$ are located within the sets $\mathcal{A}$, $\mathcal{A}_u$ and $\mathcal{A}_f$ respectively. We have to emphasize that for the nonphase–minimal plants the structure of the controller and of the overall adaptive system becomes more complicated.

There appears an additional necessity in the factorization of the polynomials $\hat{P}_u(q)$ and $\hat{P}_f(q)$ [20, 35]. The traditional basic principle of designing an adaptive system consists in eliminating the uncertainty by learning the plant through an identification of its parameters in a normal operating mode and further on, by tuning the controller due to the information obtained. The presence of an additional adaptation loop that leads to a variation of the parameters of the controller, makes the overall adaptive system highly nonlinear [35]. The system may then allow complicated nonlinear effects (bifurcation, chaotic motions, etc.). These phenomena are started to be investigated [37]. Much work is devoted to the theoretical justification of the convergence of the algorithms [1, 37]. It is usually assumed here that the plant parameters are invariant. A considerably smaller number of papers consider time–dependent plants which are estimated through on-line procedures so that the controller parameters are also time–dependent [38, 39]. The traditional adaptive schemes are therefore based on learning about the plant and on using the methods and techniques developed for completely known plants.

5 Robustly Stable Systems.

Let us abandon the idea of identifying an interval plant but rather try to ensure the operation of the interval system of control by steering the whole family of plants.

The system equations for the interval plant (2.1) have the form (1.13)

$$(Q(q)R(q) + qP_u(q)P_y(q))y(n) = \tag{5.1}$$

$$= qP_u(q)P_r(q)r(n) + R(q)P_f(q)f(n).$$

Here $Q(q)$, $P_u(q)$, $P_f(q)$ are interval polynomials while the other polynomials are of fixed parameters.

The characteristic polynomial

$$G(q) = Q(q)R(q) + qP_u(q)P_y(q) \tag{5.2}$$

is the sum of the products of interval polynomials and fixed polynomials.

The modality of a robust system, i.e. the location of the roots of $G(q)$ determines its dynamic properties. The problem which arises is to ensure the stability and the modality of a set of systems, determines by its characteristic polynomial (5.2). Usually for this aim one may use various sufficient conditions of stability, related to the requirement of minor variations of the frequency curve of the plant and therefore of minor variations of its parameters [39, 12]. A closely related topic is the H^∞ – control theory [14, 15, 40, 41]. These sufficient conditions may often lead to very strong requirements. For the solution of this problem it may be worth to use the theory of robust stability, namely the stability of a family of linear plants. There are many investigations which develop and prove some generalizations of the basic result of V.L.Kharitonov [42]. Most of these are related to continuous–time systems. The efforts of applying this theory to discrete systems were less successful [43]. There were also hardly any effective methods for investigating robust stability of feedback systems in either continuous or in discrete versions. Recently it seem that an important role in robust stability is attributed to frequency–domain methods [44, 45].

We now come up with an overview of the main results in the theory of frequency domain methods of investigating robust stability for the design of robust control systems. Define the characteristic polynomial (5.2), due to (3.2), as

$$G(q) = G^0(q) + R(q)\delta Q(q) + qP_y(q)\delta P_u(q), \tag{5.3}$$

where

$$G^0(q) = Q^0(q)R^0(q) + qP_u^0(q)P_y^0(q) \tag{5.4}$$

is a fixed polynomial which is the characteristic polynomial of the nominal system. We will refer to G^0 as the nominal characteristic polynomial.

The nominal system will be stable if and only if the characteristic polynomial $G^0(q)$ is nonzero within the unit circle $|q| \leq 1$, i.e., when

$$G^0(q) \neq 0, \ |q| \leq 1 . \tag{5.5}$$

Let us formulate a stability criterion for the nominal system that does not depend on the degree of $G(q)$ [46].

For the stability of the nominal system it is necessary and sufficient that the godograph

$$G^0(e^{-j\omega}) = \text{Re } G^0(e^{-j\omega}) + j \text{ Im } G^0(e^{-j\omega}) \tag{5.6}$$

would not surround the origin (fig. 6).

The control system for the interval plant will be robustly stable if and only if the family of polynomials $G(q)$ will be nonzero in the unit circle $|q| \leq 1$, i.e. when

$$G(q) \neq 0, \ |q| \leq 1 \tag{5.7}$$

for all $\delta Q(q) \in \mathcal{A}$, $\delta P_u(q) \in \mathcal{A}_u$.

For the formulation of robust stability criterion [45] we rewrite $R(e^{-j\omega})$ and $P_y(e^{-j\omega})$ in the form

$$R(e^{-j\omega}) = \rho(\omega)e^{j\psi(\omega)}, \ P_y(e^{-j\omega}) = \rho_y(\omega)e^{j\psi_y(\omega)} \tag{5.8}$$

and introduce the functions

$$\eta(\omega,\varphi) = \max_{\delta Q \in \mathcal{A}} \text{Re } \delta Q(e^{-j\omega})e^{j[\psi(\omega)-\varphi]}, \tag{5.9}$$

$$\eta_u(\omega,\varphi) = \max_{\delta P_u \in \mathcal{A}_u} \text{Re } \delta P_u(e^{-j\omega})e^{j[\psi_y(\omega)-\varphi]}$$

and

$$\tau(\omega) = \max_{\varphi} \frac{\text{Re } G^0(e^{-j\omega})e^{-j\omega}}{\rho(\omega)\eta(\omega,\varphi) + \rho_y(\omega)\eta_y(\omega,\varphi)} \tag{5.10}$$

where the maximum is taken over those $\varphi \in [0, 2\pi]$, for which the denominator $\tau(\omega)$ of (5.10) is nonzero. The criterion for robust stability may then be formulated as follows: *For the robust stability of the system it is necessary and sufficient that the nominal system would be stable and that the condition*

$$\tau(\omega) > 1, \ 0 \leq \omega < 2\pi \ . \tag{5.11}$$

would be fulfilled.

Geometrically this means that the godograph $G^0(e^{-j\omega})$ does not surround the origin (fig. 6) while the curve $\tau(\omega)$ in the polar coordinate system (τ, ω) does not surround and does not intersect the unit circle with center at the origin [44].

Let us introduce the modified godograph

$$\tilde{\tau}(\omega) = \frac{G^0(e^{-j\omega})}{|G^0(e^{-j\omega})|}\tau(\omega) \ , \tag{5.12}$$

then the formulation of the robust stability criterion is simplified. Namely, for the robust stability of the system it is necessary and sufficient that the modified godograph $\tilde{\tau}(\omega)$ (5.12) would not surround and would not intersect the unit circle with center at the origin (fig. 7).

Under the given criteria the control system will be robustly stable, namely, stable under all the possible variations of the parameters within the given sets. These criteria could be specified for various particular cases of interval, ellipsoidal, octaedric constraints on the plant parameters [43-45]. We however omit the details.

The robust modality of a discrete system will be defined as the property of the zeros of its characteristic polynomial to belong to a set that lies beyond a contour L_s that surrounds a unit circle. Examples of such contours L_s are given in [44]. In general they are described by the equation

$$q = \vartheta_s(\omega)e^{-j\alpha_s(\omega)}, \ 0 \leq \omega < 2\pi \ . \tag{5.13}$$

then the criterion for robust modality could be formulated similar to the criterion of robust stability in which $q = e^{-j\omega}$ should be substituted by the expression for q (5.14). Denote

$$R(\vartheta_s(\omega)e^{-j\alpha_s(\omega)}) = \rho_s(\omega)e^{j\psi_s(\omega)}, \tag{5.14}$$

$$P_y(\vartheta_s(\omega)e^{-j\alpha_s(\omega)}) = \rho_{ys}(\omega)e^{j\psi_{ys}(\omega)}$$

and introduce the functions

$$\eta_s(\omega, \varphi) = \max_{\delta Q \in A} \mathrm{Re}\ \delta Q(\vartheta_s(\omega)e^{-j\alpha_s(\omega)})e^{j[\psi_s(\omega)-\varphi]}, \tag{5.15}$$

$$\eta_{ys}(\omega, \varphi) = \max_{\delta P_u \in A_u} \mathrm{Re}\ \delta P_u(\vartheta_{ys}(\omega)e^{-j\alpha_{ys}(\omega)})e^{j[\psi_{ys}(\omega)-\varphi]}$$

and

$$\tau_s(\omega) = \max_{\varphi} \frac{\operatorname{Re} G^0(\vartheta_s(\omega)e^{j\alpha_s(\omega)})e^{-j\varphi}}{\rho(\omega)\vartheta_s(\omega)\eta_s(\omega,\varphi) + \rho_y(\omega)\vartheta_{ys}(\omega)\eta_{ys}(\omega,\varphi)} \tag{5.16}$$

where the maximum is still taken over all $\varphi \in [0, 2\pi]$ for which the denominator of (5.16) is nonzero.

Introduce a modified godograph

$$\tilde{\tau}_s(\omega) = \frac{G^0(e^{-j\omega})}{|G^0(e^{-j\omega})|}\tau_s(\omega) \ . \tag{5.17}$$

Then the criterion for robust modality may be formulated as follows. For a robust modality it is necessary and sufficient that a modified godograph $\tilde{\tau}_s(\omega)$ (5.17) would not surround and would not intersect the unit circle with center in the origin. The application of this criterion does not require a separate verification of the modality of the nominated system. The criteria for robust stability also could be specified for various particular cases. Control systems for interval plants that have the property of robust stability and modality are further referred to as robustly stable and robustly modal control systems.

Of course an appropriate quality of the process in a robust control system could be ensured under typical external perturbations or under the absence of these. Contrary to traditional adaptive systems the parameters of robust control systems are not fixed but may vary within preassigned sets without violating stability and modality. However even under constant but unknown plant parameters an optimality in a conventional minmax sense cannot be achieved.

An important problem in the theory of stable and robust modal control systems is to establish the relation between robust stability and modality and the structural control system.

6 Robust Nominal Systems under Deterministic Disturbances.

The interval plant determines a family of plants rather than a single plant. Let us select a nominal plant from this variety. It is described by the equation

$$Q^0(q)y(n) = qP_u^0(q)u(n). \tag{6.1}$$

For the nominal plant one may determine a controller

$$R^0(q)u(n) = P_r^0(q)r(n) - P_y^0(q)y(n) \tag{6.2}$$

on the basis of polynomial equations

$$Q^0(q)R^0(q) + qP_u^0(q)P_y^0(q) = T(q)G^{\text{ref}}(q) \tag{6.3}$$

and

$$P_u^0(q)P_r(q) = T(q)H^{\text{ref}}(q). \tag{6.4}$$

The nominal control system (6.1), (6.2) is described by the equation

$$(Q^0(q)R^0(q) + qP_u^0(q)P_y^0(q))y(n) = \tag{6.5}$$

$$= qP_u^0(q)P_r^0(q)r(n)$$

and satisfies given requirements, determined by the specific form of the nominal equations (6.3), (6.4).

Consider an interval plant (2.1)

$$Q(q)y(n) = qP_u(q)u(n) + P_f(q)f(n). \tag{6.6}$$

We will say that the control system for this plant is robustly nominal if the processes in this plant are determined by the equation of a nominal system (6.5).

For synthesizing a controller for a robustly nominal system, we introduce a model [35]

$$\hat{y}(n) = [1 - Q^0(q)]y(n) + qP_u^0(q)u(n). \tag{6.7}$$

The bias

$$\mathcal{E}(n) = y(n) - \hat{y}(n) \tag{6.8}$$

is equal to

$$\mathcal{E}(n) = Q^0(q)y(n) - qP_u^0(q)u(n) \tag{6.9}$$

as follows from (6.7).

However as pointed out in Section 2, the interval plant is equivalent to the nominal plant subjected to a generalized disturbance (2.4), (2.6)

$$Q^0(q)y(n) = qP_u^0(q)u(n) + \varphi(n). \tag{6.10}$$

From (6.8) and (6.9) it follows that

$$\mathcal{E}(n) = \varphi(n). \tag{6.11}$$

Therefore, the bias between the outputs $y(n)$ of the interval plant and $\hat{y}(n)$ of the nominal plant is equal to the generalized disturbance $\varphi(n)$.

Consider a controller whose equation differs from (6.2) by a term that depends on the bias

$$R^0(q)u(n) = P_r^0(q)r(n) - P_y^0(q)y(n) - A(q)\mathcal{E}(n), \qquad (6.12)$$

where $A(q)$ is a polynomial unknown at the moment.

Eliminating the control $u(n)$ from the equations (6.9) and (6.12), we come to the equation of the closed system, due to (6.11),

$$(Q^0(q)R^0(q) + qP_u^0(q)P_y^0(q))y(n) = \qquad (6.13)$$

$$= qP_u^0(q)P_r^0(q)r(n) + (R^0(q) - qP_u^0(q)A(q))\varphi(n).$$

If we select $A(q)$ so that

$$R^0(q) - qP_u^0(q)A(q) = (1 - qD(q))B(q), \qquad (6.14)$$

where $D(q)$ is the predictor polynomial (2.15), and $B(q)$ is a certain auxiliary polynomial, then, due to (2.17) the last term in (6.13) turn to be zero.

Indeed, in view of (2.17), we have

$$(1 - qD(q))B(q)\varphi(n) = B(q)(1 - qD(q))\varphi(n) = 0 \qquad (6.15)$$

and from (6.13) we come to the equation of the nominal system (6.5) (in which, we emphasize, there are no disturbances). Rewriting (6.14) as

$$qP_u^0(q)A(q) + (1 - qD(q))B(q) = R^0(q), \qquad (6.16)$$

we come to a polynomial equation that determines the polynomial $A(q)$ and the auxiliary polynomial $B(q)$.

Thus, equations (6.6), (6.12), (6.9) and (6.16) define a robustly nominal system. Being robust, this system should, of course, retain the stability property over the whole range of plants that belong to the interval plant. The generalized disturbance $\varphi(n)$ (6.11) will then contain the perturbations $\varphi_p(n)$ (2.7) whose absolute values are bounded. For defining some conditions for robustness, we have to find the characteristic polynomial of the robust nominal system described by equations (6.6), (6.12), (6.9). Making use of the representation of the polynomials in the form of a sum of nominal polynomials and variations (2.2) it is not difficult to see that the characteristic polynomial is equal to

$$G(q) = G^0(q) + (R^0(q) - qP_u^0(q)A(q))\delta Q(q) +$$

$$+ q(P_y^0(q) + Q^0(q)A(q))\delta P_u(q), \qquad (6.17)$$

where

$$G^0(q) = Q^0(q)R^0(q) + qP_u^0(q)P_y^0(q) \tag{6.18}$$

is the characteristic polynomial for the nominal system.

The characteristic polynomial $G(q)$ must satisfy the conditions for robust stability. The criteria for robust stability are given in Section 3. These criteria satisfy not only the conditions of stability but also those of realizability of robust nominal systems.

Let us now formulate the principle of constructing robust nominal systems under regular deterministic disturbances.

If the interval characteristic polynomial

$$G(q) = G^0(q) + (R^0(q) - qP_u^0(q)A(q))\delta Q(q) +$$

$$+q(P_y^0(q) + Q^0(q)A(q))\delta P_u(q)$$

is stable for all $\delta Q(q) \in \mathcal{A}$, $\delta P_u(q) \in \mathcal{A}_u$ and $\delta P_f(q) \in \mathcal{A}_f$ and there exists a predictor polynomial $D(q)$ (2.15) such that

$$(1 - qD(q))\varphi(n) = 0,$$

then for the interval plant

$$Q(q)y(n) = qP_u(q)u(n) + P_f(q)f(n)$$

the controller

$$R^0(q)u(n) = P_r^0(q)r(n) - P_y^0(q)y(n) - A(q)\mathcal{E}(n),$$

where

$$\mathcal{E}(n) = Q^0(q)y(n) - qP_u^0(q)u(n)$$

and $A(q)$ the solution to the polynomial equation

$$qP_u^0(q)A(q) + (1 - qD(q))B(q) = R^0$$

will determine a robust nominal system.

The block–scheme of a robust nominal system is given in fig. 8. It contains a nominal model that produces the bias $\mathcal{E}(n)$. In traditional adaptive systems the parameters of the nominal model are fixed in contrast with those of the tuned model.

Consider some useful particular cases. Assume that

$$R^0(q) = P_u^0(q). \tag{6.19}$$

In this case the numerator of the transfer function of the phase–minimal plant is cancelled out with the denominator of the transfer function for the controller [2, 36]. After a substitution (6.19) in the polynomial equation (6.16) we conclude that $B(q) = P_u^0(q)$ and after cancelling out the term $P_u^0(q)$ and performing some transformations we come to

$$A(q) = D(q). \tag{6.20}$$

Thus, under condition (6.19) the necessity in solving the polynomial equation disappears. A block–scheme of such a robustly nominal system coincides with the scheme in fig. 8 if $R^0(q)$ is substituted by $P_u^0(q)$ and $A(q)$ by $D(q)$. This particular case was considered in [46]. Assume that

$$A(q) = A_1(q)R^0(q). \tag{6.21}$$

In this case $B(q) = B_1(q)R^0(q)$ and after cancelling $R^0(q)$ we come to a polynomial equation of type

$$qP_u^0(q)A_1(q) + (1 - qD(q))B_1(q) = 1. \tag{6.22}$$

The block–scheme of this robust nominal system is shown in fig. 9. It differs from the one given in fig. 8 by having an autonomous loop for compensating the generalized disturbance.

A robust nominal scheme is closely related to systems with an internal model [12]. This internal model has a two-dimensional input [35]. It has an impulse reaction that is finite in time.

From equation (6.13) under conditions (6.14), (6.15) we arrive at the equation of a closed–loop nominal system

$$(Q^0(q)R^0(q) + qP_u^0(q)P_y^0(q))y(n) = \tag{6.23}$$

$$qP_u^0(q)P_r^0(q)r(n),$$

or, having in view the notations (1.3), (1.7), (1.8)

$$y(n) = \frac{W_0(q)W_r(q)}{1 - W_0(q)W_r(q)}r(n)\ . \tag{6.24}$$

A structural scheme of the nominal system is given in fig. 10. The processes in a robustly nominal system (fig. 8, fig. 9) coincide (with the accuracy of having some irregular components) with the processes in a nominal system which has neither any input nor any parametric disturbances, fig. 10.

7 Robust Nominal Systems Under Stochastic Disturbances.

In this case the equation of the nominal plant has the form

$$Q^0(q)y(n) = qP_u^0(q)u(n) + \xi(n), \tag{7.1}$$

where $\xi(n)$ satisfies conditions (2.24). For a nominal plant a synthesizing controller

$$R^0(q)u(n) = P_r^0(q)r(n) - P_y^0(q)y(n) \tag{7.2}$$

is designed due to the polynomial equations

$$Q^0(q)R^0(q) + qP_u^0(q)P_y^0(q) = T(q)G^{\text{ref}}(q), \tag{7.3}$$

$$P_u^0(q)P_r^0(q) = T(q)H^{\text{ref}}(q). \tag{7.4}$$

A nominal control system (7.1), (7.2) is described by the equation

$$(Q^0(q)R^0(q) + qP_u^0(q)P_y^0(q))y(n) = \tag{7.5}$$

$$= qP_u^0(q)P_r^0(q)r(n) + R^0(q)\xi(n)$$

and satisfies given requirements, defined by specific form of the polynomial equations (7.3), (7.4). Thus if $R^0(q) = P_u^0(q)$, then $T(q) = T_1(q)P_u^0(q)$, and after cancelling $P_u^0(q)$, we come to the polynomial equations

$$Q^0(q) + qP_y(q) = T_1(q)G^{\text{ref}}(q), \tag{7.6}$$

$$P_r^0(q) = T_1(q)H^{\text{ref}}(q), \tag{7.7}$$

which, as it is known [1, 3, 4], define the optimal nominal system, in which the error $e(n) = y(n) - y^0(n)$ has a minimal variance.

Consider an interval plant

$$Q(q)y(n) = qP_u(q)u(n) + P_f(q)f(n), \tag{7.8}$$

where now $f(n)$ is an AR–disturbance (2.25). We will say that a control system for this interval plant is a robust nominal system if it is described by an equation that differs from the equation for the nominal system (7.5) only by a term that depends on a stochastic disturbance $\xi(n)$.

Consider a controller similar to (6.12)

$$R^0(q)u(n) = P_r^0(q)r(n) - P_y^0(q)y(n) - \frac{A(q)}{C(q)}\mathcal{E}(n), \tag{7.9}$$

where the polynomials $A(q)$ and $C(q)$ are unknown at the moment and $\mathcal{E}(n)$ is still determined from (6.9). From equations (7.8), (7.9) in view of (6.11) we determine the equation for the closed loop system

$$(Q^0(q)R^0(q) + qP_u^0(q)P_y^0(q))y(n) = \tag{7.10}$$

$$qP_u^0(q)P_r^0(q)r(n) + \frac{R^0(q) - qP_u^0(q)A(q)}{C(q)}\varphi(n) .$$

Let us find the polynomials $A(q)$ and $C(q)$ from a condition similar to (6.14)

$$R^0(q) - qP_u^0(q)A(q) = (1 - qD(q))B(q), \tag{7.11}$$

where $D(q)$ is the predictor polynomial (2.15) and $B(q)$ is an auxiliary polynomial. Here

$$(1 - qD(q))B(q)\varphi(n) = B(q)(1 - qD(q))\varphi(n) =$$

$$= B(q)(1 - qD_1(q))\xi(n). \tag{7.12}$$

Also select

$$C(q) = 1 - \beta qD_1(q), \tag{7.13}$$

where β is a term close but not equal to unity, i.e. $\beta \lesssim 1$. Taking into account (7.11) and (7.12), equation (7.10) has the form

$$(Q^0(q)R^0(q) + qP_u^0(q)P_y^0(q))y(n) = \tag{7.14}$$

$$= qP_u^0(q)P_r^0(q)r(n) + B(q)\xi_\beta(n),$$

where

$$\xi_\beta(n) = \frac{1 - qD_1(q)}{1 - \beta qD_1(q)}\xi(n) \tag{7.15}$$

is a stochastic disturbance with variance

$$\sigma_\beta^2 = M\{\xi_\beta^2(n)\} = \sigma^2 + (1 - \beta^2)M_0. \tag{7.16}$$

With β close to unity, σ_β^2 is close to σ^2. Comparing the equations of a robust nominal system (7.14) with those for a nominal system (7.5) we conclude that they differ from each other through the terms $B(q)\xi_\beta(n)$ and $R^0(q)\xi(n)$ that depend on a random disturbance $\xi(n)$. From (7.11) and (7.13) we derive a polynomial equation that determines $A(q)$ and $B(q)$

$$qP_u^0(q)A(q) + (1 - qD(q))B(q) = (1 - \beta qD_1(q))R^0(q). \tag{7.17}$$

A characteristic polynomial for a robust nominal system described by equations (7.8), (7.9), (6.9) has the form

$$G(q) = G^0(q) + (C(q)R^0(q) - qO_u^0(q)A(q))\delta Q(q)+$$

$$+ q(C(q)P_y(q) + Q^0(q)A(q))\delta P_u(q), \tag{7.18}$$

and

$$G^0(q) = (Q^0(q)R^0(q) + qP_u^0)(q)P_y^0(q))C(q) \tag{7.19}$$

is the characteristic polynomial of the nominal system. The characteristic polynomial $G(q)$ must satisfy the condition for robust stability.

Let us formulate a principle for designing robust nominal systems under stochastic disturbances. If the interval characteristic polynomial

$$G(q) = G^0(q) + (C(q)R^0(q) - qP_u^0(q)A(q)\delta Q(q)+$$

$$+q(C(q)P_y^0(q) + Q^0(q)A(q))\delta P_u(q)$$

is stable and there exist predictor polynomials $D(q)$ and $D_1(q)$ such that

$$(1 - qD(q))\varphi(n) = (1 - qD_1(q))\xi(n)$$

for all $\delta Q(q) \in \mathcal{A}$, $\delta P_u(q) \in \mathcal{A}_u$, $\delta P_f(q) \in \mathcal{A}_f$, then for the interval plant

$$Q(q)y(n) = qP_u((q)u(n) + P_f(q)f(n)$$

the controller

$$R^0(q)u(n) = P_r^0(q)r(n) - P_y(q)y(n) - \frac{A(q)}{C(q)}\mathcal{E}(n),$$

where

$$\mathcal{E}(n) = Q^0(q)y(n) - qP_u^0(q)u(n)$$

and the polynomial

$$C(q) = 1 - \beta q D_1(q), \ \beta \lesssim 1,$$

while the polynomial $A(q)$ is the solution of the polynomial equation

$$qP_u^0(q)A(q) + (1 - qD(q))B(q) = (q - \beta q D_1(q))R^0(q),$$

does determine a robustly nominal system under a stochastic disturbance.

The block–scheme for a robust nominal system is shown in fig. 11. Particularly, with

$$R^0(q) = P_u^0(q) \tag{7.20}$$

it follows from the polynomial equation (7.17) that $B(q) = P_u^0(q)$, and after cancelling $P_u^0(q)$ we have

$$A(q) = D(q) - \beta D_1(q). \tag{7.21}$$

In this case the block–scheme of a robust nominal system coincides with the block–scheme shown in fig. 11 if $A(q)$ is substituted by $D(q)-\beta D_1(q)$.

If we assume that

$$A(q) = A_1(q)R^0(q), \tag{7.22}$$

then, due to (7.17), $B(q) = B_1(q)R^0(q)$, and after cancelling $R^0(q)$ we obtain, instead of (7.17), a polynomial equation

$$qP_u^0(q)A_1(q) + (1 - qD(q))B_1(q) = 1 - \beta q D_1(q). \tag{7.23}$$

The block–scheme with an autonomous compensation loop is shown in fig. 12.

8 Adaptive Robustly Nominal Systems.

In the case when the application of predictor polynomials of the simplest form $D(q) = 1$ or $D(q) = 2 - q$ is not effective, there arises a necessity in the estimation of the coefficients of the predictor polynomial through the observation of the bias $\mathcal{E}(n)$.

Since the bias is equal to the generalized disturbance (6.11)

$$\mathcal{E}(n) = \varphi(n), \tag{8.1}$$

it follows from (2.16) that

$$\mathcal{E}(n) = D(q)\mathcal{E}(n-1), \tag{8.2}$$

where $D(q)$ is the predictor polynomial with unknown vector of coefficients

$$d^* = \operatorname{coeff} D(q). \tag{8.3}$$

For estimating these coefficients one may use recurrence identification algorithms of type (4.7). Now however, these algorithms will identify the generalized disturbance instead of the plant itself. For the polynomial $\hat{D}(q) = D(q,\theta)$ with arbitrary coefficients the estimate of the bias is equal to

$$\hat{\mathcal{E}}(n) = \hat{d}(q)\mathcal{E}(n-1) = D(q,\theta)\mathcal{E}(n-1), \tag{8.4}$$

while the estimate of the parameter d^* on the basis of the observations

$$x_{\mathcal{E}}(n) = (\mathcal{E}(n-1), \mathcal{E}(n-2), ..., \mathcal{E}(n-M))^T \tag{8.5}$$

is realized by recurrence algorithms similar to (4.7)

$$d(n) = d(n-1) + \Gamma(n)[\mathcal{E}(n) - \hat{\mathcal{E}}(n)]x_{\mathcal{E}}(n), \tag{8.6}$$

or

$$d(n) = d(n-1) + \Gamma(n))(\mathcal{E}(n) - D(q, d(n-1))\mathcal{E}(n-1)x_{\mathcal{E}}(n), \tag{8.7}$$

where $\Gamma(n)$ is the gain matrix, the examples of which were given in Section 4. The estimates of the predictor polynomial $\hat{D}(q)$ allow to find the polynomials $\hat{A}(q)$ and $\hat{B}(q)$ from the polynomial equation (6.16)

$$qP_u(q)\hat{A}(q) + (1 - q\hat{D}(q))\hat{B}(q) = R^0(q). \tag{8.8}$$

This operation is repeated in each stage. With n increasing the estimates of the predictor polynomial $\hat{D}(q)$ tend to $D(q)$ and therefore $\hat{A}(q)$ tends to $A(q)$. The block–scheme of an adaptive robustly nominal system is given in fig. 13.

Under stochastic input disturbances instead of (8.2) we have, in view of (8.1)

$$\mathcal{E}(n) = D(q)\mathcal{E}(n-1) - D_1(q)\xi(n-1) + \xi(n). \tag{8.9}$$

This is an equation of autoregression and of a sliding mean (ARSM).

Let us denote the vector of the unknown coefficients as

$$d^* = \operatorname{coeff} D(q),\ d_1^* = \operatorname{coeff} D_1(q), \tag{8.10}$$

and introduce the vector of all coefficients as

$$d = (d^*,\ d_1^*) = \operatorname{coeff}(D(q),\ D_1(q)). \tag{8.11}$$

The estimate of these coefficients through observations

$$x_{\mathcal{E}}(n) = (\mathcal{E}(n-1), \mathcal{E}(n-2), ..., \mathcal{E}(n-M)), \tag{8.12}$$

$$\mathcal{E}(n-1) - \hat{\mathcal{E}}(n-1), ..., \mathcal{E}(n-M_1) - \hat{\mathcal{E}}(n-M_1))^T$$

is given by recurrence algorithms similar to (4.21)

$$d(n) = d(n-1) + \Gamma(n)(\mathcal{E}(n) - \hat{\mathcal{E}}(n))x_{\mathcal{E}}(n),$$

or

$$d(n) = d(n-1) + \Gamma(n)[\mathcal{E}(n) - D(q, d(n-1))\mathcal{E}(n-1) -$$
$$- D_1(q, d^1(n-1))(\mathcal{E}(n-1) - \hat{\mathcal{E}}(n-1))]x_{\mathcal{E}}(n). \tag{8.13}$$

Here $\Gamma(n)$ is the gain matrix, the examples of which were given above.

The estimates of the predictor polynomials $\hat{D}(q)$ and $\hat{D}_1(q)$ allow to determine the estimates for the polynomial $\hat{A}(q)$ and $\hat{B}(q)$ from the equation (7.17)

$$qP_u^0(q)\hat{A}(q) + (1 - q\hat{D}(q))\hat{B}(q) = \qquad (8.14)$$
$$= (1 - \beta q\hat{D}(q))R^0(q).$$

Under the known convergence conditions with n increasing, the values $\hat{A}(q)$ and $\hat{B}(q)$ tend to $A(q)$ and $B(q)$.

The block–scheme of an adaptive robustly nominal system under stochastic disturbances is shown in fig. 14.

After the end of the adaptation process, the adaptively robust nominal systems settle at the processes that correspond to those in the nominal control systems. In the absence of parametric disturbances the adaptively robust nominal systems reduce to adaptively invariant systems [34].

9 Conclusion.

Interval dynamic plants are those that operate under bounded uncertainty. The processing of this "a priori" information within the identification process through a tunable model and through respective parameter estimation algorithms does not allow to substantially improve the identification process (e.g. to increase the rate of identification). Therefore traditional adaptive schemes do not yield any advantages here. Moreover, their complexity and high nonlinearity which are due to variations of the parameters of the controller, do lead to a complex system behaviour (bifurcations, chaotic motions) which is not yet well studied by now.

It is certain however, that traditional adaptive systems have their specific range of applications, where the mentioned phenomena could be surpassed. Here it is important that they keep the appropriate asymptotic properties.

The conditions for the realizability of traditional adaptive schemes is the stability of the basic loop and the convergence of the adaptation algorithms. A large number of papers is devoted to the proof of the algorithm convergence.

The simplest structures are those of robustly stable control systems that have a fixed controller for controlling the interval plant. However, the properties of these robustly stable systems vary with the variations of the plant parameters. At the best they can only allow a proper quality of the control process (like stability, degree of stability, degree of oscillation etc.) The conditions for the realizability of robustly stable systems is robust stability and robust modality, determined by the characteristic polynomial $G(q)$ for the closed–loop system. This polynomial depends on both the fixed nominal polynomial and also on the parameter variations for the interval plant. In traditional adaptive systems as well as in robustly stable systems it is usually assumed that external perturbations are either absent or that they are of a standard type (jumps, harmonic, independent, stochastic). The quality of these systems under other types of disturbances, e.g. under persistent regular disturbances is substantially worse. In robustly nominal systems we eliminate these limitations as those caused by the nonphase–minimality, i.e., of the instability of the plant.

The principal condition for the realizability of robustly nominal systems is their robust stability, determined by the characteristic polynomial $G(q)$ of the closed–loop system that depends on both the nominal characteristic polynomial and also on the parameter variations for the interval plant, as well as on the predictor polynomial for the bias. In contrast with robustly stable systems, the robust stability of the characteristic polynomial $G(q)$ determines not only the stability of the robustly nominal system, but also its realizability and nonsensibility to variations of the parameters of the interval plant. Thus the implementation of simple frequency domain criteria becomes very important for investigating robust stability.

For synthesizing robust nominal systems it is important to select a nominal plant and therefore, a nominal model. This selection is not yet formalized. In the case when the information on the bias is insufficient for defining the predictor polynomial $D(q)$, one arrives at the problem of identifying this bias i.e., of estimating the coefficients of the predictor polynomial.

The identification of the bias may of course turn to be simpler than the identification of the plant itself. This fact simplifies the derivation and justification of adaptive identification algorithms and leads to robust adaptive control systems.

References

[1] Goodwin G.C., Sin K.S. *Adaptive filtering prediction and control.* New Jersey: Prentice Hall, 1984.

[2] Tsypkin Ya.Z., Aved'yan E.D. *Discrete adaptive systems of deterministic plants control.* Itogi Nauki i Techniki, Ser. Techn. Kibernetika. Moscow: VINITI, 1985, V. 18, pp. 45-78. (in Russian).

[3] Aström K.J. *Introduction to stochastic control theory.* New York: Academic Press, 1970.

[4] Tsypkin Ya.Z., Kelmans G.K. *Discrete adaptive control systems.* Itogi Nauki i Techniki, Ser. Techn. Kibernetika. Moscow: VINITI, 1984, V. 17, pp. 3-73. (in Russian).

[5] Basar T. *Theory of stochastic dynamic noncooperative games.* Oxford: basil Blackwell, 1990.

[6] Kuntsevitch V.M., Lychak M. *Guaranteed estimates. Adaption and robustness in control systems.* Lecture Notes in Control and Information Sciences. V. 169, Berlin: Springer Verlag, 1992.

[7] Voronov A.A., Rutkovsky V.Y. *State-of-the-art and prospects of adaptive systems* Automatica. The Journal of IFAC. 1984, V. 20, N 5, pp.547–557.

[8] Aström K.J., Wittenmark B. *Adaptive control systems.* Addison–Wesley, Reading, Mass., 1990.

[9] Tsypkin Ya.Z. *Adaptation and learning in automatic systems.* New York: Academic Press, 1971.

[10] Landau Y.D. *Adaptive control. The model reference approach.* New York: Marcel Dekker Inc., 1979.

[11] Ortega R., Tang Y. *Robustness of adaptive controllers – a survey.* Automatica. 1989. V 25, N 5, pp. 651–677.

[12] Morari M., Zafiriorou E. *Robust process control.* New Jersey: Prentice Hall, 1989.

[13] Basar T. *Game theory and* H^∞*–optimal control: the discrete–time case.* Proceedings of the 1990 International Conference on New Trends in Communication Control and Signal Processing. Ankara, Turkey, July, 1990, pp.669–686, Elsevier.

[14] Francis B.A. *A course in* H^∞*–control theory.* Lecture Notes in Control and Information Sciences. V. 88, Berlin: Springer Verlag, 1987.

[15] Francis B.A., Melton J.W., Zames G. H^∞*–optimal feedback controllers for linear multivariable systems.* IEEE Trans. Autom. Control. 1984. V. 29, N 10, pp.888–899.

[16] Tsypkin Ya.Z. *Sampling systems theory and its applications.* Oxford: Pergamon Press, vol. 1,2, 1964.

[17] Kuo B.C. *Digital control systems.* New York: Halt Rinehart and Winston Inc., 1980.

[18] Volgin L.N. *Optimal discrete control of dynamic systems.* Moscow: Nauka, 1986. (in Russian).

[19] Tsypkin Ya.Z. *Frequency modality criteria of linear discrete systems.* Automatica. 1990, N 3, pp. 3–7.

[20] Tsypkin Ya.Z. *Synthesis of optimal systems for non-minimum phase plants.* International Journal of System Science. 1992, V. 23, N 2, pp. 291–296.

[21] Guerrieri J. *Methods of introductions functional relations automatical on different analysers.* Thesis MIT, 1932.

[22] Shannon C. *Mathematical theory of the differential analyzer.* Journal of Mathematics and Physics. 1941, V. 20, N 4, pp. 337.

[23] Kulebakin B.S. *On behavior of constantly disturbed automatization linear systems.* Doklady AN SSSR, V. 68, N 5, pp. 73–79.

[24] Johnson C.D. *Accomodation of external disturbances in linear regulator and servomechanism problems.* IEEE Trans. Autom. Control. 1971. V. AC-16, N 6, pp. 635–644.

[25] Johnson C.D. *Theory of disturbance–accomodating controllers.* In "Advances in Control and Dynamic Systems". C.T. Leondes (ed.), V. 12, N 7, New York: Academic Press. 1976.

[26] Davisson E.J. *The output control at linear time invariant systems with unmeasurable arbitrary disturbances.* IEEE Trans. Autom. Control. 1972. V. AC-17. N 5, pp. 621–630.

[27] Wonham W.M. *Linear multivariable control: a geometric approach.* Berlin: Springer Verlag, 1979.

[28] Francis B.A., Wonham W.M. *THe internal model principle for linear multivariable regulators.* Applied Mathematics and Optimization, 1975, V. 22, N 5, pp. 170–194.

[29] Francis B.A., Wonham W.M. *The internal model principle of control theory.* Automatica, 1976, V. 12, N 5, pp. 457–465.

[30] Gonzalez D.R., Autsaklis P.J. *Internal models in regulation, stabilizing and tracking.* International Journal of Control, 1991, V. 53, N 2, pp. 411–430.

[31] Nikolskii V.A., Sevastjanov I.P. *$K(E)$-transformation of sampled functions in the problem of discrete systems research.* Avtomatika i elektromekhanika. Moscow: Nauka, 1973, pp. 30–36. (in Russian).

[32] Johnson C.D. *Discrete-time disturbance accomodating control theory with applications to missile digital control.* J. Guidance and Control. 1981, V. 4, N 2, pp. 116–125.

[33] Ulanov G.M. *Dynamic accuracy and disturbances compensation in automatic control systems.* Moscow: Mashinostroenie, 1971. (in Russian).

[34] Tsypkin Ya.Z. *Adaptive-invariant discrete control systems.* In "Foundations of adaptive control". Ed. P.V. Kokotovich. Lecture Notes in Control and Information Science. V. 160, Berlin: Springer Verlag, 1991. pp. 239–268.

[35] Zypkin Ja.S. (Tsypkin Ya.Z.) *Grunlagen der informationellen Theorie der Identification.* Berlin: VEB Verlag Technik, 1987.

[36] Aström K.J., Wittenmark B. *Computer controlled systems. Theory and design.* New Jersey: Prentice Hall, 1984.

[37] Anderson B.D.O., Bitmead R.R. et al. *Stability of adaptive systems passivity and averaging analysis.* Cambridge, Massachusetts, London: MIT Press, 1986.

[38] Anderson B.D.O., Johnstone R.M. *Adaptive systems and time varying plants.* International Journal of Control. 1983. V. 37, N 4, pp. 367–377.

[39] Horowitz I.M. *Synthesis of feedback systems.* New York: Academic Press, 1963.

[40] Francis B., Doyle J. *Linear control theory with an H^∞-optimality criterion.* SIAM Journal of Control. 1987. V. 25, N4, pp. 815–844.

[41] McFarlane D.C., Glover K. *Robust controller design using normalized coprime factor plant descriptions.* Lecture Notes in Control and Information Sciences. V. 138, Berlin: Springer Verlag, 1990.

[42] Kharitonov V.L. *Asymptotic stability of systems family of differential equations.* Differential Equations, 1978, V. 14, N 11, pp. 2086–2088. (in Russian).

[43] Jury E.I. *Robustness of discrete systems. A Survey.* Avtomatika i telemekhanika, 1990, N 5, pp. 3–28.

[44] Tsypkin Ya.Z., Polyak B.T. *Frequency criterion of robust modality of linear discrete systems.* Avtomatika, Kiev, 1990, N 3, pp. 3–9. (in Russian).

[45] Polyak B.T., Tsypkin Ya.Z. *Robust stability of discrete linear systems.* Soviet Physics Doklady, 1991, V. 36(2), pp. 111–113.

[46] Tsypkin Ya.Z. *Robust adaptive control systems.* Soviet Physics Doklady, 1990, V. 35, N 12, pp. 1013–1014.

Institute for Control Problems
Russian Academy of Sciences
Profsoyuznaya 65, Moscow 117342, Russia

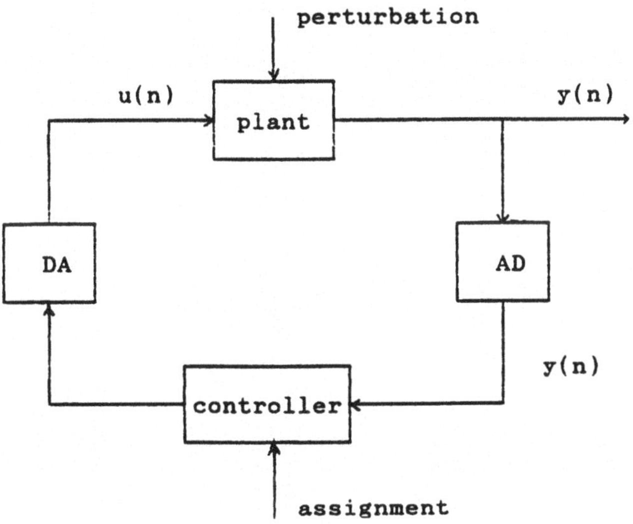

Figure 1.

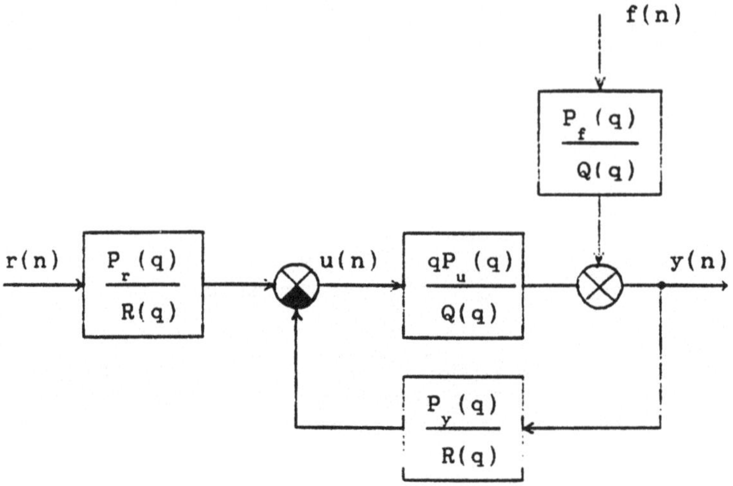

Figure 2a

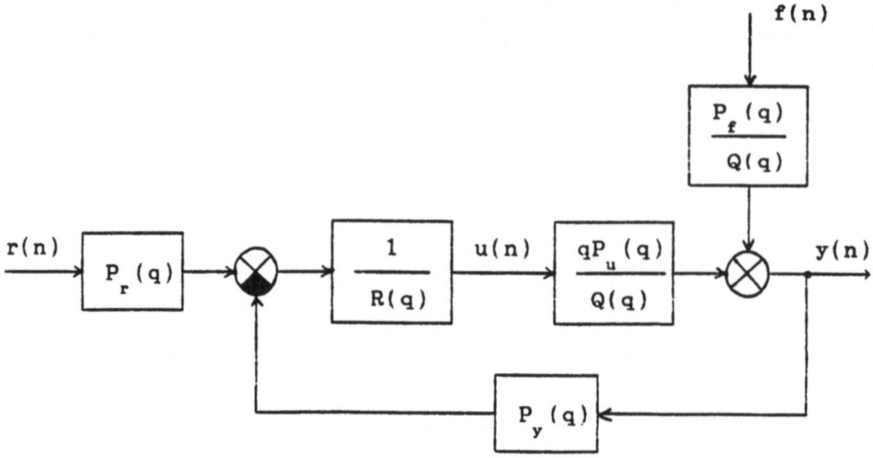

Figure 2b

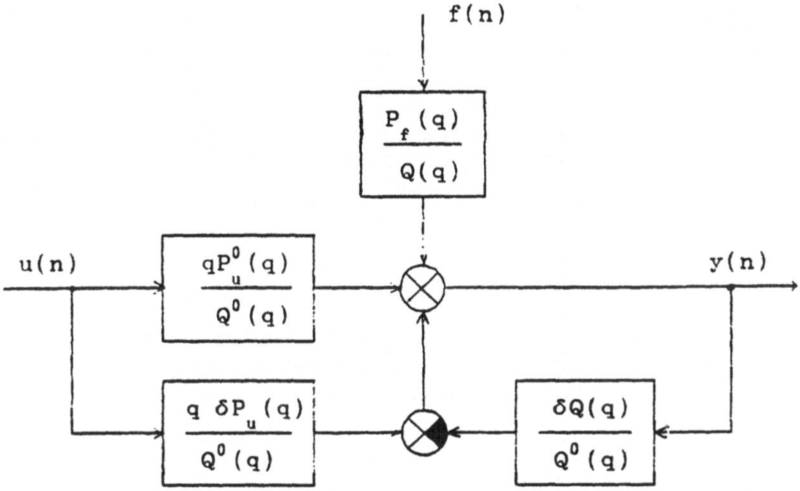

Figure 3.

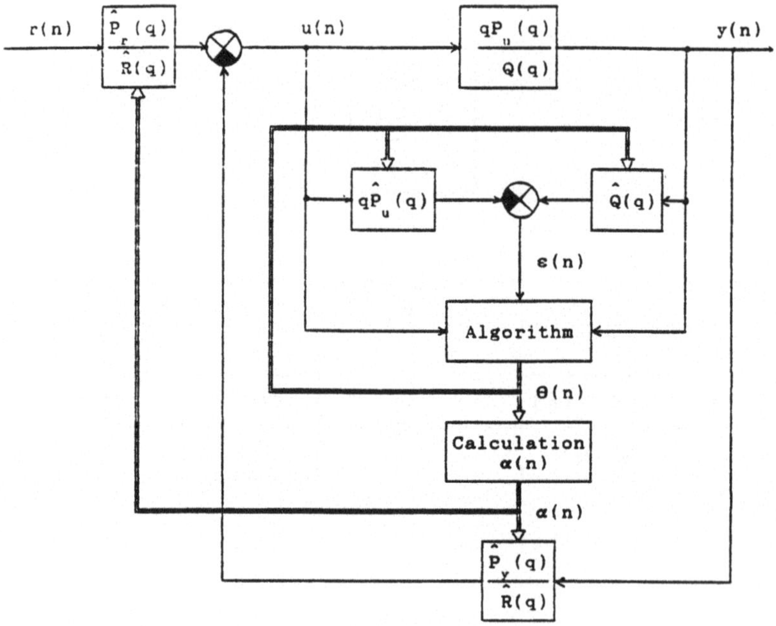
r(n)
$\hat{P}_r(q)$
$\hat{R}(q)$
u(n)
$qP_u(q)$
Q(q)
y(n)
$q\hat{P}_u(q)$
$\hat{Q}(q)$
ε(n)
Algorithm
θ(n)
Calculation
α(n)
α(n)
$\hat{P}_y(q)$
$\hat{R}(q)$

Figure 4.

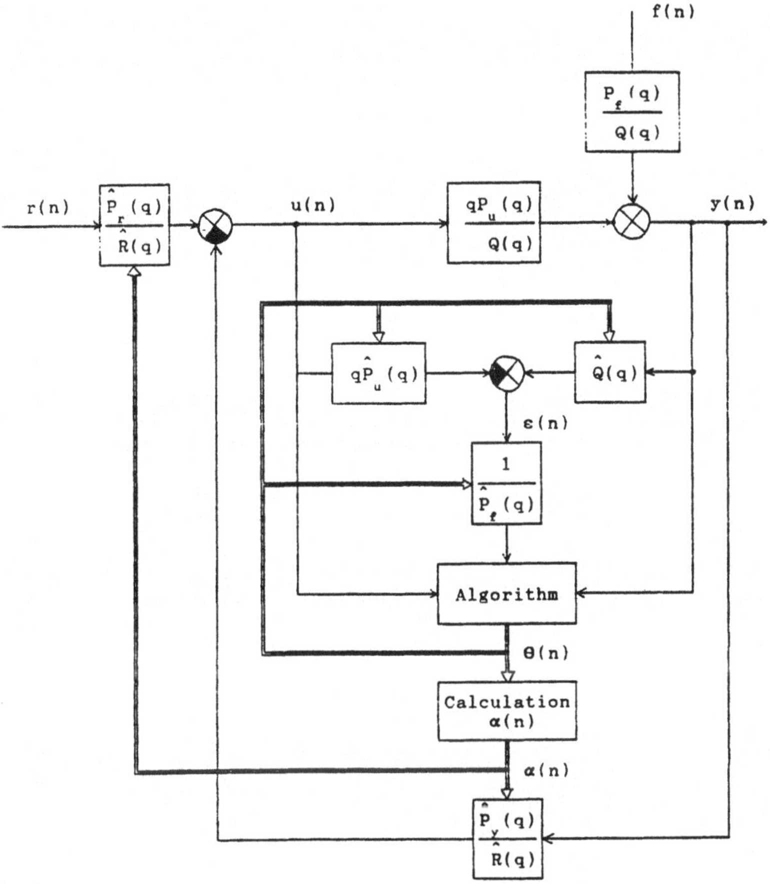
f(n)
Pf(q)
Q(q)
r(n)
Pr(q)
R(q)
u(n)
qPu(q)
Q(q)
y(n)
qPu(q)
Q(q)
ε(n)
1
Pf(q)
Algorithm
θ(n)
Calculation
α(n)
α(n)
Py(q)
R(q)

Figure 5.

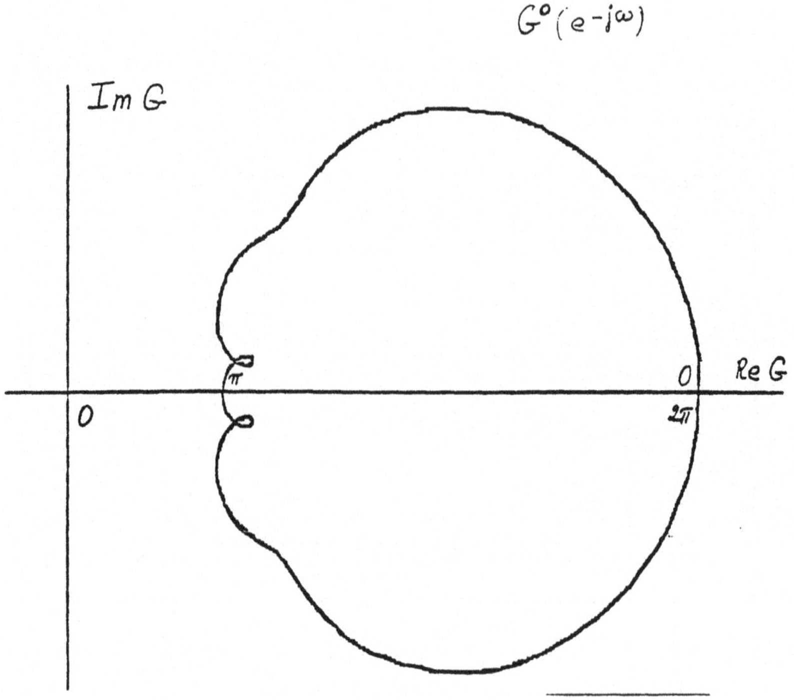

Figure 6.

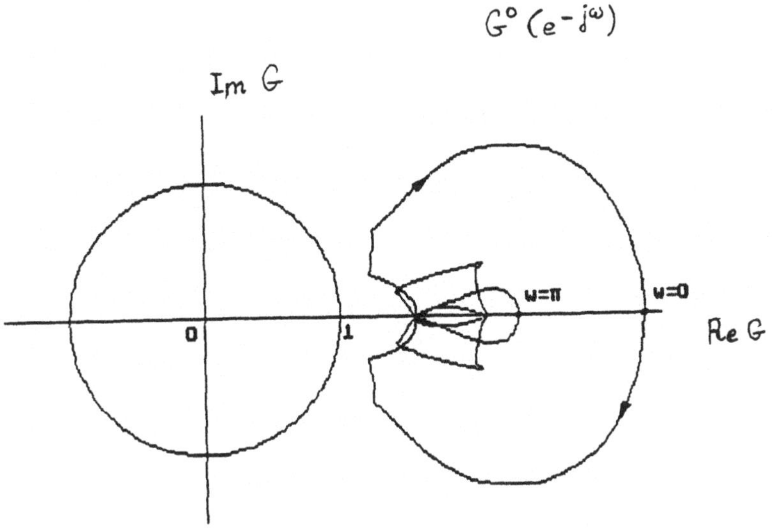

Figure 7.

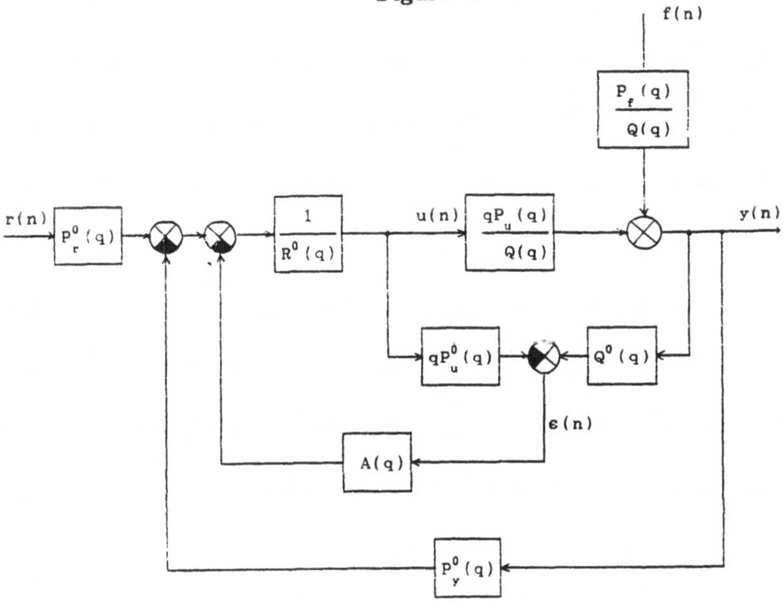

Figure 8.

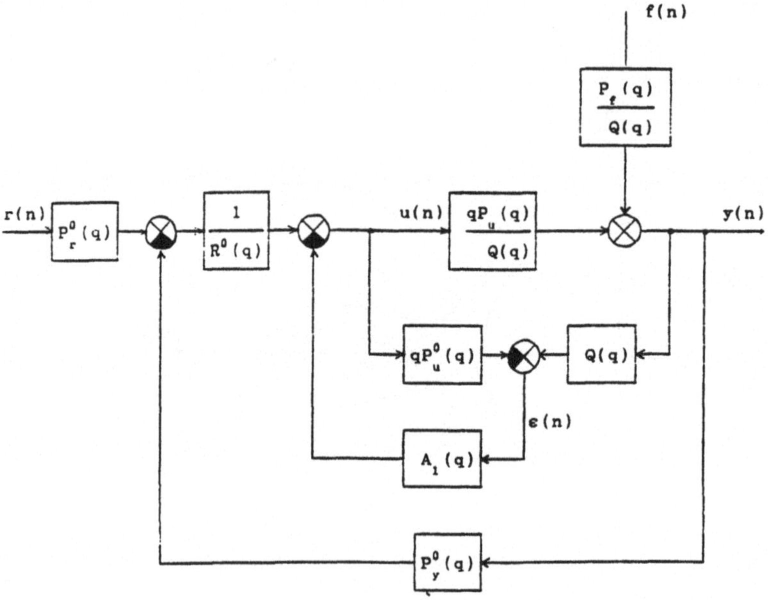

Figure 9.

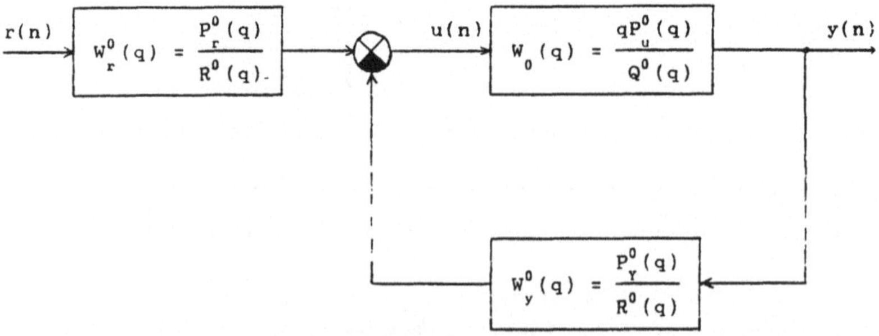

Figure 10.

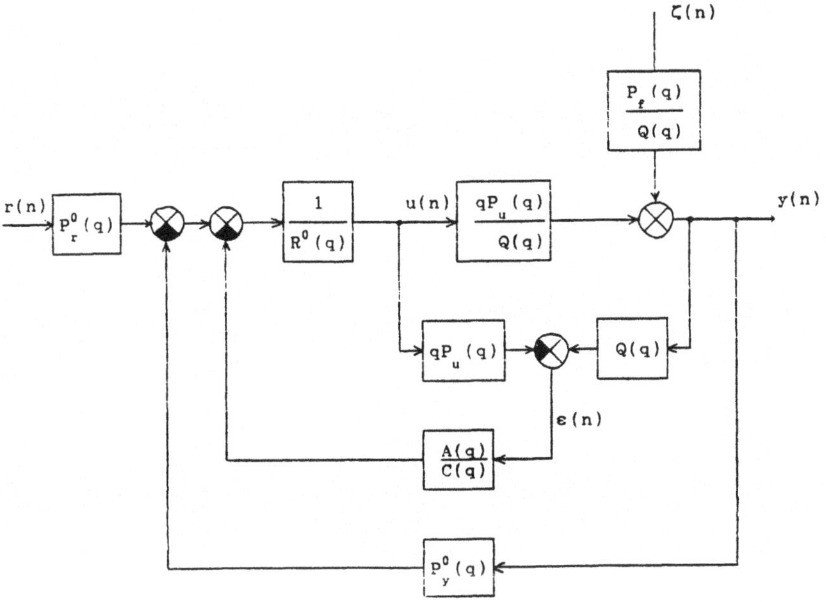

Figure 11.

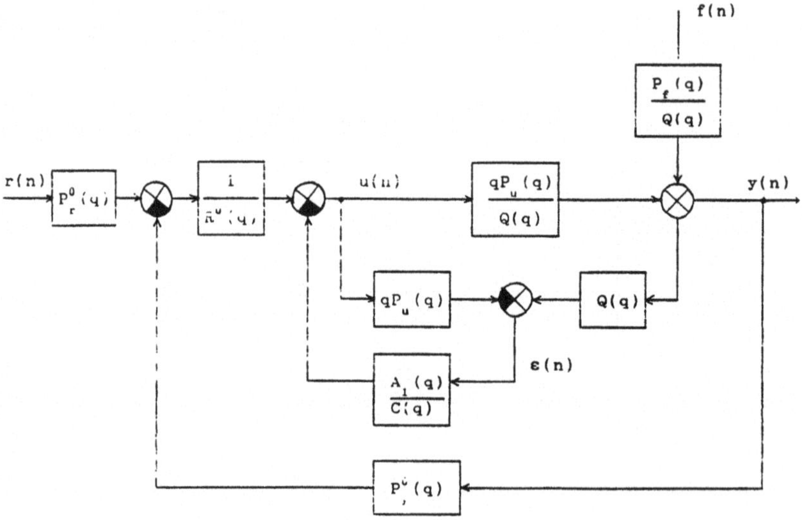

Figure 12.

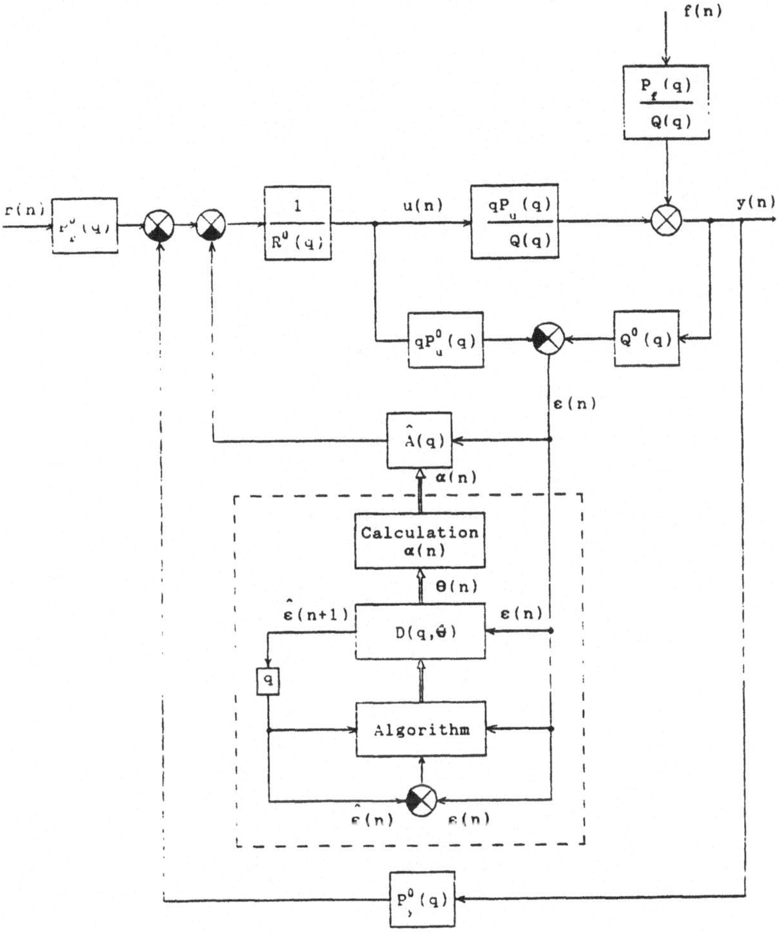
f(n)
P_f(q) / Q(q)
r(n)
1 / R^0(q)
u(n)
qP_u(q) / Q(q)
y(n)
qP_u^0(q)
Q^0(q)
ε(n)
Â(q)
α(n)
Calculation α(n)
θ(n)
ε̂(n+1)
D(q,θ̂)
ε(n)
q
Algorithm
ε̂(n)
ε(n)
P_y^0(q)

Figure 13.

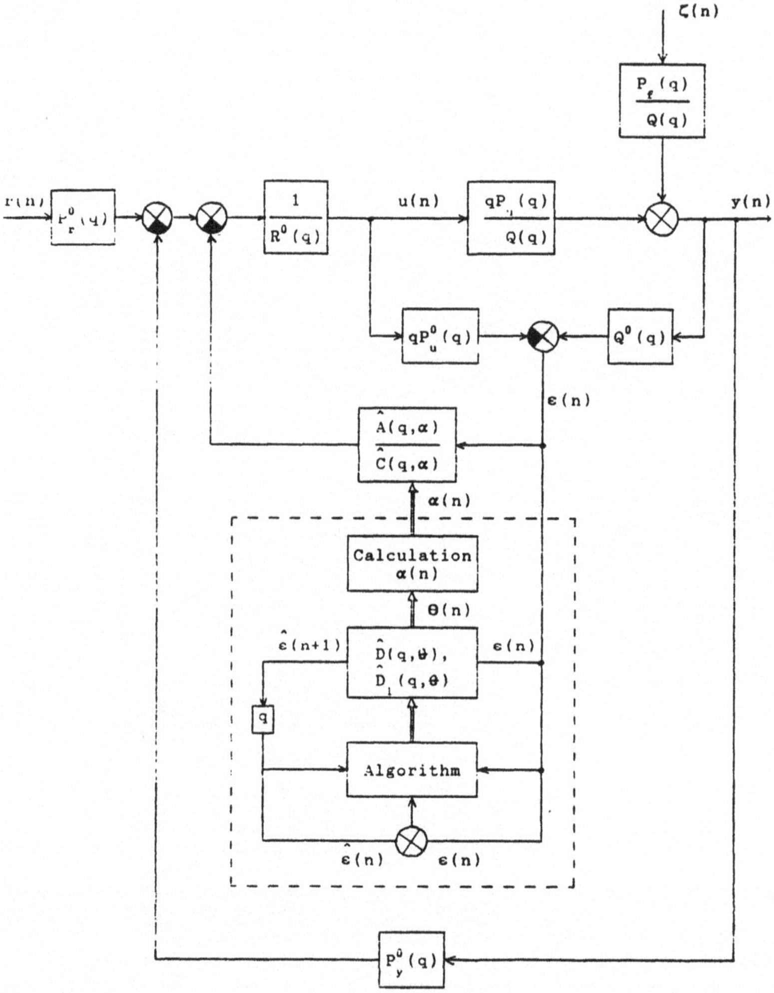
ζ(n)
P_f(q) / Q(q)
r(n)
P_r^0(q)
1 / R^0(q)
u(n)
qP_u(q) / Q(q)
y(n)
qP_u^0(q)
Q^0(q)
ε(n)
Â(q,α) / Ĉ(q,α)
α(n)
Calculation α(n)
θ(n)
ε̂(n+1)
D̂(q,θ), D̂_1(q,θ)
ε(n)
q
Algorithm
ε̂(n)
ε(n)
P_y^0(q)

Figure 14.

Progress in Systems and Control Theory

Series Editor

Christopher I. Byrnes
Department of Systems Science and Mathematics
Washington University
Campus P.O. 1040
One Brookings Drive
St. Louis, MO 63130-4899

Progress in Systems and Control Theory is designed for the publication of workshops and conference proceedings, sponsored by various research centers in all areas of systems and control theory, and lecture notes arising from ongoing research in theory and applications control.

We encourage preparation of manuscripts in such forms as LATEX or AMS TEX for delivery in camera-ready copy which leads to rapid publication, or in electronic form for interfacing with laser printers.

Proposals should be sent directly to the editor or to: Birkhäuser Boston, 675 Massachusetts Avenue, Cambridge, MA 02139, U.S.A.

PSCT1 Computation and Control: Proceedings of the Bozeman Conference, Bozeman, Montana, 1988
K. Bowers and J. Lund

PSCT2 Perspectives in Control Theory: Proceedings of the Sielpia Conference, Sielpia, Poland, 1988
B. Jakubczyk, K. Malanowski, and W. Respondek

PSCT3 Realization and Modelling in System Theory: Proceedings of the International Symposium MTNS-89, Volume 1
M. A. Kaashoek, J. H. van Schuppen, and A. C. M. Ran

PSCT4 Robust Control of Linear Systems and Nonlinear Control: Proceedings of the International Symposium MTNS, Volume II
M. A. Kaashoek, J. H. van Schuppen, and A. C. M. Ran

PSCT5 Signal Processing, Scattering and Operator Theory, and Numerical Methods: Proceedings of the International Symposium MTNS-89, Volume III
M. A. Kaashoek, J. H. van Schuppen, and A. C. M. Ran

PSCT6 Control of Uncertain Systems: Proceedings of an International Workshop, Bremen, West Germany, June 1989
D. Hinrichsen and B. Mårtensson

PSCT7 New Trends in Systems Theory: Proceedings of the Università di Genova - The Ohio State University Joint Conference, July 9-11, 1990
G. Conte, A. M. Perdon, and B. Wyman

PSCT8 Analysis of Controlled Dynamical Systems: Proceedings of a Conference in Lyon, France, July 1990
B. Bonnard, B. Bride, J. P. Gauthier and I. Kupka

PSCT9 Nonlinear Synthesis: Proceedings of a IIASA Workshop held in Sopron, Hungary, June 1989
Christopher I. Byrnes and Alexander Kurzhanski

PSCT10 Modeling, Estimation and Control of Systems with Uncertainty: Proceedings of a Conference held in Sopron, Hungary, September 1990
Giovanni B. Di Masi, Andrea Gombani and Alexander B. Kurzhanski

PSCT 11 Computation and Control II: Proceedings of the Second Conference at Bozeman, Montana, August 1990
Kenneth Bowers and John Lund

PSCT 12 Systems, Models, and Feedback: Theory and Applications. Proceedings of a U.S.-Italy Workshop in honor of Professor Antonio Ruberti, held in Capri, Italy, 15-17, June 1992
A. Isidori and T.J. Tarn

PSCT 13 Discrete Event Systems: Modelling and Control. Proceedings of a Joint Workshop held in Prague, Czechoslovakia, August 1992
S. Balemi, P. Kozak, and R. Smedinga

PSCT 14 Essays on Control: Perspectives in the Theory and its Applications
H. L. Trentelman and J. C. Willems

PSCT 15 Computation and Control III: Proceedings of the Third Bozeman Conference, Bozeman, Montana, August 1992
K. L. Bowers and J. Lund

PSCT 16 Set–valued Analysis and Differential Inclusions, A Collection of Papers from a Workshop held in Pamporovo, Bulgaria, September 1990
A. B. Kurzhanski and Vladimir M. Veliov

PSCT 17 Advances in Nonlinear Dynamics and Control: A Report from Russia
A. B. Kurzhanski

The manufacturer's authorised representative in the EU is Springer Nature Customer Service Centre GmbH, Europaplatz 3, 69115 Heidelberg, Germany. If you have any concerns regarding our products, please contact ProductSafety@springernature.com

Printed and bound by CPI Group (UK) Ltd, Croydon, CR0 4YY
29/04/2026
02099472-0004